LUMINESCENCE

edited by
ADAM M. GILMORE

LUMINESCENCE
The Instrumental Key
to the Future of Nanotechnology

Pan Stanford Publishing

Published by

Pan Stanford Publishing Pte. Ltd.
Penthouse Level, Suntec Tower 3
8 Temasek Boulevard
Singapore 038988

Email: editorial@panstanford.com
Web: www.panstanford.com

British Library Cataloguing-in-Publication Data
A catalogue record for this book is available from the British Library.

Luminescence: The Instrumental Key to the Future of Nanotechnology
Copyright © 2014 by Pan Stanford Publishing Pte. Ltd.
All rights reserved. This book, or parts thereof, may not be reproduced in any form or by any means, electronic or mechanical, including photocopying, recording or any information storage and retrieval system now known or to be invented, without written permission from the publisher.

For photocopying of material in this volume, please pay a copying fee through the Copyright Clearance Center, Inc., 222 Rosewood Drive, Danvers, MA 01923, USA. In this case permission to photocopy is not required from the publisher.

ISBN 978-981-4241-95-3 (Hardcover)
ISBN 978-981-4267-72-4 (eBook)

Printed in the USA

Contents

Preface   xiii

1. **Important Spectral and Polarized Properties of Semiconducting SWNT Photoluminescence**   1
   Shigeo Maruyama and Yuhei Miyauchi
   1.1 Important Spectral Features   1
   1.2 Phonon Sideband in Absorption   4
   1.3 Various Sidebands in Emission   6
   1.4 Cross-Polarized Absorption   11
   1.5 Transverse Quasi-Dark Excitons   15

2. **Advanced Aspects of Photoluminescence Instrumentation for Carbon Nanotubes**   23
   Said Kazaoui, Y. Futami, Konstantin Iakoubovskii, and Nobutsugu Minami
   2.1 Introduction   24
   2.2 CNT Thin-Film Fabrication Methods   25
   2.3 NIR-PL-Mapping Instruments   26
       2.3.1 Scanning-Type NIR-PL-Mapping Instrument   26
       2.3.2 FT-IR-Type NIR-PL-Mapping Instrument   28
   2.4 Outlook   31

3. **Developments in Catalytic Methodology for (n,m) Selective Synthesis of SWNTs**   35
   Yuan Chen, Bo Wang, Yanhui Yang, and Qiang Wang
   3.1 Introduction   36
   3.2 Effective Catalysts for (n,m) Selective Synthesis   37
   3.3 Growth Parameters Influencing (n,m) Selectivity   41
       3.3.1 Temperature   42
       3.3.2 Catalyst Particle   43
       3.3.3 Carbon Precursor   45
       3.3.4 "Clone" SWNTs   46

3.4	Fundamental Understanding of (n,m) Selectivity		46
3.5	Characterization Methodology for (n,m) Abundance Evaluation		51
3.6	Conclusions and Outlook		54

4. Single-Walled Carbon Nanotube Thin-Film Electronics — 61
Husnu Emrah Unalan and Manish Chhowalla

4.1	Introduction		62
4.2	Purification and Dispersion of SWNTs		65
4.3	Thin-Film Deposition Processes		69
4.4	Optoelectronic Properties of SWNTs		74
4.5	SWNT Functionalization Treatments		81
4.6	Applications and Devices		84
	4.6.1	Photovoltaic Devices	84
	4.6.2	Light-Emitting Diodes	87
	4.6.3	Supercapacitors and Batteries	87
	4.6.4	Sensors	90
	4.6.5	Electromagnetic Interference Shielding	92
	4.6.6	IR Properties and Applications	92
	4.6.7	Thin-Film Transistors	93
	4.6.8	Other Devices	98
4.7	Conclusions and Outlook		101

5. Single-Walled Carbon Nanotube-Based Solution-Processed Organic Optoelectronic Devices — 115
Ming Shao and Bin Hu

5.1	Introduction	115
5.2	Effects of SWCNTs on the Electroluminescent Performance of Organic Light-Emitting Diodes	116
5.3	CNT Effect on Photovoltaic Response in Conjugated Polymers	125

6. Exciton Energy Transfer in Carbon Nanotubes Probed by Photoluminescence — 133
Ping Heng Tan, Tawfique Hasan, Francesco Bonaccorso, and Andrea C. Ferrari

6.1	Introduction	133

6.2		The Photoluminescence Spectrum of Nanotube Bundles	134
6.3		Mechanism and Efficiency of EET in Nanotube Bundles	141
6.4		How to Distinguish EET-Induced Features from Other Sidebands in the PL Spectrum?	144
6.5		Relaxation Pathways of Excitons in Nanotube Bundles	148
6.6		How to Detect Bundles and Probe Their Concentration?	150
6.7		Exploiting EET for Photonic and Optoelectronic Applications	154
6.8		Conclusions	155

7. Advances in Dispersal Agents and Methodology for SWNT Analysis — 163

Tsuyohiko Fujigaya and Naotoshi Nakashima

7.1		Introduction	163
7.2		Characterization of Dispersion States	164
7.3		Solubilization by Dispersal Agents	165
	7.3.1	Surfactants	165
	7.3.2	Polycyclic Aromatic Compounds	169
	7.3.3	Porphyrins	173
	7.3.4	DNA	177
	7.3.5	Condensation Polymers	181
7.4		Nanotube/Polymer Composites	182
	7.4.1	Curable Monomers and Nanoimprinting	182
	7.4.2	Nanotube/Polymer Gel for NIR-Responsive Materials	183
	7.4.3	Conductive Nanotube Honeycomb Film	186
7.5		Summary	189

8. Time Domain Luminescence Instrumentation — 203

Graham Hungerford, Kulwinder Sagoo, and David McLoskey

8.1	Introduction	204
8.2	Overview	206
8.3	Light Sources	207

		8.3.1	Flashlamps	207
		8.3.2	Dye Laser Systems	209
		8.3.3	LEDs and Laser Diodes	210
		8.3.4	Femtosecond Lasers	212
		8.3.5	Supercontinuum Lasers	213
		8.3.6	Sources for Longer-Lived Decays	214
	8.4	Detectors		216
		8.4.1	Photomultiplier Tubes	217
		8.4.2	Microchannel Plate Detectors	219
		8.4.3	Avalanche Photodiodes	219
	8.5	Data Acquisition Electronics		220
		8.5.1	TCSPC Electronics	220
		8.5.2	Longer Timescale Measurements	223
	8.6	Time-Resolved Measurement System Considerations		224
	8.7	Summary		225

9. Key Approaches to Linking Nanoparticle Metrology and Photoluminescence — 229

Yu Chen, Jan Karolin, and David J. S. Birch

	9.1	Introduction		230
	9.2	Fluorescence Anisotropy Theory		234
	9.3	Experimental		236
		9.3.1	Instrumentation	236
		9.3.2	Choice of Dyes and Nanoparticles and Sample Preparation	238
	9.4	Results and Discussions		244
		9.4.1	Ludox Labeled with Extrinsic Probes	244
		9.4.2	Fluorescence from Au Nanoparticles	248
		9.4.3	Size-Dependent Fluorescence	251
	9.5	Conclusions		253

10. Nanometer-Scale Measurements Using FRET and FLIM Microscopy — 259

Margarida Barroso, Yuansheng Sun, Horst Wallrabe, and Ammasi Periasamy

	10.1	Introduction	260
	10.2	FRET Microscopy	261
	10.3	Choosing FRET Pairs	265

	10.4	Organic Dye Donor–Acceptor FRET Pair: AF488–AF555	267
		10.4.1 Filter-Based FRET Microscopy	267
	10.5	FP Donor–Acceptor FRET Pair: mTFP-mKO2	272
		10.5.1 Spectral FRET Microscopy	273
		10.5.2 FLIM-FRET Microscopy	276
	10.6	QD–Organic Dye FRET Pairs: QD566–AF568 and QD580–AF594	278
		10.6.1 Application of QDs as Donor Molecules in FRET Pairs	279
		10.6.2 Filter-Based and Spectral FRET Confocal Microscopy of QD566–AF568 and QD580–AF594	281
	10.7	Conclusions and Outlook	285
11.	**Cancer Detection and Biosensing Applications with Quantum Dots**		**291**
	Ken-Tye Yong		
	11.1	Introduction	292
	11.2	Preparation of Quantum Dots with the Hot Colloidal Synthesis Method	294
	11.3	Types of Quantum Dots Available for Biomedical and Cancer Applications	296
		11.3.1 CdSe/ZnS Core-Shell Quantum Dots	297
		11.3.2 CdTe/ZnS Core-Shell Quantum Dots	297
		11.3.3 InP/ZnS Core-Shell Quantum Dots	298
		11.3.4 PbS Quantum Dots	299
		11.3.5 Type II CdTe/CdSe Core-Shell Quantum Dots	299
		11.3.6 Silicon Quantum Dots	300
		11.3.7 Other Types of Quantum Dots	301
		11.3.8 CdSe/CdS/ZnS Quantum Rods	302
	11.4	Preparation of Water-Dispersible Quantum Dots	302
	11.5	Preparation of Bioconjugated Quantum Dots	304
	11.6	Bioconjugated Quantum Dots and Quantum Rods for in vitro Cancer Imaging and Sensing	305
	11.7	Multifunctional Quantum Dots and Quantum Rods for in vivo Cancer Targeting and Imaging	308

11.8	The Risk and Benefits of Using Functionalized Quantum Dots for Biomedical Health Care	312
11.9	Conclusions and Outlook	313

12. Zinc Oxide Nanoparticles in Biosensing Applications — 323

Linda Y. L. Wu

12.1	Introduction	324
12.2	Particle Size Control through Chemical Synthesis and Surface Modifications	326
12.3	Bandgap Modification for Visible Emission	328
	12.3.1 Photoluminescence Spectra of Pure and Doped ZnO	329
	12.3.2 Quantum Yield of Pure and Doped ZnO Colloids	330
12.4	Bioimaging Using ZnO Nanocrystals	332
	12.4.1 In vitro Bioimaging on Human and Animal Cells	332
	12.4.2 Bioimaging on a Plant System	334
	12.4.3 In vivo Bioimaging in a Rat Model	334
12.5	Cytotoxicity Tests	336
12.6	Conclusions and Outlook	339

13. Use of QDOT Photoluminescence for Codification and Authentication Purposes — 343

Shoude Chang

13.1	Introduction	344
13.2	QDOTs Used as Information Carriers	346
13.3	Information Encoding	348
13.4	Information Retrieval	355
13.5	Applications	358
	13.5.1 Anticounterfeiting	358
	13.5.2 Friend/Enemy Discrimination	360
13.6	Conclusions and Outlook	363

14. Characterization Approaches for Blue and White Phosphorescent OLEDs — 367

Brian W. D'Andrade

14.1	Introduction	368
14.2	Blue Electrophosphorescence	368

		14.2.1	Device Architecture and Energy Transfer	369
		14.2.2	Identifying High-Triplet-Energy Host Materials	370
		14.2.3	A General Route to Deep-Blue Electrophosphorescence	371
		14.2.4	Application in White OLEDs	372
	14.3	White Organic Light-Emitting Device		373
		14.3.1	Optical Characterization and Device Efficiency	374
		14.3.2	Characterization of Organic Semiconductor Materials	377

Index   383

Preface

Without question the subject of nanomaterial properties and applications represents one of the critical revolutions in modern human technology. In this light, luminescence—the instrumental key to the future of nanotechnology—highlights the key historical, contemporary, and anticipated future developments in instrumentation and experimental methodology relating to photo- and electroluminescent properties of nanomaterials. Each of the 14 chapters systematically documents the historical and current status of a key aspect of the field and concludes with the authors' views of their perceived future directions.

The volume is organized into two sections, the first dealing with the eminent subject of single-walled carbon nanotubes and the second dealing with methods and applications relating primarily to other quantum-confined nanoparticles and special aspects of time- and wavelength-resolved photoluminescence spectroscopy. The invited chapters are authored by a handpicked selection of world leaders in the field, including key, promising young researchers responsible for early, seminal contributions in the areas of single-walled carbon nanotubes, nanocrystals, and organic electroluminescent materials and applications inter alia. The book encompasses the nanoscale semiconductor field by amalgamating contributions from a broad multidisciplinary background, including applications of carbon nanotube photo- and electrophysics, materials authentification and codification, energy conservation, materials performance enhancements, electronic circuitry, video display technology, lighting, photovoltaics, quantum computing, computing memory, chemo- and biosensors, pharmaceuticals, and medical diagnostics, including key breakthroughs relating to cancer detection and treatment. Because the book's contents are composed to encompass the recent past, current, and obvious predicted developments in the future state of the art, the book should be a long-lasting, valuable resource for students and advanced researchers in academic and industrial positions in the field of nanomaterial luminescent properties and applications.

The editor gratefully acknowledges the contributing authors for their care and patience in preparing their chapters' material in a thorough and timely manner and the publisher and staff at Pan Stanford Publishing. I especially thank Stanford Chong for approaching me with the concept after attending one of my invited talks on the subject, and also for strongly supporting my efforts in organizing and completing the project. Special thanks are also offered to everyone who participated in reviewing the chapter materials for accuracy and scientific value as well as to my colleagues at HORIBA Instruments Inc. for their role in supporting and critiquing the project.

In conclusion, and despite the many challenges associated with the editor's and authors' schedules and commitments, I feel proud to present this book because I feel confident it will be highly valued for its documentation of the important field of nanomaterial luminescence characterization and practical applications thereof.

Adam M. Gilmore
Autumn 2013

Chapter 1

Important Spectral and Polarized Properties of Semiconducting SWNT Photoluminescence

Shigeo Maruyama[a] and Yuhei Miyauchi[b,c]
[a]*Department of Mechanical Engineering, The University of Tokyo, 7-3-1 Hongo, Bunkyo-ku, Tokyo 113-8656, Japan*
[b]*Institute of Advanced Energy, Kyoto University, Uji, Kyoto 611-0011, Japan*
[c]*Japan Science and Technology Agency, PRESTO, 4-1-8 Honcho Kawaguchi, Saitama 332-0012, Japan*
maruyama@photon.t.u-tokyo.ac.jp, ym2316@columbia.edu

1.1 Important Spectral Features

Photoluminescence (PL) of single-walled (carbon) nanotubes (SWNTs)[1] has been intensively studied for the optical characterization of SWNTs. By plotting PL emission intensities as a function of emission and excitation photon energy, Bachilo et al.[2] obtained a two-dimensional (2D) map of relative emission intensities. Hereafter, we refer to such a plot of photoluminescence excitation (PLE) and emission spectra as a "PL map." Figure 1.1 shows typical PL maps for SWNTs synthesized using different methods. SWNTs were

---

Luminescence: The Instrumental Key to the Future of Nanotechnology
Edited by Adam M. Gilmore
Copyright © 2014 Pan Stanford Publishing Pte. Ltd.
ISBN 978-981-4241-95-3 (Hardcover), 978-981-4267-72-4 (eBook)
www.panstanford.com

dispersed in deuterium oxide (D$_2$O) with a surfactant. A major peak in a PL map corresponds to the excitation transition energy of the second subband (E_{22}) and the photon emission energy of the first subband (E_{11}) of a specific SWNT structure defined by chiral indices (n,m).[3] The clear deference between these two PL maps reflects the difference of chirality and diameter distribution of SWNTs in different samples. Theoretical studies and recent experiments have clarified that these optical transitions are dominated by strongly correlated electron–hole (e–h) states in the form of excitons.[4–9] Since the pairs of E_{11} and E_{22} energies depend on the nanotube structure, we can separately measure a PLE spectrum from specific (n,m) SWNTs as a cross section of the PL map at an energy corresponding to the emission of the relevant SWNTs. This is useful to study optical properties of different types of SWNTs separately. In addition, as far as semiconducting SWNTs are concerned, such PL mapping is one of the most promising approaches for the determination of the structure distribution in a bulk SWNT sample if complementally combined with other optical spectroscopy such as optical absorption and Raman spectroscopy. Hence, PL spectroscopy is a powerful tool not only for investigating physical properties of SWNTs but also advancing toward development of complete synthesis or separation methods of SWNTs with only one chiral structure, which is one of the ultimate dreams in the nanoscience and technology field.

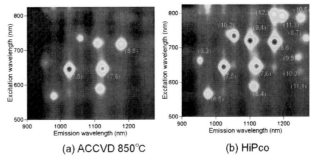

Figure 1.1 PLE map of different samples. (a) ACCVD SWNTs and (b) HiPco SWNTs dispersed in D$_2$O using a surfactant (SDS[1]).[45] ACCVD SWNTs were produced on zeolite support particles under the conditions of 850°C and 10 min for the CVD temperature and reaction time, respectively. *Abbreviations*: CVD, chemical vapor deposition; CCVD, catalytic chemical vapor deposition; ACCVD, alcohol catalytic chemical vapor deposition; SDS, sodium dodecyl sulfate.

In a PL map, one can generally find some peaks other than bright PL peaks assigned to E_{22} absorption and E_{11} emission of particular nanotube species allowed for incident light polarized parallel to the nanotube axis. Figure 1.2 shows the typical spectral features of (7,5) SWNTs. First, one can find a broad absorption peak in the shorter excitation wavelength side. Second, one can also find a small absorption peak in the longer emission wavelength side. In addition, for a fixed excitation wavelength, there is a weak emission peak (or peaks) in the longer emission wavelength side. The physical origins of these peaks have been recently demonstrated as, respectively, excitonic phonon sideband in absorption[10-13], absorption peak for perpendicularly polarized light to the nanotube axis[14,15], excitonic phonon sideband in emission[16,17], and/or emission from triplet exciton states.[18,19] Recently, there also have been reports on the direct observation of nominally dark excitons for cross-polarized excitons to the nanotube axis (transverse excitons).[20] Details of these important features will be discussed in the following sections.

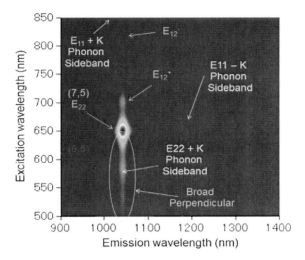

Figure 1.2 Spectral features observed in a PL map of a sample with virtually only (7,5) SWNTs. SWNTs were individually dispersed in toluene using a fluorene polymer (PFO).[46] *Abbreviation*: PFO, poly [9,9-dioctylfluorenyl-2,7-diyl].

Recent development on synthesis techniques of a single nanotube suspended over pillars or trenches enabled one to measure PL from a "single" SWNT.[21] By single-nanotube spectroscopy, further details in the PL and PLE spectra of SWNTs, such as excited exciton states[22], effect of surrounding materials on exciton energies (environmental effect)[23–26], temperature dependence of PL homogeneous line widths[27], and symmetry-induced dark excitons[28,29], have been studied. The observation of excited exciton states gave information about the exciton binding energies in clean, intrinsic SWNTs.[22] Since the surrounding materials change the Coulomb interaction strength through the dielectric screening effect, exciton energies depend on surrounding materials around SWNTs.[23–26] The temperature dependence of the homogeneous line width has shown almost linear behavior in the low temperature range below 100 K, which suggests that the exciton dephasing is dominated by the interaction between the exciton and the phonon mode with very low energy under lower excitation conditions.[27] Observation of parity-induced dark excitons has been achieved through "brightening" of dark excitons due to symmetry breaking[30] by the Aharonov–Bohm effect[31], which revealed the energy difference of bright and dark excitons about a few to several millielectron volts, depending on the nanotube diameter.[28,29]

1.2 Phonon Sideband in Absorption

Absorption peaks in PLE spectra about 200 meV above the main absorption/emission peaks of E_{11}[10–13] were attributed to a the phonon-assisted excitonic absorption and recombination process.[32,33] Theoretical prediction of the excitonic phonon sideband shape[33] was in good agreement with the PL spectrum from an individual nanotube[11] and isolated SWNTs in surfactant suspension.[12] These experimental observations of sideband features[10–12] were mainly for the excitation energy range close to the E_{11} transition energy, and the observed sidebands are attributed to excitation to the exciton-phonon bound state. However, the interpretations in these studies were based on only a peak position and line shape analysis. The direct experimental verification of the origin of the absorption sideband peaks for both E_{11} and E_{22} was achieved by isotope study using SWNTs made of the carbon-13 isotope.[13]

Figure 1.3 compares PLE spectra of normal SWNTs and carbon-13 isotope SWNTs (SW^{13}CNTs). PLE spectra in Fig. 1.3 (right panel) correspond to vertical cuts of the PL maps (left panel) at the emission energy indicated by solid lines. Each PLE spectrum is normalized by the E_{22} peak intensity for comparison. It can be seen that the PLE spectra exhibit a sideband 0.2–0.3 eV above the E_{11} and E_{22} main absorption peaks. Since the emission energies of these peaks were almost identical with that at the E_{22} absorption peak, these peaks are also attributed to emission from (7,5) (or (6,5)) SWNTs. It can be seen from Fig. 1.3 that the energy difference between the sideband peaks about 0.2 eV above and the main E_{22} peaks are reduced considerably for the PLE spectra of SW^{13}CNTs. If these peaks are excitonic phonon sidebands, the amount of the isotope shift is expected to be consistent with the value estimated from the difference of phonon energies. According to theoretical prediction by Perebveinos et al.,[33] the optical phonons near the K point of the graphene Brillouin zone (in-plane TO[16,34]) have stronger exciton-phonon coupling and dominantly contribute to the sideband. Assuming K-momentum phonons are dominant, the isotope shift of the phonon energy is estimated as about 7 meV for considering the square root of the mass ratio. Since the energy difference of ~7 meV is in good agreement with the observed energy shift of ~6–10 meV shown in Fig. 1.3, the observed sideband peaks are verified to be excitonic phonon sidebands of the E_{11} and E_{22} main absorption peak due to strong exciton-phonon coupling to the K-momentum phonons.

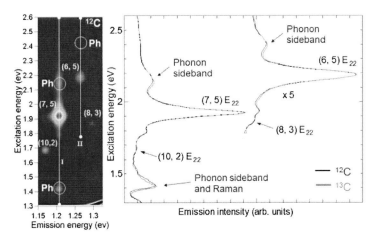

Figure 1.3 Phonon sideband for E_{11} and E_{22} absorption of (6,5) and (7,5) SWNTs.[13]

It is seen that the energy differences of these phonon sidebands from the main absorption peaks, ~0.2–0.3 eV, are considerably larger than the K-momentum optical phonon energy. Here, it should be noted that the energy difference from E_{22} is remarkably similar to the excitonic phonon sideband predicted in Ref. 33. This larger energy difference of phonon sideband peaks from E_{11} or E_{22} peaks than optical phonon energies can be interpreted by considering phonon-assisted excitation to "dark" (dipole-forbidden) exciton bands by K-momomtum phonons.[33] Figure 1.4 shows a schematic diagram of bright and dark excitons in SWNTs.[17] Energy corresponding to circle A in Fig. 1.4 indicates the excitation energy of absorption phonon sidebands. Considering the contribution of dark excitons with a finite momentum q_0 (point X in Fig. 1.4), the energy contribution from optical absorption required to make a finite-q_0 exciton is the sum of the energy of a dark exciton and the phonon (or phonons) satisfying energy momentum conservation. When the dark excitons that dominate the sideband have energies larger than the optically active E_{ii} exciton, the energy difference between E_{ii} and the phonon sideband peak can exceed the phonon energy. The amount of contribution of the dark exciton band is roughly estimated as about 30–40 meV for (7,5) nanotubes by subtracting phonon energy from the observed energy difference between the E_{ii} main absorption peaks and the sideband peaks.[13] Recently, energies of K-momentum dark excitons have been confirmed by observation of phonon sidebands in emission spectra.[16,17] This topic will be discussed in the next section.

1.3 Various Sidebands in Emission

In Fig. 1.2, one can find a small sideband feature denoted as "E_{11}-K phonon sideband." This sideband was initially reported and explained as "deep dark excitonic states" by Kiowski et al.[35] They ruled out the possibility of a phonon-dissipating emission mechanism on the basis of their experimental observation that the energy separation between the E_{11} level and the sideband (δ, see Fig. 1.4) is remarkably dependent on the SWNT diameter (d). On the other hand, Torrens et al.[16] recently proposed that this sideband originates from dipole-forbidden "dark" excitons coupled with K-point phonons (i.e., the mechanism that was ruled out in Ref. 35) on the basis of their experimental measurement of (6,5) SWNTs. Apparently, the dependence of δ on d is a key issue to resolve this conflict, which is essential for correctly understanding the properties of the weak photoemission found below the E_{11} states.

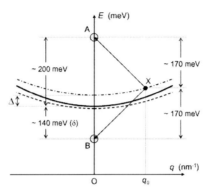

Figure 1.4 Schematic diagram showing the four lowest-energy singlet exciton states in zigzag SWNTs.[17] The abscissa denotes the exciton momentum. The solid and dashed curves denote bright and dark exciton bands with zero angular momentum, respectively. The dot-dashed curve denotes the dark exciton band with nonzero angular momentum (K-momentum).

Recently, Murakami et al.[17] reported that there is no diameter dependence on this sideband. Figure 1.5 compares PL maps of different SWNT samples with virtually only one (n,m) structure in each sample. Preparation techniques of these samples are presented in Ref. 17. Panels (d–f) correspond to the same as panels (a–c), but the intensities are presented in logarithmic scale. Panels (d–f) clearly show the existence of sidebands (indicated by the arrows) associated with the PL features of (6,5), (7,5), and (10,5) SWNTs, respectively.

Figure 1.6 shows PL spectra measured from the three different samples obtained by resonantly exciting E_{22} levels of the dominant SWNT species with excitation wavelengths 570, 655, and 800 nm, respectively. The abscissa denotes energy relative to the PL emission maxima from respective E_{11} levels. Here, spectral decomposition analysis was performed to obtain the positions and widths of the observed peaks. Specifically, these spectra were fitted with an E_{11} PL peak, a shoulder at ~45 meV below the E_{11} level and the PL sideband shown in Fig. 1.5. The shoulder peak at ~45 meV below the E_{11} was also reported in Ref. 35 and attributed to another deep dark state.

8 | *Important Spectral and Polarized Properties of Semiconducting SWNT PL*

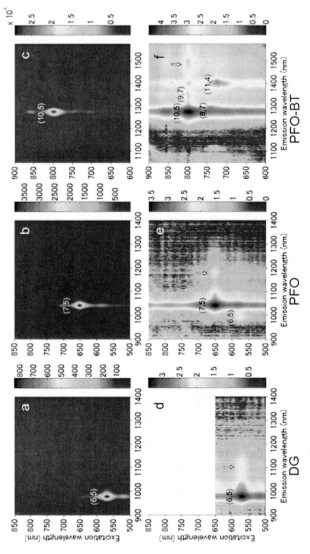

Figure 1.5 Contour maps of PLE intensities measured from the (a, d) DG, (b, e) PFO, and (c, f) PFO-BT samples, which are dominated by (6, 5), (7, 5), and (10, 5) SWNTs, respectively.[17] Intensity scale bars are attached to the right of each panel, where the unit is the photon count. Panels (d–f) are the same as panels (a–c), but the intensities are shown in log scale. Arrows indicate the locations of the sidebands. *Abbreviations:* DG, density gradient; PFO-BT, poly[(9,9-dioctylfluorenyl-2,7-diyl)-alt-co-(1,4-benzo-2,10,3-thiadiazole)].

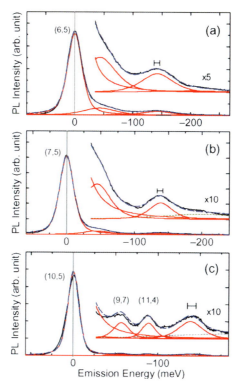

Figure 1.6 PL spectra measured for (6,5), (7,5), and (10,5) SWNTs, obtained by resonantly exciting the E_{22} levels of the dominant SWNTs with excitation wavelengths of 570, 655, and 800 nm, respectively.[17] The abscissa represents the energy relative to the E_{11} levels of (6,5), (7,5), and (10,5) SWNTs. The original spectra, shown by black solid curves, were decomposed into multiple peaks, and the summation is shown by blue dashed curves. The gray short-dash curves in (b) and (c) are spectral baselines assumed in the analysis. The uncertainties in the peak-maximum positions of the PL sidebands are indicated by the horizontal bars.

On the other hand, recently, Galland et al.[36] have attributed this peak to the radial breathing mode (RBM)-related features. As for Fig. 1.6c, additional peaks for (9,7) and (11,4) tubes were taken into account. A 50% Lorentzian+50% Gaussian line shape was assumed for all the peaks. In Fig. 1.6a–c, the blue dash curves show the summation of all the components that overlap the experimental spectra shown by the black solid curves. Uncertainties or the possible ranges of the peak maximum position of the sidebands obtained from the decomposition analysis are indicated by horizontal bars. Figure 1.6 clearly shows that the observed δ is independent of d (~140 meV, in the range of 0.75 < d < 1.05 nm) within the uncertainties indicated in Fig. 1.6a–c. The independence of δ on d suggests an involvement of phonons in the origin of these sidebands, particularly the involvement of higher-frequency optical phonons that arise from the phonon dispersion relations of graphite,[34] and hence is essentially independent of d.

Although the above emission sideband peaks whose energy difference from the main E_{11} emission peaks do not depend on the diameter can be attributed to the phonon sideband, Harutyunyan et al.[18] have recently reported that low-energy emission bands can be created in the spectra of SWNTs by intense pulsed-laser excitation. The new emission bands appeared about 190 meV below the main E_{11} emission peak of (6,4) SWNTs. This larger energy difference from the main E_{11} emission peak cannot be explained by the phonon-assisted mechanism discussed above. From the results of static and time-resolved PL measurements, they attributed this new satellite emission bands to PL from different nominally dark excitons that are "brightened" because of a defect-induced mixing of states with different spins, that is, intersystem crossing between singlet and triplet exciton states. Here, the dependence of the energy difference between this new satellite emission peaks and the main E_{11} peak on the nanotube diameter d is again a key issue. Matsunaga et al.[19] have recently reported that the weak PL peak ~130–140 meV lower than the E_{11} peak diminishes at low temperatures and that the energy separation only weakly depends on the tube diameter, which supports the interpretation that the weak low-energy peak is the phonon sideband of the K-momentum dark exciton states, as discussed above. In addition to the (emission) phonon sideband, they also found that an additional low-energy PL peak appears under intense pulsed-laser irradiation, the intensity of which increases with decreasing temperature, and the energy separation depends strongly on the tube diameter. Their results suggest that there can be two kind of origins for satellite peaks around 100–200 meV below the E_{11} peak—one is the K-momentum phonon sideband, and another is the spin-induced nominally dark excitons (triplet exciton states). From the above discussion, the apparently complicated diameter dependence of the emission sideband features reported in the pioneering work by Kiowski et al.[35] might be attributed to incomplete separation of both sideband features originating from the two different mechanisms (emission from K-momentum dark states and triplet dark states). Since intrinsic SWNTs do not have the mechanisms for spin-flip processes, one can expect only the K-momentum phonon sideband peaks for intrinsic SWNTs.

1.4 Cross-Polarized Absorption

PLE studies of SWNTs have mainly focused on the optical transitions for incident light parallel to the SWNT axis. However, there are also important absorption peaks for cross-polarized light to the nanotube axis. Figure 1.7a,b schematically show the selection rules for incident light polarized parallel and perpendicular to the SWNT axis.[31,37,38] In the case of parallel polarization, optical absorption between subbands with the same quasi-angular momentum is allowed ($\Delta\mu = 0$ or E_{ii} transitions), while $\Delta\mu = \pm 1$ transitions are allowed for perpendicular polarization (μ denotes the subband (cutting line) index in the 2D Brillouin zone of graphite). We refer to $\Delta\mu = \pm 1$ transitions between the first and second subbands as E_{12} and E_{21} transitions. Since the emission is primarily caused by recombination of electrons and holes within the first conduction and valence subbands with the same index μ ($\Delta\mu = 0$, E_{11} transitions), the emission dipole is considered to be parallel to the SWNT axis. Note that the distinct cross-polarized excitation peaks can be observed, although the cross-polarized absorption has been considered to be strongly suppressed and cannot have a peak structure due to the induced self-consistent local field (depolarization effect) within single-particle theory.[31] This is one of the most important consequences and signature of the strong Coulomb interaction and excitons in SWNTs.

Optical absorption by two different dipoles being normal to each other followed by the same E_{11} emission process can be separately probed by polarized PLE spectroscopy on randomly oriented ensemble SWNTs through anisotropy analysis.[14] The use of an ensemble sample enables one to probe very weak cross-polarized absorption peaks clearly even without a strong light source, such as a wavelength variable laser. Figure 1.7c shows the schematic diagram for the measurement of PL anisotropies of randomly oriented

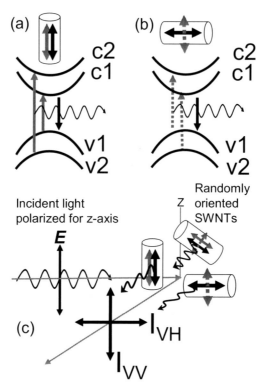

Figure 1.7 Schematics of optical transitions in SWNTs corresponding to (a) collinear and (b) perpendicular absorption and emission dipoles.[14] Solid and dotted arrows indicate $\Delta\mu = 0$ and $\Delta\mu = \pm 1$ transitions, respectively. (c) Schematic diagram for the measurement of PL anisotropies of randomly oriented SWNTs.

SWNTs. Because a PL emission signal is enhanced when directions of an emission dipole and the emission polarizer are the same, the direction of emission polarizer determines the mainly probed SWNTs. In the I_{VV} spectra, the PL emission excited by collinear absorption— $\Delta\mu = 0$ absorption (e.g., E_{22} absorption) → $\Delta\mu = 0$ E_{11} emission—is enhanced, whereas emission excited by perpendicular absorption— $\Delta\mu = \pm 1$ absorption (e.g., E_{12} absorption) → $\Delta\mu = 0$ E_{11} emission—is enhanced in the I_{VH} spectra. Polarized PLE spectra are obtained with the emission polarizer oriented parallel to (I_{VV}) or perpendicular to (I_{VH}) the direction of the vertically polarized excitation, and the pure component for parallel ($I_{||}$) and perpendicular ($I_\perp$) dipoles relative to the nanotube axis were obtained using the relationships[14,39] $I_{||}$ = $[(r_{exp} - r_\perp)/(r_{||} - r_\perp)](I_{VV} + 2I_{VH})$ and $I_\perp = [(r_{||} - r_{exp})/(r_{||} - r_\perp)](I_{VV} + 2I_{VH})$, where $r_{exp} \equiv (I_{VV} - I_{VH})/(I_{VV} + 2I_{VH})$ is the anisotropy and $r_{||}$ and $r_\perp$ are the maximum and minimum anisotropies for parallel and perpendicular dipoles, which are related by $r_\perp = -0.5 r_{||}$.[39] Details of the experimental technique are presented in Ref. 14. The validity of the spectra obtained by this technique has been confirmed by direct observation of $I_{||}$ and $I_\perp$ spectra by single-nanotube spectroscopy[15] and polarized PLE measurements on ensemble SWNTs dispersed in toluene using an organic dispersing agent.[40]

Figure 1.8a,b shows the decomposed PL maps. In Fig. 1.8c,d the decomposed PLE spectra of (7,5) and (6,5) SWNTs are presented. For the spectra in Fig. 1.8, a value of $r_{||} = 0.3$ was determined from the maximum value of observed anisotropy for SWNTs in the lower excitation energy range without any peak structure.[14] While the main peaks corresponding to $\Delta\mu = 0$ transitions are seen in the $I_{||}$ spectra in Fig. 1.8a,c, it can be clearly seen that the $I_\perp$ spectra in Fig. 1.8b,d exhibit two distinct peaks below the respective E_{22} peaks and there exist small, but nonzero intensity tails above the distinct peaks in the $I_\perp$ maps. The lowest $I_\perp$ peaks are relatively sharp, while the second $I_\perp$ peaks are broader. Here it was confirmed that the ambiguity of $r_{||}$ around 0.3 does not substantially change the peak positions of the lowest $I_\perp$ peaks.

One can expect the degeneracy of E_{12} and E_{21} and only one peak around $\sim(E_{11} + E_{22})/2$ assuming e–h symmetry in SWNTs. Furthermore, within single-particle theory, one can expect that the E_{12} and E_{21} transitions drop almost in the middle of E_{11} and E_{22}, that is, $\sim(E_{11} + E_{22})/2$.[41] However, the experimental results showed pairs of peaks for cross-polarized absorption bands and the peak energies were considerably blue-shifted from $(E_{11} + E_{22})/2$. The prediction from a simple theory cannot reproduce the experimental results. These discrepancy were initially explained as follows:[14] the pairs of cross-polarized absorption peaks were simply attributed to the e–h asymmetry in SWNTs (i.e., degeneracy lifting between E_{12} and E_{21}) or excited exciton states for cross-polarized (transverse) excitons. The higher cross-polarized absorption energies than the expectation from single-particle theory were attributed to smaller exciton binding energies for transverse excitons than excitons polarized parallel to the nanotube axis (longitudinal excitons).

14 | Important Spectral and Polarized Properties of Semiconducting SWNT PL

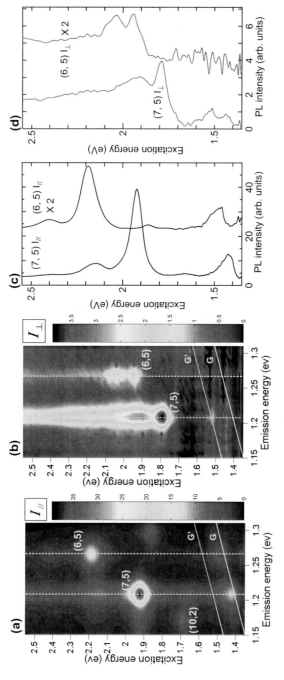

Figure 1.8 Decomposed PL maps for (a) collinear ($I_{\parallel}$) and (b) perpendicular ($I_{\perp}$) dipoles.[14] Dotted lines in (a) and (b) indicate the emission energies of respective SWNTs. Solid lines indicate the position of Raman lines for G and G' bands.[1] Peaks for $I_{\perp}$ spectra are indicated by arrows. Decomposed PLE spectra of (6,5) and (7,5) SWNTs for (c) collinear ($I_{\parallel}$) and (d) perpendicular ($I_{\perp}$) dipoles. The PLE spectra for (6,5) SWNTs are magnified (× 2) and offset for comparison in (c) and (d).

Recent theoretical studies by Uryu and Ando,[42,43] Zhao and Mazumdar,[7] and Wang et al.[44] that included electron–electron (e–e) and electron–hole (e–h) Coulomb interactions have shown that the transverse excitons can have finite oscillator strength due to strong excitonic effects and their excitation energies are considerably blue-shifted from $(E_{11} + E_{22})/2$. As for this point, the initial interpretation is virtually correct. However, Wang et al.[44] have reported that only one distinct peak for the cross-polarized absorption is expected because E_{12} and E_{21} excitons interact with each other and one bright and one dark exciton states are created. Considering the mixing of E_{12} and E_{21} excitons, we cannot attribute the pair of peaks simply to the E_{12} and E_{21} excitons. The difference of peak shapes between the two peaks may be a key to address this issue. To clarify this issue, further theoretical studies including the e–h asymmetry, intervalley mixing, and electron–phonon interactions might be needed.

1.5 Transverse Quasi-Dark Excitons

As discussed in the previous sections, the optical transitions in semiconducting SWNTs are dominated by strongly bound e–h states called excitons.[4–8] Figure 1.9a compares the selection rules for excitation polarized parallel (longitudinal) and perpendicular (transverse) to the nanotube axis. Longitudinal excitons consist of electrons and holes in the same one-dimensional (1D) subband ($E_{11}, E_{22}, \ldots$). On the other hand, transverse excitons[7,42] have quasi-angular momentum connecting the electron and hole states across these subbands (E_{12} and E_{21}). As shown in Fig. 1.1b, the degenerate exciton states near the K and K' points in momentum space are theoretically predicted to yield optically active (bright) and inactive (dark) exciton states through the intervalley Coulomb interaction for both longitudinal and transverse excitons.[7] Only dark states for longitudinal excitons[28–30] have been experimentally confirmed using the Aharonov–Bohm effect in strong magnetic fields. As for the transverse excitons, if the intrinsic e–h asymmetry in SWNTs is large enough to give considerable degeneracy lifting between E_{12} and E_{21} excitons, one can expect that the transverse dark excitons acquire finite oscillator strength even without a magnetic field. Hereafter, we refer to these nominally dark excitons as transverse quasi-dark excitons.

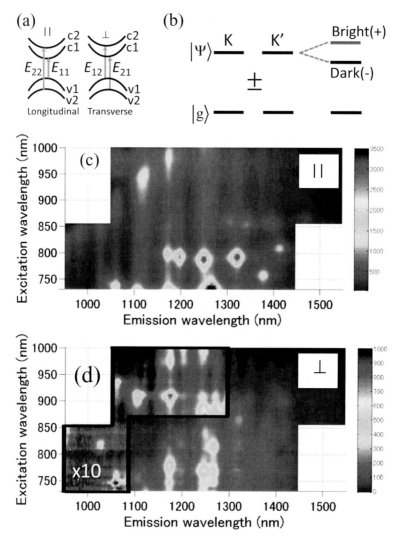

Figure 1.9 Schematic diagram of (a) the selection rules for incident light polarized parallel (||) and perpendicular (⊥) to the nanotube axis and (b) intervalley mixing of K and K' excitons.[20] The exciton wavefunctions are even and odd superpositions of those near the K and K' points in momentum space. This superposition gives the bright and dark states for longitudinal and transverse excitons. PLE maps for excitations polarized (c) parallel and (d) perpendicular to the nanotube axis. In (d), the PL intensities in the region surrounded by solid lines have been magnified 10 times.

Recently, we have demonstrated the first direct observation of transverse quasi-dark excitons using polarized PLE spectroscopy.[20] The intrinsic e–h asymmetry lifts the degeneracy of the transverse exciton wavefunctions at two equivalent K and K' valleys in momentum space, which gives finite oscillator strength to transverse dark exciton states. The experimental procedure and analysis are almost the same as presented in the previous section, and only differences were the use of wavelength variable laser as a light source and the use of HiPco SWNTs as a sample. The use of laser and ensemble SWNTs enablea us to obtain very weak quasi-dark exciton absorption signals with high sensitivity.

Figure 1.9c,d shows PLE maps for excitations parallel (longitudinal excitons) and perpendicular (transverse excitons) to the nanotube axis. For perpendicular excitation, the observed PLE peak positions were completely different from those for parallel excitation, as discussed in the previous section. Both peak positions and PLE spectral shapes of the dominant peaks for parallel and perpendicular incident light are consistent with those in previous measurements.[2,14,15] For parallel excitation, near-infrared PL due to the e–h recombination of longitudinal excitons within the first subband E_{11} by excitation within the second subband E_{22} are observed. On the other hand, E_{11} PL by excitation of transverse excitons between the first and second subbands (E_{12} and E_{21}) is observed for perpendicular excitation. The distinct transverse exciton peaks have other broad absorption peaks and intensity tails to the high-energy side, as has been reported previously.[14,15] Superposition of the E_{12} and E_{21} exciton wavefunctions in the K and K' valleys could produce the bright and dark transverse exciton states, as introduced above. Hereafter, we refer to the bright transverse exciton states as $E_{12}^{(+)}$.

On the low-energy side of $E_{12}^{(+)}$ we found weak but distinct novel absorption peaks approximately 200–300 meV below the $E_{12}^{(+)}$ absorption peaks, as shown in the outlined region in Fig. 1.9d. We attribute these small absorption peaks below $E_{12}^{(+)}$ to exciton absorption by quasi-dark excitons that acquire finite oscillator strength due to the degeneracy lifting of E_{12} and E_{21} excitons originating from the intrinsic e–h asymmetry in SWNTs. We hereafter refer to these exciton transitions as $E_{12}^{(-)}$.

Figure 1.10a shows the observed excitation energies for longitudinal and transverse excitons. $E_{12}^{(+)}$ (triangles) and $E_{12}^{(-)}$ (squares) were observed between E_{11} and E_{22} (circles). $E_{12}^{(+)}$ are

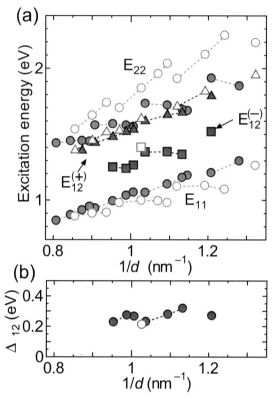

Figure 1.10 (a) Excitation energy plot for E_{11} and E_{22} (circles), $E_{12}^{(+)}$ (triangles), and $E_{12}^{(-)}$ (squares) exciton states as a function of the inverse diameter.[20] $E_{12}^{(+)}$ for (7,5) SWNTs were taken from Ref. 14. E_{11} and/or E_{22} for small-diameter SWNTs were obtained with Xe lamp excitation using a 5 nm slit width (excitation wavelength below 730 nm). (b) Energy difference between $E_{12}^{(+)}$ and $E_{12}^{(-)}$ plotted as a function of the inverse diameter. In (a) and (b), filled and open marks correspond to type I ($2n + m$ mod $3 = 1$) and type II SWNTs ($2n + m$ mod $3 = 2$), respectively.

close to E_{22}, while $E_{12}^{(-)}$ are relatively close to E_{11}. Note that phonon-related features such as phonon sidebands and Raman scatterings can be ruled out as an explanation for the $E_{12}^{(-)}$ peaks because the $E_{12}^{(-)}$ peaks do not have a constant energy difference from E_{11}, $E_{12}^{(+)}$, and E_{22}. Figure 1.10b shows the diameter dependence of the energy difference Δ_{12} between $E_{12}^{(+)}$ and $E_{12}^{(-)}$ ($\Delta_{12} \equiv E_{12}^{(+)} - E_{12}^{(-)}$). Δ_{12} is approximately 200–300 meV for the observed nanotube species and depends on the specific nanotube structure. The magnitude of the bright-dark energy splitting for transverse excitons is much larger than that of longitudinal excitons, as is consistent with previous theoretical predictions.[7,43] On the basis of the comparison between the experimental and theoretical results on the transverse quasi-dark excitons, we can evaluate the degree of e–h asymmetry corresponding to each (n,m) SWNT with specific chiral structures. Details of the chirality dependent e–h asymmetry in SWNTs will be discussed elsewhere.[20]

References

1. M. J. O'Connell et al., *Science*, **297**, 593 (2002).
2. S. M. Bachilo, M. S. Strano, C. Kittrell, R. H. Hauge, R. E. Smalley, R. B. Weisman, *Science*, **298**, 2361 (2002).
3. R. Saito, G. Dresselhaus, M. S. Dresselhaus, *Physical Properties of Carbon Nanotubes*, London: Imperial College Press, 1998.
4. T. Ando, *J. Phys. Soc. Jpn.*, **66**, 1066 (1997).
5. C. D. Spataru, S. Ismail-Beigi, L. X. Benedict, S. G. Louie, *Phys. Rev. Lett.*, **92**, 077402 (2004).
6. V. Perebeinos, J. Tersoff, P. Avouris, *Phys. Rev. Lett.*, **92**, 257402 (2004).
7. H. Zhao, S. Mazumdar, *Phys. Rev. Lett.*, **93**, 157402 (2004).
8. F. Wang, G. Dukovic, L. E. Brus, T. F. Heinz, *Science*, **308**, 838 (2005).
9. J. Maultzsch et al., *Phys. Rev. B*, **72**, 241402 (2005).
10. S. G. Chou et al., *Phys. Rev. Lett.*, **94**, 127402 (2005).
11. H. Htoon, M. J. O'Connell, S. K. Doorn, V. I. Klimov, *Phys. Rev. Lett.*, **94**, 127403 (2005).
12. F. Plentz, H. B. Ribeiro, A. Jorio, M. S. Strano, M. A. Pimenta, *Phys. Rev. Lett.*, **95**, 247401 (2005).
13. Y. Miyauchi, S. Maruyama, *Phys. Rev. B*, **74**, 035415 (2006).
14. Y. Miyauchi, M. Oba, S. Maruyama, *Phys. Rev. B*, **74**, 205440 (2006).
15. J. Lefebvre, P. Finnie, *Phys. Rev. Lett.*, **98**, 167406 (2007).
16. O. N. Torrens, M. Zheng, J. M. Kikkawa, *Phys. Rev. Lett.*, **101**, 157401 (2008).
17. Y. Murakami, B. Lu, S. Kazaoui, N. Minami, T. Okubo, S. Maruyama, *Phys. Rev. B*, **79**, 195407 (2009).

18. H. Harutyunyan, T. Gokus, A. A. Green, M. C. Hersam, M. Allegrini, A. Hartschuh, *Nano Lett.*, **9**, 2010 (2009).
19. R. Matsunaga, K. Matsuda, Y. Kanemitsu, *Phys. Rev. B*, **81**, 033401 (2010).
20. Y. Miyauchi, H. Ajiki, S. Maruyama, arXiv, 0906.3381 (2009).
21. J. Lefebvre, J. M. Fraser, P. Finnie, Y. Homma, *Phys. Rev. B*, **69**, 075403 (2004).
22. J. Lefebvre, P. Finnie, *Nano Lett.*, **8**, 1890 (2008).
23. S. Chiashi, S. Watanabe, T. Hanashima, Y. Homma, *Nano Lett.*, **8**, 3097 (2008).
24. Y. Miyauchi, R. Saito, K. Sato, Y. Ohno, S. Iwasaki, T. Mizutani, J. Jiang, S. Maruyama, *Chem. Phys. Lett.*, **442**, 394 (2007).
25. Y. Ohno, S. Iwasaki, Y. Murakami, S. Kishimoto, S. Maruyama, T. Mizutani, *Phys. Rev. B*, **73**, 235427 (2006).
26. F. Wang et al., *Phys. Rev. Lett.*, **96**, 167401 (2006).
27. K. Matsuda, T. Inoue, Y. Murakami, S. Maruyama, Y. Kanemitsu, *Phys. Rev. B*, **77**, 033406 (2008).
28. R. Matsunaga, K. Matsuda, Y. Kanemitsu, *Phys. Rev. Lett.*, **101**, 147404 (2008).
29. A. Srivastava, H. Htoon, V. I. Klimov, J. Kono, *Phys. Rev. Lett.*, **101**, 087402 (2008).
30. J. Shaver, J. Kono, O. Portugall, V. Krstic, G. L. J. A. Rikken, Y. Miyauchi, S. Maruyama, V. Perebeinos, *Nano Lett.*, **7**, 1851 (2007).
31. H. Ajiki, T. Ando, *J. Phys. Soc. Jpn.*, **62**, 1255 (1993).
32. J. Jiang, R. Saito, A. Gruneis, S. G. Chou, G. G. Samsonidze, A. Jorio, G. Dresselhaus, M. S. Dresselhaus, *Phys. Rev. B*, **71**, 045417 (2005).
33. V. Perebeinos, J. Tersoff, P. Avouris, *Phys. Rev. Lett.*, **94**, 027402 (2005).
34. R. Saito, A. Jorio, A. G. Souza Filho, G. Dresselhaus, M. S. Dresselhaus, M. A. Pimenta, *Phys. Rev. Lett.*, **88**, 027401 (2002).
35. O. Kiowski, K. Arnold, S. Lebedkin, F. Hennrich, M. M. Kappes, *Phys. Rev. Lett.*, **99**, 237402 (2007).
36. C. Galland, A. Högele, H. E. Türeci, A. Imamoğlu, *Phys. Rev. Lett.*, **101**, 067402 (2008).
37. I. Božović, N. Božović, M. Damnjanović, *Phys. Rev. B*, **62**, 6971 (2000).
38. A. Grüneis, R. Saito, G. G. Samsonidze, T. Kimura, M. A. Pimenta, A. Jorio, A. G. S. Filho, G. Dresselhaus, M. S. Dresselhaus, *Phys. Rev. B*, **67**, 165402 (2003).

39. J. R. Lakowicz, *Principles of Fluorescence Spectroscopy*, New York: Plenum, 1999.
40. S. Lebedkin, F. Hennrich, O. Kiowski, M. M. Kappes, *Phys. Rev. B*, **77**, 165429 (2008).
41. A. Grüneis, R. Saito, J. Jiang, G. G. Samsonidze, M. A. Pimenta, A. Jorio, A. G. Souza Filho, G. Dresselhaus, M. S. Dresselhaus, *Chem. Phys. Lett.*, **387**, 301 (2004).
42. S. Uryu, T. Ando, *Phys. Rev. B*, **74**, 155411 (2006).
43. S. Uryu, T. Ando, *Phys. Rev. B*, **76**, 115420 (2007).
44. Z. Wang, H. Zhao, S. Mazumdar, *Phys. Rev. B*, **76**, 115431 (2007).
45. S. Maruyama, Y. Miyauchi, Y. Murakami, S. Chiashi, *New J. Phys.*, **5**, 149 (2003).
46. A. Nish, J.-Y. Hwang, J. Doig, R. J. Nicholas, *Nat. Nano.*, **2**, 640 (2007).

Chapter 2

Advanced Aspects of Photoluminescence Instrumentation for Carbon Nanotubes

Said Kazaoui, Y. Futami, Konstantin Iakoubovskii,* and Nobutsugu Minami

National Institute of Advance Industrial Science and Technology (AIST), 1-1-1 Higashi, Tsukuba, Ibaraki 305-8565, Japan
s-kazaoui@aist.go.jp

In this chapter, we describe an advanced technique for measuring the photoluminescence (PL) maps of semiconducting carbon nanotube (CNT) thin films in the near-infrared (NIR) range, typically from 800 to 2,300 nm. This was made possible by developing both NIR-PL instruments as well as CNT thin-film fabrication methods. Two types of NIR-PL instruments were implemented, (1) a scanning type, consisting of a single element InGaAs detector connected to a lock-in amplifier attached to a 0.12 m monochromator (800–2,200 nm), and (2) a Fourier transform infrared (FT-IR) type, including a single element InGaAs detector (range 900–2,300 nm, resolution 16 cm^{-1}).

---

*Current affiliation: National Institute for Materials Science (NIMS), 1-2-1, Sengen, Tsukuba 305-0047, Japan

---

Luminescence: The Instrumental Key to the Future of Nanotechnology
Edited by Adam M. Gilmore
Copyright © 2014 Pan Stanford Publishing Pte. Ltd.
ISBN 978-981-4241-95-3 (Hardcover), 978-981-4267-72-4 (eBook)
www.panstanford.com

The light source was either a cw tunable Ti:sapphire laser (695–1,055 nm range) or a Xe lamp coupled to a 0.3 m monochromator (300–1,600 nm range). The CNT thin films were prepared by drop-casting, spin-coating, or dip-coating aqueous solutions of CNT dispersed with surfactant/gelatin, gelatin, or cellulose derivatives (carboxymethylcellulose, in particular). It is important to stress that the films were dry by removing traces of water, which otherwise would strongly absorb in the NIR range. Our technique was applied to a wide range of single- and double-wall CNTs, unveiling chirality and diameter distributions that have so far been inaccessible by commercially available systems.

2.1 Introduction

Carbon nanotubes (CNTs) are fascinating materials whose electronic properties (semiconducting or metallic) strongly depend on their chiral indexes specified by a pair of integers (n,m).[1,2]

Currently, CNTs are produced by various growth techniques, such as arc discharge, laser ablation, catalyst-assisted chemical vapor deposition (CVD; supergrowth), high-pressure CO gas decomposition with Fe catalyst (HiPCO), and direct injection pyrolytic synthesis (DIPS).[1-3] All these techniques produce CNTs with a rather different distribution of diameters and chiral indexes (n,m). It is essential to monitor these distributions and then to feed this information back to synthesize specific types of CNTs. Moreover, physical and chemical treatments carried out to tune the electronic properties, to purify, or to sort the CNTs strongly depend on their chiral indexes.[2] Therefore, there is a strong demand for an easily accessible technique for monitoring these distributions and their changes. For instance, optical absorption and Raman and photoluminescence (PL) spectroscopies have been extensively utilized to fulfill that demand.[1-6] Note that the PL techniques can only probe semiconducting CNTs, whereas the optical absorption and Raman spectroscopies provide information on both semiconducting and metallic CNTs. In this chapter, we will focus on an advanced aspect of PL instrumentation for probing the broader chiral index distribution of semiconducting CNT thin films.

PL instruments are already commercially available, such as those produced by Horiba and Shimadzu, but their wavelength ranges are quite limited. Therefore, we have developed two types of near-infrared (NIR)-PL instruments, namely, scanning and Fourier transform infrared (FT-IR).[7] Using a detector with an extended IR sensitivity made it possible to measure a much broader wavelength range than that accessible by previous systems. The light source is either a continuous-wave (cw) tunable Ti:sapphire laser (695–1,055 nm range) or a Xe lamp coupled to a 0.3 m monochromator (300–1,600 nm range), as described in Section 2.3.

It is worth noticing that PL measurements are often performed on CNTs dispersed in aqueous or organic solutions, which strongly absorb light in the NIR range. In addition, most of the expected applications, such as optical saturable absorber, field effect transistor, and gas sensor, require thin films or dense networks of CNTs.[8–10] In this context, we have fabricated solvent-free CNT thin films (dense CNT networks), eliminating the disturbance due to solvent, as described in Section 2.2.

2.2 CNT Thin-Film Fabrication Methods

CNT powders were dispersed in 1% deuterium oxide (D_2O) solutions of gelatin, sodium carboxymethylcellulose (CMC), sodium dodecyl sulfonate (SDS), or sodium dodecylbenzene sulfonate (SDBS) (Fig. 2.1) using a tip sonifier (20 kHz, 100 W, for 15 min). Several other dispersants have been tried as well, including polymers—poly(vinyl-alcohol) and hydroxyethylcellulose—and surfactants (sodium cholate, Triton-X). Next, the dispersions were ultracentrifuged for five hours at 150,000 g (45,000 rpm), and the upper 80% of the supernatant was collected.

Aqueous solution	Thin film

Figure 2.1 CNT dispersed in SDBS aqueous solution and thin film on a glass substrate.

The thus-prepared CNT supernatant solutions were utilized to fabricate thin films by drop-cast, spin-coating, or dip-coating methods. Samples were subsequently dried in air or in a controlled environment to completely remove residual traces of the solvent. In this procedure, gelatin and CMC were chosen because they are very efficient dispersants and have good film-forming properties for CNTs.[11,12] In addition, gelatin and CMC prevent the aggregation (rebundling) of CNTs upon drying.

We applied this sample preparation technique to single-wall (carbon) nanotubes (SWNTs) grown by the arc-discharge method (purchased from Carbolex Company) and the DIPS technique at the AIST, as well as double-wall nanotubes (DWNTs) produced by the CVD technique.

2.3 NIR-PL-Mapping Instruments

2.3.1 Scanning-Type NIR-PL-Mapping Instrument

Figure 2.2 shows the block diagram of our scanning-type NIR-PL instrument. Basically, the PL emission of CNTs, stimulated by a cw Ti:sapphire laser (700–1,000 nm, Spectra-Physics 3900S pumped by a 5-W Millenia pro 3s) is dispersed by a monochromator (700–2,500 nm, grating 600 g/mm, 1.6 µm Blaze, Shimadzu SPG-120IR) and is collected by a single-element InGaAs detector (800–2,200 nm, Hamamatsu G7754-01).

Further technical details are as follows:

(i) To improve the signal-to-noise ratio (SNR), PL emission is modulated with a mechanical chopper (typically in the range ~33–93 Hz) with the detector connected to a lock-in amplifier (NF-5600A). This greatly improves the SNR, which is otherwise poor for NIR detectors due to a large dark current.

(ii) To record the PL map at constant excitation intensity, a computer-controlled neutral density (ND) filter was utilized. The computer receives a signal from a photodiode and sends commands to a step motor. The motor rotates a silica disk, which has a gradient metal coating and acts as a variable ND filter. This implementation is essential because the intensity of the light source varies with wavelength and to some extent with time.

Our scanning-type NIR-PL instrument is computer controlled using a LabView computer program. The cw Ti:sapphire laser is equipped with a micrometer OptoMikeE (OMEC-2BG), which allows for step-by-step wavelength scanning controlled by the computer. Several optical filters are utilized to reject the laser light (in the range 690–1,050 nm) and to cut the high-order diffracted light due to the monochromator/grating system (low-wavelength cut filters at 900, 1,400 nm). All optical lenses are NIR grades. Note that the spectral response of the system (monochromator, grating, detector) can be calibrated using a standard black-body source.

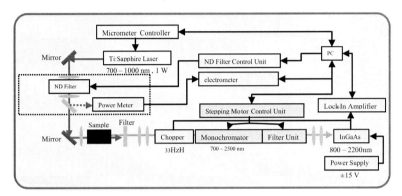

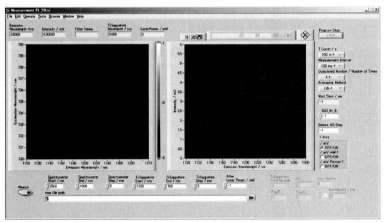

Figure 2.2 Block diagram and control panel of the scanning-type NIR-PL-mapping setup.

Figure 2.3 shows an example of an NIR-PL map of SWNTs Carbolex/SDBS/gelatin thin films recorded using our scanning-type PL instrument. The chiral indexes of the most well-resolved PL peaks are indicated on the basis of the standard data of Weisman and Bachilo.[5,6]

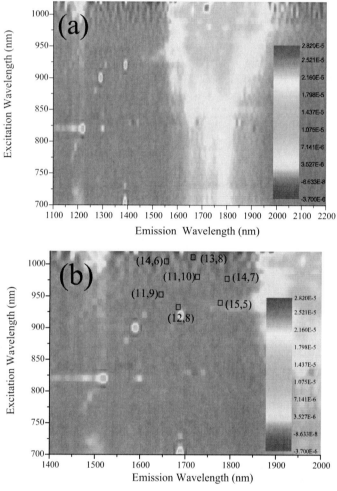

Figure 2.3 (a) The entire PL map of a Carbolex/SDBS/gelatin thin film using the NIR-PL scanning-type instrument, and (b) the chiral index (n,m) assignment of the PL peaks.

2.3.2 FT-IR-Type NIR-PL-Mapping Instrument

Most of the discussed problems are avoided in the FT-IR detection technique.[13] Our PL detection system is based on a commercial JASCO FT-IR-800 spectrometer equipped with a nitrogen-cooled IR detector (Hamamatsu G7754-01) and a broadband CaF2 beam splitter.

The FT-IR detection technique has several interesting advantages:

(i) The whole PL spectrum is sent, without the signal attenuating dispersion, to the detector, thus greatly increasing the SNR.
(ii) The lack of second-order (vs. wavelength) signals facilitates correct, automatic measurements of wide-range spectra (no need to insert order-selecting filters).
(iii) High spectral resolution (<1 cm^{-1}) is easily achieved even with the simplest FT-IR spectrometers.
(iv) The low-frequency noise components, especially problematic for extended IR detectors, do not contribute to the FT-IR signal.
(v) The absence of the dispersive element extends the maximal scanning range. However, the absence of the dispersive element also requires much better rejection of the excitation light. Consequently, a tunable laser, rather than a common lamp monochromator tandem, is best suited for PL excitation in the FT-IR PL technique.

One possible issue with the FT-IR technique is that the scanning speed of the moving mirror (one component of the interferometer)[13] can exceed the response time of the detector/amplificator unit (here typically 100 Hz), which might distort the interferogram and its Fourier transform (here the PL signal).

The FT-IR technique has already been applied to detect the PL of CNTs, but it was limited to 1.7 µm by the spectral response of the Ge detector.[14] We have also used an FT-Raman spectroscope to measure the PL emission, but it uses only a single excitation wavelength (at 1,064 nm provided by a cw Nd:yttrium aluminum garnet [YAG] laser, a nitrogen-cooled InGaAs detector, and a JEOL JRS-FT 6500N FT-Raman spectrometer).

Figure 2.4 presents the PL maps of Carbolex/SDBS/gelatin thin films recorded using our FT-IR-type PL-mapping instrument. The similarity in the position and width of the PL peaks between the results obtained using the FT-IR type (Fig. 2.4) and the scanning type (Fig. 2.3) confirms the reproducibility and reliability of these measurement systems.

Next, we shall give two more examples to illustrate that our FT-IR PL instrument is excellent for recording the PL map over a large spectral range. The first example (Fig. 2.5) shows that the DIPS/

SDBS/gelatin thin film is characterized by a large distribution of the chiral index. At a lower wavelength (bottom-left corner) the PL peaks are well resolved, whereas at a long wavelength (top-right corner) the PL peaks are generally broader. The latter feature is intrinsic to the sample and is not due to the spectral resolution of our instrument (which is typically ~16 cm^{-1}). Note the presence of well-resolved, high-order Raman modes on the top-right corner of Fig. 2.5 (the shaded area in the top-right corner of Fig. 2.5 is cut due to the laser rejection filter).

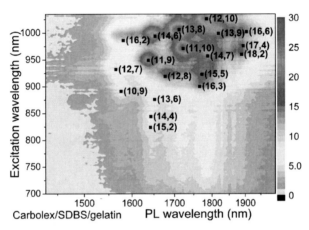

Figure 2.4 PL map of a Carbolex/SDBS/gelatin thin film recorded with an FT-IR-type instrument.

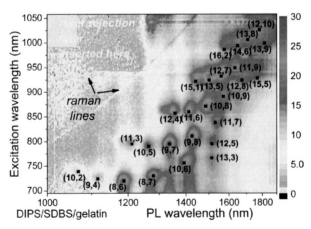

Figure 2.5 PL map of a DIPS/SDBS/gelatin thin film recorded using an FT-IR-type instrument.

The second example (Fig. 2.6) shows that the PL map for a DWNT thin film, prepared from a DWNT/SDBS/gelatin solution, is characterized by an unprecedented "bimodal" distribution. There are strong PL peaks at short (bottom-left corner) and long (top-right corner) excitation and emission wavelengths, but the emission is weak in the intermediate range. These PL peaks are assigned to the inner and outer shells of DWNTs, respectively, using the widely accepted data of Weisman and Bachilo (shifted by a few percent points due to the matrix effect). The unlabeled oval-shaped signals around the top-left corner could be assigned to phonon-assisted excitation features.[15] Observation of PL from both the inner and outer tubes in the DWNTs studied here suggests that interaction between them is weak (for more elaborated discussions see Ref. 16).

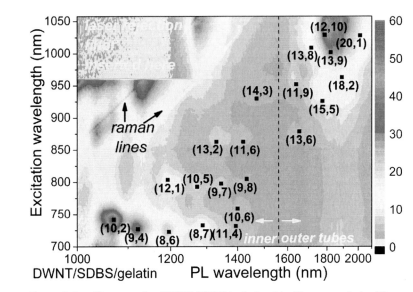

Figure 2.6 PL map of a DWNT/SDBS/gelatin thin film recorded with an FT-IR-type instrument.

2.4 Outlook

Development of novel NIR-PL instruments and methods to prepare thin film allows the PL mapping of semiconducting CNTs over a broad range, typically from 0.8 to 2.3 µm. These techniques have been applied to a wide range of single-wall and double-wall CNTs, unveiling chirality and diameter distributions that have so far escaped elucidation, despite its importance. This achievement opens possibilities for further studies, such as detecting the effect of external perturbations (chemical and electrochemical doping, gas exposure, pressure, etc.) on the inner and outer shells of DWNTs.[16]

Our instruments are still under development. In the future, we need to combine some of the advantages of the scanning-type and FT-IR-type instruments in a same setup. The FT-IR technique is very powerful and versatile to record not only PL but also the Raman and optical absorption with the same instrument. We hope that such instruments could become commercially available in the near future.

Acknowledgments

We thank Dr. T. Okazaki (AIST) for his valuable discussions. K. I. thanks the JSPS Postdoctoral Fellowship Program.

References

1. R. Saito, G. Dresselhaus, M. S. Dresselhaus, eds., *Physical Properties of Carbon Nanotubes*, Singapore: Imperial College Press, 1998.
2. A. Jorio, M. S. Dresselhaus, G. Dresselhaus, eds., *Carbon Nanotubes, Topics in Applied Physics 111*, Berlin, Heidelberg: Springer-Verlag, 2008.
3. T. Saito, W.-C. Xu, S. Ohshima, H. Ago, M. Yumura, S. Iijima, *J. Phys. Chem. B*, **110**, 5849 (2006).
4. H. Kataura et al., *Synth. Met.*, **103**, 2555 (1999).
5. S. M. Bachilo, S. M. Strano, C. Kittrell, R. H. Hauge, R. E. Smalley, R. B. Weisman, *Science*, **298**, 2361 (2002).
6. R. B. Weisman, M. S. Bachilo, *Nano Lett.*, **3**, 1235 (2003).
7. K. Iakoubovskii, N. Minami, S. Kazaoui, T. Ueno, Y. Miyata, K. Yanagi, H. Kataura, S. Ohshima, T. Saito, *J. Phys. Chem. B*, **110**, 17420 (2006).
8. T. R. Schibli, K. Minoshima, H. Kataura, E. Itoga, N. Minami, K. Miyashita, S. Kazaoui, M. Tokumoto, Y. Sakakibara, *Opt. Express*, **13**, 8025 (2005).
9. N. Izard, S. Kazaoui, K. Hata, T. Okazaki, T. Saito, S. Iijima, N. Minami, *Appl. Phys. Lett.*, **92**, 243112 (2008).
10. A. Karthigeyan, N. Minami, K. Iakoubovskii, *Jpn. J. Appl. Phys.*, **47**(9), 7440 (2008).

11. Y. Kim, N. Minami, S. Kazaoui, *Appl. Phys. Lett.*, **86**, 73103 (2005).
12. N. Minami, Y. Kim, K. Miyashita, S. Kazaoui, B. Nalini, *Appl. Phys. Lett.*, **88**, 93123 (2006).
13. P. R. Griffiths, J. De Haseth, *Fourier Transform Infrared* Spectrometry, Wiley-Interscience, 1986.
14. S. Lebedkin, K. Arnold, F. Henrich, R. Krupke, B. Renker, M. M. Kappes, *New. J. Phys.*, **5**, 140 (2003).
15. S. G. Chou, F. Plentz, J. Jiang, R. Saito, D. Nezich, H. B. Ribeiro, A. Jorio, M. A. Pimenta, Ge. G. Samsonidze, A. P. Santos, M. Zheng, G. B. Onoa, E. D. Semke, G. Dresselhaus, M. S. Dresselhaus, *Phys. Rev. Lett.*, **94**, 127402 (2005).
16. K. Iakoubovskii, N. Minami, T. Ueno, S. Kazaoui, H. Kataura, *J. Phys. Chem. C*, **112**, 11194 (2008).

Chapter 3

Developments in Catalytic Methodology for (*n,m*) Selective Synthesis of SWNTs

Yuan Chen, Bo Wang, Yanhui Yang, and Qiang Wang
School of Chemical and Biomedical Engineering, Nanyang Technological University, Singapore 637459, Singapore
chenyuan@ntu.edu.sg

Single-walled (carbon) nanotubes (SWNTs) are hollow carbon cylinders with one atomic-layer thick wall. Many potential applications of SWNTs rely on their unique chiral (*n,m*) structures. SWNT samples with well-defined (*n,m*) species are highly desired. However, current synthesis methods usually produce samples with random (*n,m*) distributions. In this review, we focus on (*n,m*) selective synthesis of SWNTs by highlighting (1) catalysts that are selective toward particular (*n,m*) species, (2) growth parameters that are capable of influencing the (*n,m*) selectivity, (3) current fundamental understandings of the (*n,m*) selectivity from both experimental and theoretical aspects, and (4) characterization methodologies that can evaluate the (*n,m*) abundance of bulk SWNT samples. This review aims to assist future research on (*n,m*) selective synthesis of SWNTs.

Luminescence: The Instrumental Key to the Future of Nanotechnology
Edited by Adam M. Gilmore
Copyright © 2014 Pan Stanford Publishing Pte. Ltd.
ISBN 978-981-4241-95-3 (Hardcover), 978-981-4267-72-4 (eBook)
www.panstanford.com

3.1 Introduction

Single-walled (carbon) nanotubes (SWNTs) are hollow carbon cylinders with one atomic-layer thick wall. They are promising for many potential applications due to their extraordinary electronic, thermal, and mechanical properties.[1] These properties rely on their unique atomic chiral structures. Depending on the direction in which a graphene sheet is rolled into a tube, as illustrated by a chiral index (n,m) in Fig. 3.1, approximately one-third of SWNTs are metallic and two-thirds of SWNTs are semiconducting with different bandgaps.[2]

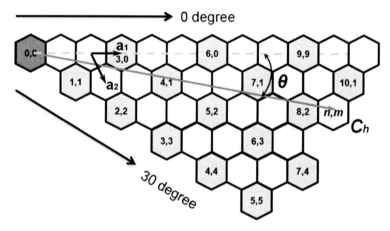

Figure 3.1 The (n,m) structure of an SWNT can be considered as a graphene sheet rolled into a cylinder at a specific direction. Related to an origin point (0,0), each point on the graphene sheet (the center of each hexagon where a carbon atom locates) can be identified by a pair of integers (n,m) ($0 \leq |m| \leq n$). The rolling direction from the origin point toward any other points on the graphene sheet can be defined by a chiral vector, $C_h = n\mathbf{a}_1 + m\mathbf{a}_2$, where ($\mathbf{a}_1,\mathbf{a}_2$) is a pair of lattice unit vectors of the graphene sheet. θ is considered as the chiral angle. Tubes are known as armchair nanotubes when $n = m$ and zigzag nanotubes when $m = 0$. SWNTs can also be categorized into two groups, metallic (tubes satisfying the condition $n - m = 3j$, where j is integer, labeled in grey) and semiconducting (tubes having either $n - m = 3j + 1$ or $3j + 2$, labeled in white).

Common SWNT synthesis methods include arc discharge[3], laser ablation[4], and chemical vapor deposition (CVD).[5] These methods share the same principle: atomic carbon derived from either solid carbon sources (such as graphite rods used in arc discharge and laser ablation methods) or carbon-bearing gases (such as hydrocarbons, CO, and volatile solvents in the CVD method) could organize into thermodynamically stable SWNTs upon interaction with nanoparticles. Current synthesis methods usually produce SWNTs with random (n,m) distributions; therefore, the synthesis selectivity toward a specific (n,m) structure remains very low. To realize many potential SWNT applications, bulk SWNT samples with a well-defined (n,m) structure are desired. This requires us to develop a novel catalytic methodology that can precisely control the (n,m) structure of SWNTs. Some other challenges in the field of SWNT synthesis, such as mass production and organization, can be found in a recent review.[6]

The development of postsynthetic approaches for further (n,m) selective enrichment was summarized by Hersam.[7]

In this review, we focus on the (n,m) selective synthesis of SWNTs. We aim to provide an overview focusing on (1) catalysts that are selective toward particular (n,m) species, (2) growth parameters that are capable of influencing the (n,m) selectivity, (3) current fundamental understandings of the (n,m) selectivity from both experimental and theoretical aspects, and (4) characterization methodologies that can evaluate the (n,m) abundance of bulk SWNT samples.

3.2 Effective Catalysts for (*n,m*) Selective Synthesis

Catalysts play essential roles in providing nanoparticles for SWNT formation. Their roles include (1) assisting carbon precursor decomposition to yield active atomic carbon, (2) serving as seeds to organize carbon atoms into carbon cap structures, (3) continually supplying carbon atoms into growing SWNTs, and (4) terminating SWNT growth when required conditions are met. A large variety of catalyst preparation methods have been explored, such as impregnation, coprecipitation, sol-gel process, metallic thin-film deposition, and many others.[8,9] However, most catalysts have low selectivity to SWNTs over carbonaceous impurities (e.g., multiwalled nanotubes [MWNTs], amorphous carbon, and graphite), to say nothing of selectivity toward a specific (n,m) structure. The difficulty in preparing highly selective catalysts lies in the fact that small nanoparticles (1–2 nm size range) have to be stabilized at high reaction temperatures (600–1,000°C) to initiate the growth of SWNTs. Only in a few rare cases, (n,m) selective catalysts have been reported. Here, we highlight a few selected cases that help illustrate issues that should be considered when designing more effective (n,m) selective catalysts.

The first effective (n,m) selective growth was reported by Resasco et al. using a Co and Mo bimetallic catalyst and CO as feedstock, which results in SWNTs with a narrow (n,m) distribution centered at (6,5) and (7,5) (Fig. 3.2).[10] This catalyst was synthesized by impregnating cobalt nitrate and ammonium heptamolybdate precursors with silica particles (6 nm pore diameter). The total metal loading in the catalyst was 2 wt%, with a Co:Mo molar ratio of 1:3. Before CO exposure, the catalyst was heated to 500°C in a flow of H_2. The CO disproportionation reaction used to produce SWNTs was then run in a fluidized bed reactor under pure CO of 5 atm at 750°C.[10] The success of Co-Mo catalysts comes from a synergistic effect between Co and Mo.[11,12] Resasco et al. explained that without Mo, Co reacts with the silica surface to form Co_3O_4, which converts into large particles, producing a mixture of MWNTs and carbon filaments. When Mo is present, Co^{2+} are well dispersed, which then convert into small Co clusters on the surface of molybdenum oxide, leading to SWNTs with a narrow diameter distribution at 0.74–0.81 nm.[11,12] When the diameter of SWNTs is well confined, so is its (n,m) distribution. Moreover, photoluminescence excitation (PLE) spectroscopy results evidenced that (6,5) and (7,5) accounted for 57% of the semiconducting tubes, presuming that tube abundance is proportional to its PLE intensity.[10]

Followed by Resasco's work, Maruyama et al. showed that a similar narrow (n,m) distribution can be obtained using an Fe/Co catalyst through low-temperature alcohol growth.[13] The Fe/Co catalyst supported on zeolite was prepared by impregnating iron acetate and cobalt acetate on a Y-type zeolite (HSZ-390HUA). The amounts of Fe and Co were 2.5 wt% each. The salt dissolved in ethanol (20 mL) was sonicated with zeolite power (1 g) for 10 minutes and dried for 24 hours at 80°C. The SWNT growth was carried out in 5 Torr ethanol without catalyst pre-reduction. PLE spectroscopy results indicated that a narrower (n,m) distribution was obtained

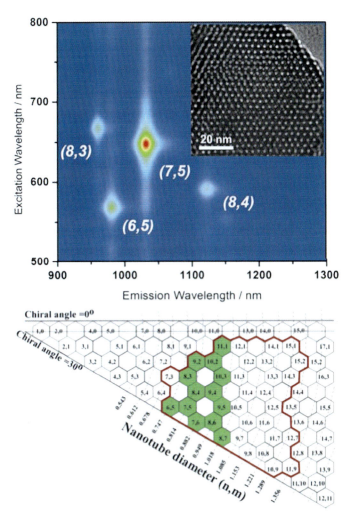

Figure 3.2 SWNTs with a narrow (n,m) distribution obtained on a Co–MCM-41 catalyst. (Top) 2D PLE intensity map as a function of the excitation and emission wavelength for a dispersed Co–MCM-41 SWNT sample in D$_2$O. Inserted TEM image shows the uniform pore structure of Co–MCM-41. (Bottom) Chirality maps of aqueous suspensions of Co–MCM-41 (green hexagons) and HiPco SWNTs (enclosed by a solid red line); chirality maps are reproduced from Ref. 23 with permission from the American Chemical Society. *Abbreviations*: 2D, two dimensional; TEM, transmission electron microscopy; D$_2$O, deuterium oxide; Co–MCM-4, cobalt–mobile crystalline material.

at a growth temperature of 650°C as compared to 750°C and 850°C. Maruyama et al. also found that high selectivity toward the higher-chiral-angle tubes close to armchair tubes existed, and this tendency was more pronounced for smaller-diameter tubes.[13] No detailed characterization on the physical and chemical properties of this Fe/Co catalyst was reported. We speculate that two factors may contribute to the narrow (n,m) distribution obtained in this process: (1) A relatively low growth temperature prevents aggressive aggregation of metal clusters, resulting in small metal clusters with a narrow size distribution. Fe and Co could easily convert into large particles under a higher reaction temperature due to the weak interaction between Fe/Co and zeolite. (2) Alcohol may play special roles during growth. Maruyama et al. proposed that alcohols produce –OH radicals, which may attack nearby carbon atoms with a dangling bond to form CO, eliminating the formation of amorphous carbon.[14] More recently, Amama et al. have proposed that hydroxyl species (such as –OH radicals derived from alcohol) absorbed on the catalyst support surface may reduce diffusion rates of metal atoms, leading to better control of the metal cluster size distribution.[15]

Meanwhile, a new catalyst, Co-incorporated MCM-41 (Co–MCM-41), was developed, and it shows high selectivity toward (7,5) tubes.[16–19] In contract to impregnation methods used in the previous two catalysts, in which metal ions likely stay near the surface of catalyst supports, Co–MCM-41 synthesis adopts a coprecipitation approach, with isolated Co ions staying in the amorphous silica matrix. Two silica sources (fumed silica and tetramethylammonium silicate) were mixed with metal ions (cobalt sulfate) in the presence of a long-alkyl-chain surfactant (trimethylalkylammonium bromide). Silica and Co ions condense together slowly around surfactant micelles over six days at pH = 11.5. After removing organic surfactants by calcination at 540°C, Co–MCM-41 evolves as an amorphous silica matrix (surface area > 1,000 m^2/g) with a uniform mesoporous structure (the full width at half-maximum of the pore size distribution curve is less than 0.2 nm).[20] In-depth in situ X-ray absorption studies on Co–MCM-41 during pre-reduction and SWNT growth evidenced that Co^{1+} intermediate species are produced after pre-reduction under H$_2$ at 500°C.[21] These Co species preserve the tetrahedral environment in the MCM-41 silica framework, and they are resistant to complete reduction to the metal in H$_2$. After exposure to CO, the Co^{1+} species possibly form Co carbonyl-like compounds with high mobility. Increasing Co mobility allows fast migration and formation of small metal clusters capable of dissociating CO and initiating the growth of SWNTs.[22] The efficient control of Co species using an MCM-41 template has produced as-synthesized SWNTs with the narrowest (n,m) distribution reported so far.[23]

Recently, Dai et al. demonstrated that an Fe and Ru bimetallic catalyst can afford SWNT growth with a narrow (n,m) distribution in methane CVD.[24] At 600°C, this catalyst produced predominantly (6,5) tubes. At 850°C, the dominant semiconducting species produced are (8,4), (7,6), and (7,5). The Fe-Ru catalyst was prepared by sonicating a mixture of silica particles with iron nitrate and ruthenium chloride hydrate in methonal for one hour. No detailed catalyst characterization was presented. It is likely that Ru could play a similar role as Mo in the Co-Mo catalyst to stabilize Fe during reduction. A lower growth temperature of 600°C might also help limit the aggregation of Fe clusters.

Despite the noticeable selectivity reported on these catalysts, their applicability in achieving selective synthesis of desired (n,m) SWNTs is still limited. All catalysts show similar selectivity toward higher-chiral-angle tubes, in particular small tubes, such as (6,5) and (7,5). No effective selectivity toward other chiral tubes has been reported so far, especially larger-diameter tubes with smaller bandgaps that would be more promising to construct high-performance transistors.[25] Developing new catalysts, which are capable of precisely controlling small metal clusters during SWNT growth, remains an unconquerable task.

3.3 Growth Parameters Influencing (*n,m*) Selectivity

SWNT synthesis is a sensitive catalytic reaction that involves several physical and chemical processes affecting the formation of SWNTs: carbon precursor decomposition, nanocluster nucleation, initial carbon network formation, continuous growth of tube structure, and termination of tube growth. In particular, current experimental results suggest that a narrow SWNT growth condition window may exist, in which the right balance is necessary among several processes, especially between carbon supply and nanocluster nucleation.[18,19] By varying growth conditions, these dynamic balances can be controlled to some extent, leading to the production of SWNTs with different (n,m) distributions. Here, we summarize several growth parameters that have been reported to influence (n,m) selectivity.

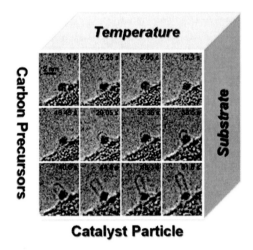

Figure 3.3 Growth parameters influencing (n,m) selectivity. TEM images are reproduced from Ref. 68 with permission from the American Chemical Society.

3.3.1 Temperature

Temperature is one of the predominate parameters that affect almost every aspects of SWNT growth. Higher temperature usually leads to a faster carbon precursor decomposition rate, more aggressive nanocluster nucleation, and less defects in SWNTs. In general, fast nanocluster nucleation under a high growth temperature would result in SWNTs with a broad diameter distribution having a much wider (n,m) distribution. SWNTs with a narrow (n,m) distribution have been produced at temperatures around 600–650°C in alcohol growth on a Co/Fe catalyst[13] and CH_4 growth on an Fe-Ru catalyst.[24] The growth temperature window is shifted up to 700–800°C with good selectivity toward (6,5) and (7,5) tubes in CO decomposition on a Co-Mo catalyst and Co–MCM-41. This temperature up-shift may be attributed to better control of Co nanoparticles in these two catalysts.[10,18] Even so, when the growth temperature is further increased to 850–900°C, a narrow (n,m) distribution disappears owing to the increase of the metal particle size.[18,26]

Generally, production of SWNTs with a narrow (n,m) distribution at a larger diameter (e.g., $d > 1.2$ nm) would require new catalysts that can stabilize larger nanoparticles at a high reaction temperature. These catalysts currently are still not available.

An intriguing case was reported by Zhang et al.[27] The temperature of an isolated individual metal particle was increased from 900°C to 950°C during growth. The diameters of one individual tube grown from this heated metal particle decreased from 1.81 nm of (14,12) to 1.73 nm of (20,3). Even though the (n,m) structure of the growing SWNT cannot be predetermined, the (n,m) can be changed by varying temperature. It is proposed that temperature variation may change the carbon solubility in the catalyst particle and the interfacial energy between the catalyst and the SWNT.[27] Higher temperature would favor the formation of smaller SWNTs with higher stain energy in the wall of the SWNT.[27] On the basis of this finding, we speculate that if a large number of individual nanoparticles can be stabilized on a surface and the nanoparticle aggregation can be eliminated, it may be possible to vary (n,m) selection easily through changing the reaction temperature.

A high growth temperature is required because carbon decomposition yielding carbon atoms for SWNT formation is more efficient at high temperature. Nevertheless, some methods have been utilized to facilitate carbon decomposition at lower temperatures. Plasma-enhanced CVD can enable SWNT growth at 600°C in 0.5 Torr CH_4 over ferritin or 0.1 nm Fe films deposited on SiO_2/Si substrates.[28] Moreover, on the basis of transport measurement on individual tube field-effect transistors (FETs), small tubes ($d \approx 1.1$ nm) show much higher semiconducting percentages than a random distribution.[28] Using density functional theory calculations, Dai et al. proposed that selectivity toward semiconducting tubes is higher because semiconducting tubes have lower formation energies compared to those metallic tubes, especially for small-diameter tubes.[29] However, no (n,m) distribution analysis was presented to evaluate any possible (n,m) selectivity.

3.3.2 Catalyst Particle

Besides temperature, the catalyst particle and carbon precursor are other two key parameters in SWNT formation. There are various factors influencing the catalyst particle and carbon precursor. For the catalyst particle, the types of elements are the first factor to be considered. Fe, Co, and Ni are the common effective metals for hydrocarbon decomposition, and so are they for SWNT growth. Co and Fe were reported to be (n,m) selective.[10,24] However, Ni was found to have a very strong affinity with carbon precursors such as CO, which impedes the control of Ni particle aggregation, resulting in SWNTs having a wide diameter distribution.[30]

Recently, SWNT growth has been found possible on various metals (e.g., Au, Ag, Cu, Pt, Pd, Mn, Mo, Cr, Sn, Mg, and Al), even semiconductor particles (e.g., Si and SiC), insulator particles (SiO_2), and diamond.[24,31–43] Although no detailed mechanism is yet available to explain the growth mechanism on these new catalyst particles, it is feasible that initial carbon precursor decomposition may happen on many materials when the temperature is sufficiently high, even in the gas phase without any catalyst. Atomic carbon species could organize on various small nanoparticles (both metal and nonmetal) to form SWNTs, if the nanoparticles have a suitable size. However, there is a big difference between "being able to grow SWNTs" and "effectively catalyzing SWNT growth," not even to mention that no specific (n,m) selectivity has been reported on any catalysts other than Co and Fe. The applicability of a new catalyst for (n,m) selective growth is still waiting to be explored.

The subtract/support is the next factor influencing the catalyst particle. For Co and Fe metal nanoparticles, catalyst support may perturb their (n,m) selectivity. Using an absorption spectrum-fitting method, Resasco et al. showed that the domain size of small metal oxides can be changed because of the different interactions between supports (MgO vs. SiO_2), leading to the growth of more small-chiral-angle tubes (8,4) on MgO supports.[26] We investigated four oxide catalyst supports—SiO_2, Al_2O_3, MgO, and TiO_2—for the feasibility of (n,m) selection control.[44] SiO_2 and MgO provide well-dispersed Co species on Mo oxides, which resulted in Co clusters in a narrow diameter range for SWNT growth. However, PLE spectroscopy analysis showed that plenty of small-chiral-angle tubes (8,3) also grow from a catalyst supported on SiO_2. The overlap of absorption peaks of (6,5) and (8,3) tubes may have complicated the absorption spectrum-fitting method. No specific selectivity toward small-chiral-angle tubes was evidenced.[44]

Growing SWNTs on flat crystal substrates, such as quartz or sapphire, can effectively control the alignment and position of

SWNTs.[45-48] It may also be possible to use the epitaxial interaction between the metal catalyst and a single-crystal substrate to influence the (n,m) selection. Ago et al. reported that aligned SWNTs grown on the A- and R-planes of sapphire using Co-Mo catalysts have narrower diameter distributions.[49] PLE measurements from as-grown SWNTs lying on a substrate demonstrated that near-zigzag SWNTs were produced on the A-plane and near- armchair SWNTs were grown on the R-plane. Ago et al. suggested that dissimilar atomic structures of the A- and R-plane sapphire surfaces induce differences in the particle morphology and orientation, thereby affecting the SWNT chirality.[49] Moreover, the epitaxial relationship due to the matching of the lattice constants of sapphire and specific SWNTs could also contribute to the crystal plane-dependent growth.[49] Another sample of selective growth on flat crystal substrates was reported by Liu et al., showing that the Y-cut quartz lattice on the surface may be a key factor in obtaining high-purity semiconducting SWNTs.[50]

3.3.3 Carbon Precursor

For the carbon precursor, its chemistry and pressure effect have showed strong impacts on the (n,m) selection in SWNT growth. Resasco et al. showed that CH_4 is less (n,m) selective than CO at 800°C on a Co-Mo catalyst because H_2, as a by-product of CH_4, can increase the rate of reduction and aggregation of Co clusters, as well as hinder the nucleation of carbon species on the surface of the cluster by decreasing the carbon surface fugacity.[26] Liu et al. reported a highly selective growth of semiconducting tubes using a mixture of methanol and ethanol.[50] They emphasized that the –OH radical from methanol can selectively etch metallic SWNTs because of their smaller ionization potential as compared to semiconducting ones.[50] We have compared the (n,m) selectivity of four representative carbon precursors—CO, ethanol, methanol, and C_2H_2 on a Co-Mo catalyst.[51] Narrowly (n,m) distributed SWNTs can only be obtained under high-pressure CO or vacuumed ethanol and methanol. Different carbon precursors have different decomposition rates, resulting in different carbon-feeding rates to Co surfaces. We proposed that because the Co clusters were continuously agglomerating owing to high-temperature sintering, other than just one (n,m) tube with global minimum stability energy on metal cluster surfaces, several local minima could coexist with small difference in stability. When different carbon-feeding rates match with these local minimum stabilities, SWNTs enriched with various (n,m) species can be produced. This was further verified by the selective growth of bulk SWNT samples enriched with three different dominant chiralities—(6,5), (7,5), and (7,6)—through adjusting the pressure of CO on Co-Mo catalysts from 2 to 18 bar.[52] SWNT growth may be initiated faster under higher CO pressure due to the higher concentration of active carbon atoms available, which leads to smaller-diameter tubes like (6,5) and (8,3).

3.3.4 "Clone" SWNTs

Other than directly (n,m) selective growth, an innovative approach is to "clone" SWNTs. Tour et al. showed that Fe particles can be anchored on the end of each SWNTs and C_2H_4 can be used as a carbon source to further increase the length of the original SWNTs.[53] The protocol involves cutting a single (n,m)-type tube into many short nanotubes, using each of those short nanotubes as a template for growing longer nanotubes of the same type, and then repeating the process.[53] It remains to be proven whether the new-grown tubes would have the same (n,m) type as the original one. A challenge for cloning from a metal particle is that the same metal particle may grow tubes with different (n,m) types, as evidenced in the temperature variation experiment.[27] A new development on this approach is to grow SWNTs via open-end growth without metal catalysts. Zhang et al. showed that short SWNTs after removal of –COOH and –OH functional groups can grow under a CH_4 and C_2H_4 mixture at 975°C, and 9% of tubes can grow longer on a SiO_2/Si substrate, while 40% tubes can grow longer on an ST-cut quartz substrate.[54]

3.4 Fundamental Understanding of (n,m) Selectivity

In short, there is still no reliable mechanism available to provide a clear path to achieve (n,m) selectivity in SWNT growth, although large amounts of efforts have been devoted on this topic. Here, we provide a general description of the current understandings based on our selection of references.

As illustrated in Fig. 3.4, various proposed mechanisms may be classified according to the different interaction scenarios among three basic components in SWNT growth: carbon, metal cluster, and substrate. According to the interaction strength between carbon and the metal cluster, one mechanism is the vapor-liquid-solid (VLS) growth mechanism, which originated from the growth of carbon fibers.[55] It considers that carbon atoms (i.e., from a hydrocarbon gas) would dissolve in liquid metal particles. When carbon concentration in liquid particles is oversaturated, carbon excess precipitates in the form of nanotubes. One question often raised to challenge the VLS growth mechanism is whether metal nanoparticles are really liquid at relative low SWNT growth temperatures. The Gibbs–Thompson model predicts that the melting temperature of the metal cluster decreases with decreasing size.[56] A few recent studies showed that the melting temperature of commonly used metals, such as Ni, Fe, and Co, in the 1.1–1.6 nm diameter range is in the 600–800°C range. Furthermore, the melting temperature of metals as well as the solubility of carbon can be influenced by carbon concentration and the substrate.[57–60] All these results strongly support the feasibility of the VLS growth mechanism. On the other hand, if metal clusters do remain as solid crystalline particles, a yarmulke mechanism[5] suggests that carbon atoms would absorb on the surface of metal particles and adsorbed carbon would agglomerate and form a carbon cap to lower the overall energy of the system. Adsorbed carbon can continually diffuse on the surface of metal particles toward the growing carbon cap. Subsequently, the cap can lift from the solid surface, forming a tube while more carbon is incorporated. Molecular dynamics simulations have demonstrated that such a mechanism is also feasible.[61–65] Between the above two mechanisms, many experimental and theoretical studies point to the possibility of fluctuating metal particles or metal carbide crystals. In such a scenario, nanoparticles are neither purely liquid nor crystalline. Carbon atoms can be incorporated into nanoclusters and cause dynamic reconstruction of nanoclusters.[64,66,67] Furthermore, considering the different interactions between the substrate and metal clusters, both tip growth (a metal cluster would detach from the substrate and sit on the tip of a growing carbon nanotube) and root growth (nanoparticles are anchored on the substrate while growing SWNTs extending out) could exist. As illustrated in Fig. 3.4, the growth of SWNTs from nonmetal clusters is also feasible, and recent experimental results have confirmed SWNT growth from SiC and Si.[39,40,42]

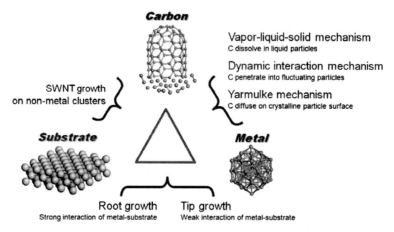

Figure 3.4 Different growth mechanisms and their correlations with interactions among carbon, metal clusters, and substrates.

SWNTs were first identified in TEM.[3,68] In situ TEM has remained the most powerful tool to reveal the details of SWNT growth.[66,67,69–74] However, as listed in Table 3.1, observations from various studies support different growth hypotheses discussed previously. The diversity in TEM results may originate from different reaction conditions employed. For instance, when the reaction temperature is relatively low, nanoparticles are more likely presented as a crystalline structure. The difference in metal adhesion energies among Ni, Co, and Fe may influence the possibility of carbide formation and carbon diffusion inside nanoparticles. In all TEM studies, the concentrations of atomic carbons on the nanoparticle surface are very low due to the low carbon precursor pressure, and this could bring out different diffusion pathways compared to realistic SWNT growth conditions. The SWNT growth mechanism could be differed when growth conditions are varied. It is possible that different proposed mechanisms can only be applied under a particular set of growth conditions.

Table 3.1 In situ TEM observations of carbon nanotube growth

Ref.	Metal/Size (nm)	Support	Carbon source	Pressure (mbar)	Temperature (°C)	Carbon nanostructure	Metal cluster structure	Metal–carbon interaction	Growth mode/rate (nm/s)
69	Ni/5–20	MgAl$_2$O$_4$	CH$_4$	2	500–540	Carbon fibers	Crystalline reshaping	Surface diffusion of C on Ni step edges	Tip/N.A.
70	Ni/1.5–3	SiO$_2$	C$_2$H$_2$	Up to 0.13	480	MWNTs or SWNTs	Melting recrystalline	C cap matching with the facets on Ni particles	Root/6–9
71	Ni/<6	MgO	C$_2$H$_2$	5.3 × 10^{-6}	650	SWNTs	Crystalline	Surface diffusion of C	Root/ 0.011–0.31
72	Fe, Co, FeCo, or Ni	Inside MWNTs	Inner wall of MWNTs	High vacuum	600 under e-beam	SWNTs MWNTs	Crystalline with carbide domains	Bulk diffusion of C, tube attached to metal	Root/84

(Continued)

Fundamental Understanding of (n,m) Selectivity | 49

Table 3.1 (Continued)

Ref.	Metal/Size (nm)	Support	Carbon source	Pressure (mbar)	Temperature (°C)	Carbon nanostructure	Metal cluster structure	Metal-carbon interaction	Growth mode/rate (nm/s)
66	Ni, Fe 5	SiO_2	C_2H_2	10	615	SWNTs MWNTs	Fluctuating crystalline	Lift-off of carbon caps, dynamic reshaping of metal	Root/N.A.
73	Ni 7–30	MgO	C_2H_2	5.3×10^{-6}	650	MWNTs	Crystalline	Surface diffusion of C	Foot and tip/0.34
67	Fe	SiO_2	C_2H_2	0.1	600	SWNTs MWNTs	Fe carbide, fluctuating	Bulk diffusion of C in Fe carbide	Root/N.A.
74	N.A.	Inside MWNTs	C from another tube	High vacuum	527–1527 under e-beam	SWNTs	N.A.	Directly incorporate into closed cap	N.A./0.36

Lastly, we would like to highlight several theoretical mechanism studies related to (n,m) selectivity of SWNTs. The early explanations of the (n,m) selectivity are based on the carbon cap structure stability and the heat of formation energies.[13,29] Balbuena et al. investigated the (n,m)-dependent aromaticity and reactivity of different carbon caps.[75,76]

Although small differences are observed among (n,m) species, considering the involvement of the metal cluster and the substrate in most of the SWNT growth processes, such mechanisms are oversimplified. Furthermore, considering the influence of the metal surface, Reich et al. proposed that particular (n,m) caps could be favored by their epitaxial relationship to the solid catalyst surface.[77,78] This mechanism implies the possibility of (n,m) selective control through matching an SWNT cap with a specific local crystalline lattice on metal surfaces. Balbuena et al. provided further studies on the interaction between carbon caps and small metal clusters, indicating that the interaction strengths are different among (n,m) species.[79–81] Reich et al. also claimed that once the (n,m) selectivity is determined during the cap nucleation stage, it grows into a unique nanotube.[77,78] However, this argument may not be correct, because studies have shown that the chirality of SWNTs can further change during the growth.[82,83] On the other hand, Ding et al. noticed that the strong adhesion between SWNTs and catalyst particles is necessary to support nanotube growth and the dissociation energy may be (n,m) dependent.[84] This finding implies that alloying metals with different adhesion strengths to SWNTs could allow (n,m)-selective growth.[84] From the viewpoint of the SWNT growth rate, Ding et al. proposed that the SWNT growth rate is proportional to the chiral angle of the tube.[85] (n,m) species with a faster growth rate result in tubes with a longer length. Through sonication cutting during the sample preparation for spectroscopic characterizations, longer tubes convert into more short fragments. The chirality selectivity observed in SWNT spectroscopic characterizations may be due to the length differences of different (n,m) species other than the real (n,m) selection.[85]

3.5 Characterization Methodology for (n,m) Abundance Evaluation

Many efforts have been devoted to obtain monodisperse SWNTs either by synthesis or by separation; all these efforts have to be based on an accurate (*n,m*) abundance evaluation methodology. However, accurately determining (*n,m*) abundance in a given sample is still a challenging task. Various topics related to SWNT characterization have been discussed in other chapters of this book. Here, we only highlight a few issues that we believe should be considered for a reliable (*n,m*) abundance evaluation.

Early experimental characterization of the (*n,m*) structure was achieved by scanning tunneling microscopy (STM).[31] TEM, electron diffraction, and field emission have been used to determine the atomic structure of individual SWNTs.[86–91] Although STM and TEM are able to determine the structure of individual tubes, the ultrahigh vacuum condition and the limited samples that can be explored prevent them being a common method for determining the (*n,m*) abundance of bulk samples. Recently, the (*n,m*) distribution of enriched SWNT dispersions was evaluated on the basis of the statistical analysis by TEM for about 40 individual tubes in each sample.[92] However, most of the SWNTs in samples are found to form bundles on the TEM specimen grids; the (*n,m*) distribution determined on a small number of individual tubes may not exactly agree with the value for bundles.[92] This TEM-based methodology needs further verification.

Resonance Raman spectroscopy (RRS)[93], ultraviolet-visible-near-infrared (UV-vis-NIR) absorption spectroscopy[94], and PLE spectroscopy[95] are currently available techniques that can be used for nondestructive (*n,m*) abundance evaluation. RRS has the advantages of being applicable to both individual tubes and tubes in bundles, detecting metallic tubes, and measuring either solid or solution samples, through the radial breathing mode (RBM) vibration of SWNTs. However, due to the strong resonance effect[96], performing a complete (*n,m*) distribution analysis would require continuous laser excitations to march the resonance energies with every (*n,m*) species. Such continuous laser excitation systems are only available in few research groups in the world, which further limit their applicability. UV-vis-NIR absorption spectroscopy measures the optical absorption of SWNTs. Both semiconducting and metallic SWNTs can be detected. This instrument is widely available. However, the deconvolution of absorption spectra of a bulk SWNT sample into distinct (*n,m*) contributions is very difficult because electron transitions of (*n,m*) tubes with different diameters highly overlap, especially for samples having a broad (*n,m*) distribution. In contrast, PLE spectroscopy provides a quick identification of the majority of semiconducting SWNTs. No spectrum deconvolution is needed, if a 2D intensity map (as shown in Fig. 3.2) is employed. The abundance of each (n,m) species can also be easily related to their PLE intensity. Attributed to these advantages, PLE spectroscopy plays a crucial role in the development of (n,m) selective growth methodologies. The first efficient (n,m) selective growth was verified through PLE spectroscopy.[10]

Most of the studies on (n,m) selective synthesis and enrichment utilize PLE spectroscopy as the primary method to evaluate the (n,m) selectivity. However, some drawbacks exist: no metallic tubes can be detected, and the current PLE detectors are insensitive to the emissions from large-diameter SWNTs. Moreover, good-quality spectra can only be obtained in both PLE spectroscopy and UV-vis-NIR absorption spectroscopy when SWNTs are dispersed individually or in very small bundles.[95] This requirement sometimes complicates the (n,m) abundance evaluation because spectroscopic results obtained after SWNT purification and dispersion may not be the same as those of as-synthesized samples.

These three spectroscopic techniques have the same issue concerning the abundance determination. Measured optical signal intensity I^{exp} can be simplified as $I^{exp} = \eta \times A \times I^{cal} \times I^{exc}$, where η is the detection efficiency of the experimental system, I^{cal} is the intrinsic photophysical parameter for a particular SWNT (n,m) structure, A is the abundance of the particular SWNT, and I^{exc} is the excitation intensity.[97] The intrinsic photophysical parameter I^{cal} can be considered as the Raman cross section for RRS, the quantum efficiency for PLE spectroscopy, and the absorption extinction coefficient for UV-vis-NIR absorption spectroscopy, respectively. Ideally, if accurate intrinsic intensity (I^{cal}) is available, the population of individual (n,m) species in a mixture can be obtained by the equation $A(n,m) = I^{exp}(n,m)/[\eta \times I^{cal}(n,m) \times I^{exc}]$. Several theoretical models have emerged to predict the intrinsic photophysical parameter I^{cal} of different (n,m) species.[98-101] These models have been applied in various (n,m) abundance evaluations.[23,51] However, their predictions have not been validated by experimental measurements because of the lack of (n,m)-resolved samples (where the abundance $A(n,m)$ is known). Some experimental results have indicated a large deviation from theoretical predictions.[92,97]

On the other hand, evaluating SWNT samples by the electronic performance of FETs composed of individual, a small bundle, or a network of SWNTs becomes popular.[25,102] This popularity relies on the fact that FETs are basic components of many potential SWNT applications and that FET performance parameters (e.g., I_{on}/I_{off} ratio and mobility) seem to be more sensitive than spectroscopic techniques in detecting the content of metallic tubes in a sample.[103]

Nevertheless, researchers should be very cautious when comparing device performance data from different research groups to draw conclusions on the metallicity purity of SWNT samples. This is because FET performance parameters are highly depending on device structures, such as the metal of electrodes, the device channel width, the channel length, the thickness of oxide layers, the contacting mode between SWNTs and electrodes (top gate or bottom gate), the annealing treatment history of devices, the number of SWNTs in the device channel, contact resistance among SWNTs, and the measurement conditions.

Overall, development of rational characterization methodologies that can assist the convergence of both theoretical models and experimental sample preparation toward the accurate determination of (n,m) abundance is highly desired.

3.6 Conclusions and Outlook

This review has focused on the (n,m) selective synthesis of SWNTs. Several catalysts, including Co-Mo, Co/Fe, Fe-Ru, and Co–MCM-41, have demonstrated similar selectivity toward higher-chiral-angle tubes, especially small-diameter tubes, such as (6,5) and (7,5). Various growth parameters, such as temperature, the metal, the substrate, the chemical structure of the carbon precursor, and pressure, can influence (n,m) selectivity to some extent. However, the overall (n,m) selectivity in synthesis strategies is still limited. The challenge relies on the fact that SWNT growth involves nanoscale dynamic interactions among carbon, the metal cluster, and the substrate at high temperature. Although many aspects of SWNT growth have been investigated from both theoretical and experimental approaches, the growth mechanism is still poorly understood. A number of characterization techniques have been applied in (n,m) abundance evaluation. Among them, PLE spectroscopy plays a crucial role in developing (n,m) selective growth methods. However, a reliable and accurate (n,m) abundance evaluation methodology is still elusive.

SWNT (n,m)-selective growth presents significant difficulties to current experimental and theoretical techniques.

Looking forward, SWNTs are one of the most promising nanoscale semiconductor materials that have the potential to bring out a nanotechnology revolution. (n,m) selective synthesis holds the key to enable many spectacular applications of SWNTs. The ultimate goal should be large-scale, economical synthesis of any desired (n,m) SWNT structures at preferred substrates with suitable density, length, and orientation. Breakthroughs in the following areas would bring us much closer to this ultimate goal: (1) novel catalysts that have better control of nanoclusters, in terms of uniform size, shape, and location on substrates at SWNT growth temperature; (2) a more realistic theoretical mechanism that is able to elucidate the complex dynamic interactions among carbon, metal clusters, and substrates; and (3) a better characterization methodology that can accurately determine the abundance of different (n,m) species in bulk samples (samples can be either individual tubes or tube bundles and either liquid dispersion or solid powers).

Acknowledgments

This work was supported by the Defense Science & Technology Agency, (MINDEF-NTUJPP/08/03), the Environment & Water Industry Development Council (0802-IRIS-12), and the National Research Foundation, Singapore (NRFCRP2-2007-02).

References

1. M. Endo, M. S. Strano, P. M. Ajayan, "Potential applications of carbon nanotubes," in J. Ado, G. Dresselhaus, M. S. Dresselhaus, eds., *Carbon Nanotubes Advanced Topics in the Synthesis, Structure, Properties and Applications*, Vol. 111, Berlin, Heidelberg, New York: Springer-Verlag, 2008, 13.
2. M. S. Dresselhaus, G. Dresselhaus, P. Avouris, eds., *Carbon Nanotubes Synthesis, Structure, Properties, and Applications*, Berlin: Springer, 2001, 447.
3. S. Iijima, T. Ichihashi, *Nature*, **6430**, 603 (1993).
4. T. Guo, P. Nikolaev, A. Thess, D. T. Colbert, R. E. Smalley, *Chem. Phys. Lett.*, **1–2**, 49 (1995).

5. H. J. Dai, A. G. Rinzler, P. Nikolaev, A. Thess, D. T. Colbert, R. E. Smalley, *Chem. Phys. Lett.*, **3–4**, 471 (1996).
6. E. Joselevich, H. Dai, J. Liu, K. Hata, A. H. Windle, "Carbon nanotube synthesis and organization," in J. Ado, G. Dresselhaus, M. S. Dresselhaus, eds., *Carbon Nanotubes Advanced Topics in the Synthesis, Structure, Properties and Applications*, Vol. 111, Berlin, Heidelberg, New York: Springer-Verlag, 2008, 101.
7. M. C. Hersam, *Nat. Nanotechnol.*, **7**, 387 (2008).
8. A. C. Dupuis, *Prog. Mater. Sci.*, **8**, 929 (2005).
9. E. Lamouroux, P. Serp, P. Kalck, *Cat. Rev.—Sci. Eng.*, **3**, 341 (2007).
10. S. M. Bachilo, L. Balzano, J. E. Herrera, F. Pompeo, D. E. Resasco, R. B. Weisman, *J. Am. Chem. Soc.*, **37**, 11186 (2003).
11. D. E. Resasco, W. E. Alvarez, F. Pompeo, L. Balzano, J. E. Herrera, B. Kitiyanan, A. Borgna, *J. Nanopart. Res.*, **1–2**, 131 (2002).
12. W. E. Alvarez, F. Pompeo, J. E. Herrera, L. Balzano, D. E. Resasco, *Chem. Mater.*, **4**, 1853 (2002).
13. Y. Miyauchi, S. Chiashi, Y. Murakami, Y. Hayashida, S. Maruyama, *Chem. Phys. Lett.*, **1–3**, 198 (2004).
14. S. Maruyama, R. Kojima, Y. Miyauchi, S. Chiashi, M. Kohno, *Chem. Phys. Lett.*, **3–4**, 229 (2002).
15. P. B. Amama, C. L. Pint, L. McJilton, S. M. Kim, E. A. Stach, P. T. Murray, R. H. Hauge, B. Maruyama, *Nano Lett.*, **1**, 44 (2009).
16. S. Lim, D. Ciuparu, C. Pak, F. Dobek, Y. Chen, D. Harding, L. Pfefferle, G. Haller, *J. Phys. Chem. B*, **40**, 11048 (2003).
17. D. Ciuparu, Y. Chen, S. Lim, G. L. Haller, L. Pfefferle, *J. Phys. Chem. B*, **2**, 503 (2004).
18. Y. Chen, D. Ciuparu, S. Lim, Y. Yang, G. L. Haller, L. Pfefferle, *J. Catal.*, **2**, 453 (2004).
19. Y. Chen, D. Ciuparu, S. Lim, Y. Yang, G. L. Haller, L. Pfefferle, *J. Catal.*, **2**, 351 (2004).
20. L. Wei, B. Wang, Q. Wang, L. J. Li, Y. H. Yang, Y. Chen, *J. Phys. Chem. C*, **45**, 17567 (2008).
21. D. Ciuparu, P. Haider, M. Fernandez-Garcia, Y. Chen, S. Lim, G. L. Haller, L. Pfefferle, *J. Phys. Chem. B*, **34**, 16332 (2005).
22. D. Ciuparu, Y. Chen, S. Lim, Y. Yang, G. L. Haller, L. Pfefferle, *J. Phys. Chem. B*, **40**, 15565 (2004).
23. Z. T. Luo, L. D. Pfefferle, G. L. Haller, F. Papadimitrakopoulos, *J. Am. Chem. Soc.*, **48**, 15511 (2006).

24. X. Li, X. Tu, S. Zaric, K. Welsher, W. S. Seo, W. Zhao, H. Dai, *J. Am. Chem. Soc.*, **51**, 15770 (2007).
25. L. Zhang, S. Zaric, X. M. Tu, X. R. Wang, W. Zhao, H. J. Dai, *J. Am. Chem. Soc.*, **8**, 2686 (2008).
26. G. Lolli, L. A. Zhang, L. Balzano, N. Sakulchaicharoen, Y. Q. Tan, D. E. Resasco, *J. Phys. Chem. B*, **5**, 2108 (2006).
27. Y. G. Yao, Q. W. Li, J. Zhang, R. Liu, L. Y. Jiao, Y. T. Zhu, Z. F. Liu, *Nat. Mater.*, **4**, 283 (2007).
28. Y. M. Li, D. Mann, M. Rolandi, W. Kim, A. Ural, S. Hung, A. Javey, J. Cao, D. W. Wang, E. Yenilmez, Q. Wang, J. F. Gibbons, Y. Nishi, H. J. Dai, *Nano Lett.*, **2**, 317 (2004).
29. Y. M. Li, S. Peng, D. Mann, J. Cao, R. Tu, K. J. Cho, H. J. Dai, *J. Phys. Chem. B*, **15**, 6968 (2005).
30. Y. Chen, D. Ciuparu, Y. Yang, S. Lim, C. Wang, G. L. Haller, L. D. Pfefferle, *Nanotechnology*, **7**, 476 (2005).
31. T. W. Odom, J. L. Huang, P. Kim, C. M. Lieber, *Nature*, **6662**, 62 (1998).
32. D. Takagi, Y. Homma, H. Hibino, S. Suzuki, Y. Kobayashi, *Nano Lett.*, **12**, 2642 (2006).
33. D. Takagi, A. Yamazaki, Y. Otsuka, H. Yoshimura, Y. Kobayashi, Y. Homma, *Chem. Phys. Lett.*, **4–6**, 213 (2007).
34. S. Bhaviripudi, E. Mile, S. A. Steiner, A. T. Zare, M. S. Dresselhaus, A. M. Belcher, J. Kong, *J. Am. Chem. Soc.*, **6**, 1516 (2007).
35. D. Yuan, L. Ding, H. Chu, Y. Feng, T. P. McNicholas, J. Liu, *Nano Lett.*, **8**, 2576 (2008).
36. K. Kobayashi, R. Kitaura, Y. Kumai, Y. Goto, S. Inagaki, H. Shinohara, *Chem. Phys. Lett.*, **4–6**, 346 (2008).
37. M. Ritschel, A. Leonhardt, D. Elefant, S. Oswald, B. Buchner, *J. Phys. Chem. C*, **24**, 8414 (2007).
38. Y. Zhang, W. Zhou, Z. Jin, L. Ding, Z. Zhang, X. Liang, Y. Li, *Chem. Mater.*, **24**, 7521 (2008).
39. B. Liu, W. Ren, L. Gao, S. Li, Q. Liu, C. Jiang, H.-M. Cheng, *J. Phys. Chem. C*, **49**, 19231 (2008).
40. D. Takagi, H. Hibino, S. Suzuki, Y. Kobayashi, Y. Homma, *Nano Lett.*, **8**, 2272 (2007).
41. D. Takagi, Y. Kobayashi, Y. Homma, *J. Am. Chem. Soc.*, **20**, 6922 (2009).
42. B. Liu, W. Ren, L. Gao, S. Li, S. Pei, C. Liu, C. Jiang, H.-M. Cheng, *J. Am. Chem. Soc.*, **6**, 2082 (2009).

43. T.-Y. Yang, W.-C. Yang, T.-C. Tseng, C.-M. Tsai, T.-R. Yew, *Appl. Phys. Lett.*, **22**, 223103 (2007).
44. B. Wang, Y. Yang, L. J. Li, Y. Chen, *J. Mater. Sci.*, **12**, 3285 (2009).
45. A. Ismach, L. Segev, E. Wachtel, E. Joselevich, *Angew. Chem., Int. Ed.*, **45**, 6140 (2004).
46. S. Han, X. L. Liu, C. W. Zhou, *J. Am. Chem. Soc.*, **15**, 5294 (2005).
47. C. Kocabas, S. H. Hur, A. Gaur, M. A. Meitl, M. Shim, J. A. Rogers, *Small*, **11**, 1110 (2005).
48. L. Ding, D. N. Yuan, J. Liu, *J. Am. Chem. Soc.*, **16**, 5428 (2008).
49. N. Ishigami, H. Ago, K. Imamoto, M. Tsuji, K. Iakoubovskii, N. Minami, *J. Am. Chem. Soc.*, **30**, 9918 (2008).
50. L. Ding, A. Tselev, J. Y. Wang, D. N. Yuan, H. B. Chu, T. P. McNicholas, Y. Li, J. Liu, *Nano Lett.*, **2**, 800 (2009).
51. B. Wang, C. H. P. Poa, L. Wei, L.-J. Li, Y. Yang, Y. Chen, *J. Am. Chem. Soc.*, **29**, 9014 (2007).
52. B. Wang, L. Wei, L. Yao, L.-J. Li, Y. Yang, Y. Chen, *J. Phys. Chem. C*, **40**, 14612 (2007).
53. R. E. Smalley, Y. B. Li, V. C. Moore, B. K. Price, R. Colorado, H. K. Schmidt, R. H. Hauge, A. R. Barron, J. M. Tour, *J. Am. Chem. Soc.*, **49**, 15824 (2006).
54. Y. G. Yao, C. Q. Feng, J. Zhang, Z. F. Liu, *Nano Lett.*, **4**, 1673 (2009).
55. R. T. K. Baker, M. A. Barber, R. J. Waite, P. S. Harris, F. S. Feates, *J. Catal.*, **1**, 51 (1972).
56. H. M. Duan, F. Ding, A. Rosen, A. R. Harutyunyan, S. Curtarolo, K. Bolton, *Chem. Phys.*, **1**, 57 (2007).
57. A. R. Harutyunyan, E. Mora, T. Tokune, K. Bolton, A. Rosen, A. Jiang, N. Awasthi, S. Curtarolo, *Appl. Phys. Lett.*, **16**, 163120 (2007).
58. A. R. Harutyunyan, N. Awasthi, A. Jiang, W. Setyawan, E. Mora, T. Tokune, K. Bolton, S. Curtarolo, *Phys. Rev. Lett.*, **19**, 195502 (2008).
59. A. R. Harutyunyan, O. A. Kuznetsov, C. J. Brooks, E. Mora, G. G. Chen, *ACS Nano*, **2**, 379 (2009).
60. E. Mora, J. M. Pigos, F. Ding, B. I. Yakobson, A. R. Harutyunyan, *J. Am. Chem. Soc.*, **36**, 11840 (2008).
61. J. Gavillet, A. Loiseau, C. Journet, F. Willaime, F. Ducastelle, J. C. Charlier, *Phys. Rev. Lett.*, **27** (2001).
62. Y. Shibuta, S. Maruyama, *Chem. Phys. Lett.*, **3-4**, 381 (2003).
63. X. Fan, R. Buczko, A. A. Puretzky, D. B. Geohegan, J. Y. Howe, S. T. Pantelides, S. J. Pennycook, *Phys. Rev. Lett.*, **14**, 145501 (2003).

References

64. H. Amara, C. Bichara, F. Ducastelle, *Phys. Rev. Lett.*, **5**, 056105 (2008).
65. J. Y. Raty, F. Gygi, G. Galli, *Phys. Rev. Lett.*, **9**, 096103 (2005).
66. S. Hofmann, R. Sharma, C. Ducati, G. Du, C. Mattevi, C. Cepek, M. Cantoro, S. Pisana, A. Parvez, F. Cervantes-Sodi, A. C. Ferrari, R. Dunin-Borkowski, S. Lizzit, L. Petaccia, A. Goldoni, J. Robertson, *Nano Lett.*, **3**, 602 (2007).
67. H. Yoshida, S. Takeda, T. Uchiyama, H. Kohno, Y. Homma, *Nano Lett.*, **7**, 2082 (2008).
68. D. S. Bethune, C. H. Kiang, M. S. Devries, G. Gorman, R. Savoy, J. Vazquez, R. Beyers, *Nature*, **6430**, 605 (1993).
69. S. Helveg, C. Lopez-Cartes, J. Sehested, P. L. Hansen, B. S. Clausen, J. R. Rostrup-Nielsen, F. Abild-Pedersen, J. K. Norskov, *Nature*, **6973**, 426 (2004).
70. R. Sharma, P. Rez, M. M. J. Treacy, S. J. Stuart, *J. Electron Microsc.*, **3**, 231 (2005).
71. M. Lin, J. P. Y. Tan, C. Boothroyd, K. P. Loh, E. S. Tok, Y. L. Foo, *Nano Lett.*, **3**, 449 (2006).
72. J. A. Rodriguez-Manzo, M. Terrones, H. Terrones, H. W. Kroto, L. T. Sun, F. Banhart, *Nat. Nanotechnol.*, **5**, 307 (2007).
73. M. Lin, J. P. Y. Tan, C. Boothroyd, K. P. Loh, E. S. Tok, Y. L. Foo, *Nano Lett.*, **8**, 2234 (2007).
74. C. Jin, K. Suenaga, S. Iijima, *ACS Nano*, **6**, 1275 (2008).
75. J. Zhao, P. B. Balbuena, *J. Phys. Chem. C*, **34**, 13175 (2008).
76. J. Zhao, P. B. Balbuena, *J. Phys. Chem. C*, **10**, 3482 (2008).
77. S. Reich, L. Li, J. Robertson, *Chem. Phys. Lett.*, **4–6**, 469 (2006).
78. S. Reich, L. Li, J. Robertson, *Phys. Rev. B: Condens. Matter*, **16**, 165423 (2005).
79. D. A. Gomez-Gualdron, P. B. Balbuena, *Nanotechnology*, **48**, 485604 (2008).
80. D. A. Gomez-Gualdron, P. B. Balbuena, *Nanotechnology*, **21**, 215601 (2009).
81. D. A. Gomez-Gualdron, P. B. Balbuena, *J. Phys. Chem. C*, **2**, 698 (2009).
82. W. M. Zhu, A. Rosen, K. Bolton, *J. Chem. Phys.*, **12**, 124708 (2008).
83. W. M. Zhu, H. M. Duan, K. Bolton, *J. Nanosci. Nanotechnol.*, **2**, 1222 (2009).
84. F. Ding, P. Larsson, J. A. Larsson, R. Ahuja, H. M. Duan, A. Rosen, K. Bolton, *Nano Lett.*, **2**, 463 (2008).

85. F. Ding, A. R. Harutyunyan, B. I. Yakobson, *Proc. Natl. Acad. Sci. U S A*, **8**, 2506 (2009).
86. L.-C. Qin, *Phys. Chem. Chem. Phys.*, **1**, 31 (2007).
87. M. Kociak, K. Suenaga, K. Hirahara, Y. Saito, T. Nakahira, S. Iijima, *Phys. Rev. Lett.*, **15**, 155501 (2002).
88. K. A. Dean, B. R. Chalamala, *J. Appl. Phys.*, **7**, 3832 (1999).
89. K. A. Dean, O. Groening, O. M. Kuttel, L. Schlapbach, *Appl. Phys. Lett.*, **18**, 2773 (1999).
90. K. A. Dean, B. R. Chalamala, *J. Vac. Sci. Technol. B*, **2**, 868 (2003).
91. M. Khazaei, K. A. Dean, A. A. Farajian, Y. Kawazoe, *J. Phys. Chem. C*, **18**, 6690 (2007).
92. Y. Sato, K. Yanagi, Y. Miyata, K. Suenaga, H. Kataura, S. Lijima, *Nano Lett.*, **10**, 3151 (2008).
93. A. Jorio, R. Saito, J. H. Hafner, C. M. Lieber, M. Hunter, T. McClure, G. Dresselhaus, M. S. Dresselhaus, *Phys. Rev. Lett.*, **6**, 1118 (2001).
94. A. Hagen, T. Hertel, *Nano Lett.*, **3**, 383 (2003).
95. S. M. Bachilo, M. S. Strano, C. Kittrell, R. H. Hauge, R. E. Smalley, R. B. Weisman, *Science*, **5602**, 2361 (2002).
96. H. Kataura, Y. Kumazawa, Y. Maniwa, I. Umezu, S. Suzuki, Y. Ohtsuka, Y. Achiba, *Synth. Met.*, **1–3**, 2555 (1999).
97. D. A. Tsyboulski, J.-D. R. Rocha, S. M. Bachilo, L. Cognet, R. B. Weisman, *Nano Lett.*, **10**, 3080 (2007).
98. V. N. Popov, L. Henrard, P. Lambin, *Nano Lett.*, **9**, 1795 (2004).
99. S. Reich, C. Thomsen, J. Robertson, *Phys. Rev. Lett.*, **7**, 077402 (2005).
100. J. Jiang, R. Saito, A. Gruneis, S. G. Chou, G. S. Ge, A. Jorio, G. Dresselhaus, M. S. Dresselhaus, *Phys. Rev. B: Condens. Matter*, **20**, 205420 (2005).
101. Y. Oyama, R. Saito, K. Sato, J. Jiang, G. G. Samsonidze, A. Gruneis, Y. Miyauchi, S. Maruyama, A. Jorio, G. Dresselhaus, M. S. Dresselhaus, *Carbon*, **5**, 873 (2006).
102. M. S. Arnold, A. A. Green, J. F. Hulvat, S. I. Stupp, M. C. Hersam, *Nat. Nanotechnol.*, **1**, 60 (2006).
103. L. Wei, C. W. Lee, L. J. Li, H. G. Sudibya, B. Wang, L. Q. Chen, P. Chen, Y. H. Yang, M. B. Chan-Park, Y. Chen, *Chem. Mater.*, **24**, 7417 (2008).

Chapter 4

Single-Walled Carbon Nanotube Thin-Film Electronics

Husnu Emrah Unalan[a] and Manish Chhowalla[b]
[a]*Department of Metallurgical and Materials Engineering, Middle East Technical University, Ankara, Turkey*
[b]*Department of Materials, Imperial College London, UK Department of Materials Science and Engineering, Rutgers University, USA*
unalan@metu.edu.tr, manish1@rci.rutgers.edu

Two-dimensional single-walled (carbon) nanotube (SWNT) networks—also referred to as thin films—for large area electronics have been a rapidly growing area of research since 2003. SWNT thin films provide an alternative to the lithographically intensive fabrication process for individual SWNT devices and can be deposited at room temperature over large areas on inexpensive substrates. SWNT thin films display unique and tunable optoelectronic properties and thus can be utilized in applications ranging from electrodes for organic solar cells and light-emitting diodes to thin-film transistors. Here we describe the different fabrication routes for the assembly of SWNT thin films primarily including solution

---

Luminescence: The Instrumental Key to the Future of Nanotechnology
Edited by Adam M. Gilmore
Copyright © 2014 Pan Stanford Publishing Pte. Ltd.
ISBN 978-981-4241-95-3 (Hardcover), 978-981-4267-72-4 (eBook)
www.panstanford.com

deposition approaches—starting from dispersion related issues—and then briefly describing dry transfer strategies. Thereafter, we concentrate on the optoelectronic properties of SWNT thin films and postdeposition strategies for engineering their optoelectronic properties. Finally, the use of SWNT thin films in various devices will be presented in detail.

4.1 Introduction

Carbon nanotubes (CNTs) are graphene sheets rolled up to form seamless cylindrical tubes.[1-3] Their ends can be capped with bisected fullerene molecules, or they can be open. Nanotubes can be tens of micrometers long and are either single walled (one shell) or multiwalled nanotubes (MWNTs) (many shells).[1-3] Double-walled (carbon) nanotubes (DWNTs), which possess characteristics of both SWNTs and MWNTs, have also been isolated and synthesized in large quantities.[4,5] It is now widely known that the electronic properties of SWNTs depend on the arrangement of carbon atoms around the circumference of the tubes. The chirality (orientation of the graphene lattice relative to the tube axis) and diameter of SWNTs determine the band structure and hence the electronic properties. As a result, SWNTs can be metallic (~0 eV bandgap) or semiconducting (0.4–0.7 eV bandgap).[1-3] The bandgap of the semiconducting SWNT is strain induced due to curvature effects, and it decreases with the SWNT diameter. One-third of SWNTs with all possible chiralities are metallic, the other two-thirds being semiconducting. [1-3]

SWNTs can be synthesized using three primary methods: arc discharge[6-8], laser ablation[9-11], and chemical vapor deposition (CVD).[12-15] The vapor-liquid-solid (VLS) growth mechanism leads to the formation of CNTs where diffusion of carbon occurs through a liquid-phase catalyst, followed by precipitation of graphitic filaments. The carbon source for nanotubes is vaporized carbon atoms from a solid target in arc discharge and laser ablation and hydrocarbon gaseous species in CVD. Arc discharge and laser ablation can be classified as high-temperature (>3,000 K near the discharge) and short-reaction-time (μs–ms) processes, whereas CVD is a moderate-temperature (700–1,400 K) and long-reaction-time (typically minutes to hours) process. Arc discharge and laser ablation produce SWNTs as powder samples in bundles, while CVD allows synthesis of SWNTs on substrates[16–19] as well as in powder form.[20,21] Additionally, CVD allows control over the diameter[22,23], length, and orientation[24,25] of nanotubes. One important point to note is that although there are promising efforts on the preferential growth of semiconducting/metallic SWNTs, it is presently not possible to grow all metallic or all semiconducting SWNTs.[23,26–28] Recently, the catalyst-conditioning ambient investigated by Harutyunyan et al. was shown to affect the relative abundance of metallic and semiconducting tubes.[29] Successful separation of SWNTs by diameter and electrical properties to enhance semiconducting/metallic behavior in device applications will be discussed in Section 4.5.

The realization of MWNTs at NEC Laboratories by Sumio Iijima in 1991[30] followed by the discovery of SWNTs in 1993[6,31] started a new era of research in nanotechnology. The unique one-dimensional structure and cylindrical symmetry of SWNTs leads to appealing mechanical and electrical properties, which have received a great deal of attention and investigation. Numerous potential applications have been proposed for CNTs due to their extraordinary characteristics. A list of applications includes but is not limited to conductive and high-strength composites, energy storage devices, sensors, field emission displays, and semiconductor devices such as field-effect transistors (FETs).

Nanoelectronics utilizing CNTs has been considered as the most promising application of nanotechnology. In principle CNTs can be useful for downsizing circuit dimensions and providing a corresponding increase in computational power (Moore's law[32]). Additional effects such as ballistic transport also make them interesting for high-frequency and interconnect applications.[33,34] The unique quantum wire-like properties make them useful in novel devices such as a spin transport medium for spintronics.[35,36] Experiments have shown that metallic SWNTs can carry currents up to 10^9 A/cm^2 (compared to $\sim 10^5$ A/cm^2 for metals), which make them particularly useful in high-power electronic circuits.[37,38] On the other hand, semiconducting SWNTs connected to two metal electrodes can function as FETs. Semiconducting SWNTs can be switched from conducting to insulating state at room temperature by modulating the gate voltage.[39] However, fabrication methods for individual SWNT devices such as transistors are not integrated and sometimes involve crude techniques such as dragging an SWNT over predefined electrodes using an atomic force microscope (AFM) tip[40,41] or drop-casting a dilute SWNT solution onto prepatterned electrodes[39] or directly growing SWNTs between electrodes.[42–44] Although outstanding and record-beating results have been obtained using individual SWNTs, several inherent variabilities in SWNT properties, such as chirality and diameter, make it challenging to fabricate reproducible and uniform devices. In addition, the limited drive current per SWNT makes it almost impossible to match the current levels required for applications such as microwave circuits or display drivers. The practical fabrication and integration challenges along with fundamental limitations such as low drive currents and inadequate density of tubes for interconnects have limited the implementation of SWNTs into electronics. Therefore, despite the intense efforts and substantial investment in CNT research over the last two decades, individual SWNT devices are not likely to be incorporated into mainstream electronics applications in the foreseeable future.

SWNT networks on the other hand offer advantages by alleviating some of the device integration challenges associated with individual SWNT devices.[45] That is, by creating stable SWNT inks, networks of SWNTs can be deposited onto substrates over large areas using solution-based techniques such as inkjet printing, spin coating, spraying, or roll-to-roll printing. Although the electronic properties of the SWNT thin films are substantially lower than those of individual nanotubes, the ease of processing and device fabrication makes them useful for large-area electronics where devices with moderate performance on inexpensive and flexible platforms are required. In addition, SWNT thin films possess unique properties, which have opened new applications. For example, SWNT thin films are transparent and conducting and can be deposited on flexible substrates such as plastics, making them useful as a potential replacement for indium tin oxide (ITO) in organic electronic devices. The recent advancements in postgrowth separation of metallic SWNTs from semiconducting ones using a centrifugal density gradient by Hersam et al.[46–49] has allowed fabrication of high-performance optoelectronic devices from SWNT networks. In this chapter, we review the progress in the field of SWNT thin-film devices. The field is relatively new, with the first report on SWNT network devices by Snow et al.[50] appearing in 2003, but has grown rapidly, in part, due to advancements from Rinzler et al.[51–53], Gruner et al.[54–56], Rogers et al.[57–60], and others.

4.2 Purification and Dispersion of SWNTs

Prior to deposition of thin films, SWNTs must be purified to remove impurities and by-products from the synthesis process and subsequently dispersed to create uniform and stable suspensions of individual nanotubes that are free of bundles. The optoelectronic properties of SWNT thin films are sensitive to the quality of the material. Despite efforts to the contrary, all SWNT growth techniques produce significant amounts of impurities such as amorphous and graphitic carbon and carbon-encapsulated catalytic metal nanoparticles. The presence of such impurities necessitates purification. It is critical to employ purification methods that only remove carbon impurities and metal catalyst particles without damaging the SWNTs. Purification specifically depends on the SWNT growth process, and the following methods are just a few of the many that have been reported to obtain high-purity SWNTs: hydrothermal method[61], gaseous or catalytic oxidation[62,63], nitric acid reflux[64,65], peroxide reflux[66], cross-flow filtration[67], chromatography[68], microwave oxidation[69], and deoxyribonucleic acid (DNA) wrapping.[70]

A common approach for the purification of CVD-grown SWNTs is to use a two-step process starting with strong oxidation followed by acid reflux. The oxidation of SWNTs starts at temperatures above 350°C.[71] However, the catalytic effect of metal catalyst particles lowers the oxidation temperature so that oxidation at 350°C leads to the destruction of SWNTs.[72] Oxidative treatment of raw SWNTs is effective in removing extraneous carbonaceous material and exposing the transition metal catalyst particles. The second step in SWNT purification is acid refluxing. Exposed transition metal particles can be dissolved in acid and removed via filtration or centrifugation. Generally, concentrated nitric (HNO_3) or hot hydrochloric (HCl) acid is used for the refluxing step. However, sidewall damage has been observed in the HNO_3-refluxed SWNT samples and therefore HCl is more commonly used. Transmission electron microscope (TEM) images of the as-received and purified high-pressure carbon monoxide conversion (HiPCO) process SWNTs following the procedure described above are shown in Figs. 4.1a and 4.1b, respectively. Long bundles of SWNTs with higher-contrast catalytic iron particles can be readily observed in Fig. 4.1a. Most of the catalytic particles have been removed through the purification process, as indicated in Fig. 4.1b.

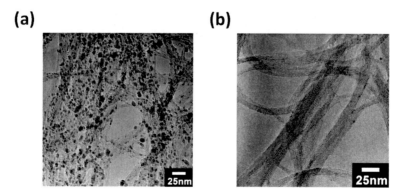

Figure 4.1 TEM images of (a) as-received and (b) purified HiPCO SWNTs.

SWNTs functionalized with carboxyl groups are obtained after the acid treatment. Functionalization imparts new electrical and mechanical properties that are different from those of pristine nanotubes.[73,74] New characteristics arise from the attachment of chemical species to the sidewalls or ends of the nanotubes. Functionalization can be therefore used to tailor the interactions of CNTs with other entities, such as a solvent[75], host matrix[76], or other nanotubes.[77]

The realization of many of the proposed applications of SWNTs is hampered by difficulties encountered in their processing and manipulation. One of the biggest problems is the tendency of SWNTs to form bundles or ropes. This bundling perturbs the electronic structure of the SWNTs and thus complicates attempts to separate them by size or type or to use them as individual macromolecular species. Research has been focused on the dispersion of SWNTs in solvents through oxidation, functionalization, and surfactant or polymer wrapping. Dispersion is particularly problematic since highly polarizable, smooth-sided SWNTs readily form parallel bundles or ropes with van der Waals binding energies of ~500 eV per micrometer of tube-to-tube contact.[9,78] Thus, any successful dispersion process must initially overcome this substantial van der Waals energy. Dispersion of bundles into individual SWNTs is difficult since as-grown SWNTs are insoluble in most solvents.[7] Well-dispersed SWNTs facilitate processing, particularly for thin-film deposition where transport through individual SWNTs is essential.

Two main approaches have been proposed to achieve uniform dispersion via functionalization. First is the noncovalent attachment of functional groups to the sidewalls of the SWNTs. This approach can immobilize functional molecules on the surface of the SWNTs, while maintaining their intrinsic properties. Typical noncovalent functionalization methods include formation of dispersed SWNTs in water by wrapping with various polymers such as polyvinylpyrolidone (PVP), as demonstrated by O'Connell et al.[79] An AFM image of the PVP-dispersed SWNTs on substrates is shown in Fig. 4.2a. The inset in Fig. 4.2a shows the possible wrapping arrangements. Likewise, DNA has been used to disperse SWNTs in water by wrapping the tubes via relatively weak π stacking.[80] An AFM image of the DNA-dispersed SWNTs is shown in Fig. 4.2b with a suggested DNA-binding structure. The image inset also shows picture of the starting SWNT solution and two enriched solutions using DNA wrapping. The most widely used dispersion method is encapsulation by supramolecular systems of small molecules such as surfactant micelles through ultrasonic treatments. Commonly used surfactants are sodium dodecyl sulfate (SDS),[81-84] sodium dodecylbenzene sulfonate (NaDDBS)[82,85], and Triton-X.[51,86,87] Surfactant adsorption at interfaces has been widely studied because of its importance in detergents, lubrication, and colloid stabilization. SWNTs are insoluble in most solvents and especially in water, but they have been reported to form stable suspensions of individual SWNTs when sonicated in a 1 wt% aqueous solution of SDS followed by ultracentrifugation and supernatant removal, as reported by O'Connell et al.[81] Wrapping of the SDS micelles around individual and bundles of SWNTs are shown in Fig. 4.2c, with the corresponding absorption spectra taken in solution. Noncovalent methods are usually used to disperse SWNTs into stable suspensions and do not disrupt the extended π network, causing little or no change to electronic properties of SWNTs. These dispersions, however, require the use of large amounts of surfactants that must be removed for optoelectronic applications. Bundling has been found to rapidly increase with time after sonication is terminated. Therefore, to obtain SWNT thin films with reproducible properties, it is crucial that the state of suspension be carefully monitored.[88]

Covalent functionalization on the other hand includes chemical attachment of functional groups to the sidewalls[89,90] or defect sites.[91,92] This method produces defects on the pristine structure, which can alter the intrinsic properties of the SWNTs. Covalent functionalization of SWNTs improves the interaction between sidewalls and polymers, resulting in improved stress transfer for composite applications.

Some examples of covalent functionalization of the SWNTs include fluorination[74], 1,3-dipolar addition[93], glucosamine attachment[92], and sidewall carboxylic acid attachment.[94] Alternative approaches for dispersion of SWNTs via shortening of SWNTs down to 100–300 nm by vigorous acid attack have also been explored.[95]

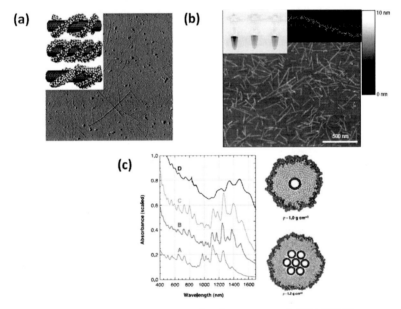

Figure 4.2 (a) AFM, 5 µm amplitude image of PVP-SWNTs on a functionalized substrate. Inset shows some possible wrapping arrangements of PVP on an (8,8) SWNT. (Reproduced with permission from Ref. 79. Copyright 2001 Elsevier.) (b) AFM image of the DNA-wrapped SWNTs on substrates following anion exchange chromatography. Inset shows photographs from starting material and two different enriched fractions and a model of the one of the several binding structures of DNA to SWNTs. (Reproduced with permission from Ref. 80. Copyright 2003 Nature Publishing Group.) (c) Absorption spectra of SDS-dispersed SWNTs in water, showing well-resolved features. Right-hand side shows cross-sectional model of individual and seven-tube bundle of SWNTs in a cylindrical SDS micelle. (Reproduced with permission from Ref. 81. Copyright 2002 American Association for the Advancement of Science [AAAS].)

It is also possible to disperse unfunctionalized SWNTs using organic solvents, but this only allows dispersions with low concentrations. Several examples of organic solvents for SWNT dispersion include *N,N*-dimethylformamide (DMF), *N,N*-dimethylacetamide (DMA), *N,N*- diethylacetamide (DEA), and *N,N*-dimethylpropanamide (DMP)[96], *N*-methyl-2-pyrrolidinone (NMP)[97,98], and 1,2-dichlorobenzene.[99] Recently some effort has been spent on SWNT dispersions in solvents containing conjugated polymers. Less amounts of conjugated polymers are required for the dispersion of the same amount of SWNTs as compared to surfactants. In addition, conjugated polymers have higher conductivity values, which can improve intertube junction resistance. This may lead to thin films with better optoelectronic properties compared to their counterparts prepared with surfactants. Examples of commonly used conjugated polymers include poly(3,4-ethylenedioxythiophene):poly(styrenesulfonate), or (PEDOT:PSS)[100,101], poly-3-hexylthiophene (P3HT)[101-104], and poly-3-octyl-thiophene (P3OT).[105] In some of these studies, samples with greater than 50 wt% SWNT loading have been achieved.

4.3 Thin-Film Deposition Processes

SWNT thin-film electronics can potentially avoid some of the problems associated with the utilization of individual SWNTs in devices.[50,51,54,55,58,84,106–109] In addition, SWNT networks have unique electrical (good performance on plastic), optical (transparent at visible wavelengths), and mechanical (bendability) properties that are difficult to achieve using conventional materials. SWNT thin films have also been proposed as an alternative to organic and amorphous semiconductors for plastic electronics. Although the mobility of the random SWNT networks is lower than that of individual nanotubes, they offer the advantages of high transparency and flexibility. Nevertheless, preparation of homogeneous and uniform networks is of paramount importance for all applications. Figure 4.3 shows SEM images of SWNT thin films in different configurations. Figures 4.3a and 4.3b show thin films of randomly oriented SWNTs of high and low density, respectively, while Figs. 4.3c and 4.3d exhibit arrays of partially and nearly perfectly aligned SWNTs, respectively.

There are three primary methods for the deposition of SWNT thin films. They can be formed either by direct growth on substrates[50, 57,111], deposition of previously grown SWNTs onto substrates from a solution[50,51,54,55,58,106–109], or transfer onto desired substrates after CVD growth.[58,107,108] The solution deposition method offers the advantage of using purified SWNTs, without amorphous carbon and catalyst particles, which is critical for device applications. In addition, various transparent and flexible substrates can be used since the deposition[55,106–108,112] is done at room temperature or at temperatures below 100°C. The SWNT density in thin films can be controlled by adjusting the solution concentration or repeating deposition steps. Different solution-based SWNT thin-film deposition methods include spin coating,[102] dip coating[84,108], solution casting and transfer printing[58,107,108,113], spray coating[55,114–116], Langmuir–Blodgett drawing[117], rod coating[118], the commonly used vacuum filtration method[51,56,109], and inkjet printing.[119,120] Commonly used substrates for the deposition of SWNT thin films include glass, quartz, silicon dioxide on silicon, poly(ethylene terephthalate) (PET), and polyimide (Kapton). These deposition methods usually lead to randomly oriented SWNTs. One exception is using gas flow to partially align the still wet SWNTs following deposition.[84,121] There are also examples of alignment using strong magnetic fields.[122,123]

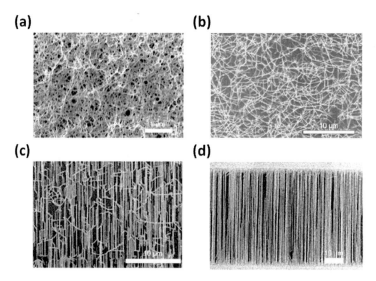

Figure 4.3 SEM images of (a) random and dense, (b) random and sparse, (c) partially aligned, and (d) perfectly aligned SWNT arrays. (Reproduced with permission from Ref. 110. Copyright 2008 Springer.)

Commonly used vacuum filtration is a reproducible method enabling deposition of uniform SWNT thin films. It is possible to deposit thin films from dilute suspensions of SWNTs filtered through mixed cellulose ester (MCE) or anodic aluminum oxide (AAO) membranes[56], which can then be transferred onto various substrates. The vacuum filtration process, first developed by Wu et al. for the deposition of transparent and conducting uniform SWNT thin films, consists of five steps: (i) preparation of dispersed suspension of purified SWNTs, (ii) vacuum filtration of the SWNT suspension onto the filtration membrane, (iii) removal of the surfactant, (iv) transfer of the SWNT film onto the substrate, and (v) dissolution of the filtration membrane with an appropriate solvent. These five steps are relevant for MCE membranes. When utilizing AAO membranes, SWNTs (following surfactant removal, step iii) are transferred to recipient substrates by transfer printing utilizing an elastomeric stamp.[124,125] The schematic of the vacuum filtration process is shown in Fig. 4.4. Dispersed SWNT solutions are carefully poured into the glass funnel to prevent bubble formation that can disrupt the film uniformity. While filtering, a vacuum of 10^{-1}–10^{-2} Torr is maintained for 15 to 60 minutes to ensure that the film is completely dry. Then the residual surfactant is removed by repeated washing with deionized (DI) water until no bubbles are apparent in the filtrate stream. The vacuum is released, and the membrane with the SWNT film is carefully removed from the filter assembly. The filter paper is then rewetted and placed on substrates with the SWNT thin film on the substrate, as indicated in the schematic. Compressive loading and heat are applied to encourage adhesion of the SWNT network to the substrate. The filter membrane is then dissolved away using successive acetone baths and rinsing with methanol so that only the SWNT thin film remains on the substrate.

Thin films prepared by vacuum filtration are homogeneous because the process is self-regulating in that as SWNTs accumulate in one region, the permeation rate in that region is lower relative to the bare regions. The higher permeation rate leads to accumulation in the uncovered regions so that uniform coverage on the membrane is achieved. Furthermore, the density of SWNTs (SWNTs/μm^2) can be varied by controlling the concentration and volume of the filtrated solution.

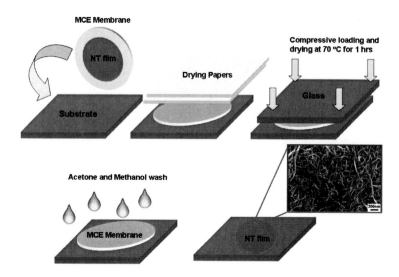

Figure 4.4 Schematic of the SWNT thin-film deposition process using vacuum filtration. The process proceeds left to right. Following filtration, the membrane is covered with a glass slide and drying papers, compressively loaded, and dried for two hours. The membrane is then removed with acetone and washed with methanol. Finally the SWNT thin films are dried with a nitrogen gun. Inset shows an SEM image for SWNT thin films.

Patterned SWNT thin films can be produced through lithographic patterning of the AAO membrane.[124] A schematic of this process and the patterned SWNT films are shown in Fig. 4.5a. In another approach, sacrificial polystyrene (PS) beads are used to pattern SWNT thin films, where the beads are mixed with surfactant-assisted SWNT dispersions prior to filtration. SEM images of the SWNT thin films before and after removal of the PS beads are shown in Fig. 4.5b. This incorporated porosity is found to change the electrolytic capacitance, comparative perfusion rates, and optical properties of the SWNT thin films.[53] Change in optical properties through incorporation of the pores is revealed in bottom images in Fig. 4.5b. Another patterning strategy is the use of sacrificial silica nanoparticles[126], SEM images of which are provided in top panel of Fig. 4.5c. The SWNT thin films are deposited onto the silica nanoparticles and subsequently removed via hydrofluoric acid treatment. SEM images of the SWNT thin films before and after removal of the silica beads for different bead sizes are shown in the bottom panel of Fig. 4.5c.

Thin-Film Deposition Processes | 73

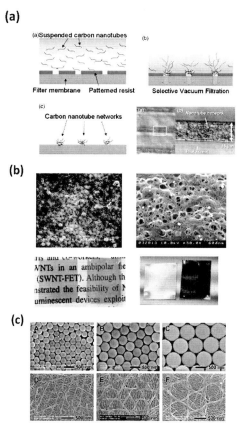

Figure 4.5 (a) Schematic diagram of the selective vacuum filtration process. SWNT suspension on the patterned filter surface, SWNTs being guided to the exposed pattern surface by low-pressure liquid siphon, and formation of the patterned SWNT thin films on the filter surface, optical microscope image of the patterned SWNT thin film. (Reproduced with permission from Ref. 124. Copyright 2007 American Institute of Physics.) (b) SEM images of the SWNT thin films before (upper left) and after (upper right) the removal of the PS beads. Bottom images compare the transparency and reflection of the films with and without porosity, showing the change in optical properties through porosity engineering. (Reproduced with permission from Ref. 53. Copyright 2009 American Chemical Society.) (c) SEM images of the silica–colloid SWNT complex crystal films before (A–C) and after etching silica spheres (D–E). Silica nanoparticles have a particle size of 300 (A), 500 (B), and 900 nm (C). (Reproduced with permission from Ref. 126. Copyright 2008 Wiley-VCH.)

4.4 Optoelectronic Properties of SWNTs

Many factors can affect the optoelectronic properties of SWNT thin films. For instance, for the same SWNT density, highly conducting, less defective, pure, and well-dispersed SWNTs possess higher conductivity at the same transparency. The doping level and the semiconducting-to-metallic SWNT ratio can also be added to the list of factors influencing the properties. Two parameters, sheet resistance and the normal incidence transmittance of the films at 550 nm, are used to assess and compare SWNT thin films. Sheet resistance is calculated using $R = R\,(W/L)$, where R, W, and L are direct current (DC) resistance, channel width, and channel length, respectively. Sheet conductance is then defined as $G = 1/R$.

The transport mechanism in SWNT networks has been described as percolation amongst randomly aligned conducting sticks. Owing to the low percolation threshold along with excellent electrical properties of SWNTs (high mobility and conductivity), low concentration (surface coverage) of SWNTs is required to achieve a highly conductive layer. The low surface coverage also limits light absorption and scattering, making these films highly transparent. Although the percolation model describes the optoelectronic properties of the SWNT thin films reasonably well, it assumes that the nanotubes are individual sticks and neglects the presence of bundles. Also, equal conductivities for the sticks are assumed, whereas the thin films consist of 33% metallic and 67% semiconducting SWNTs, which have dramatically different electrical properties. Typical conductance values for metallic SWNTs and semiconducting SWNTs are ~70 µS and 1.5 µS, respectively.[2,3]

Hu et al. measured the percolation threshold in a series of vacuum-filtrated SWNT thin films and found that the sheet conductance varies according to:[56]

$$G = C_{MET}(V_{SOL} - V_C)^n \tag{4.1}$$

where V_{SOL} is the solution volume, V_C is the critical volume for percolation, C_{MET} depends on the conductance of a single metallic SWNT, and n is a critical exponent with values 1.33 and 1.94 for two and three dimensions, respectively.[127,128] V_C however, depends on the concentration of SWNTs. The relationship between the conductivity (σ) of the network to the conductivity of the single stick (σ_{MET}) and the relative volume fraction, Φ, occupied by the sticks is described by:

$$\sigma = \sigma_{MET} (\Phi - \Phi_C)^n \quad (4.2)$$

Under the assumption of fully random networks, the percolation threshold (Φ_C) is independent of the concentration and volume of the solution and related only to the SWNT length, diameter, and degree of disorder.[129] Φ_C can be expressed as the ratio of the SWNT volume to the excluded volume. Figure 4.6a shows the sheet conductivity of SWNT thin films spanning almost seven orders of magnitude due to an enhancement in network connectivity. The continuous line in Fig. 4.6a clearly demonstrates a good agreement between the two-dimensional (2D) percolation theory and the experimental data.

Taking into account the volume and concentration of the solution, fits yield exponents ranging from 1.5 to 1.65.[56,109] Both values are close to but slightly higher than theoretical values. This is because the percolation model does not account for the fact that both metallic and semiconducting SWNTs contribute to the conduction. The resistance of metallic-metallic SWNT junctions has been shown to be less than the resistance of metallic-semiconducting junctions due to presence of Schottky barriers.[130] The probability of all metallic paths is small at low SWNT densities so that the intertube junctions limit the conductance. However, as the network density increases more pathways involving all metallic SWNTs are formed, yielding an increase in conduction with film density that is not accounted for by the standard theory. Regardless, the close correspondence of the calculations and experimental measurements provide evidence that the behavior of SWNT networks can be understood through percolating finite-sized sticks with variable intertube coupling. The conductivity values below the percolation threshold are shown in Fig. 4.6b. In this nonpercolating regime, a mean square fit shows that the data clearly follows an exponential trend instead of Eq. 4.2. In a 2D system of sticks, ($\Phi_C - \Phi$) is related to the intertube distance, Δr. Thus, the behavior below the metallic percolation threshold can be attributed to an alternative conduction mechanism such as tunneling or hopping based on exponential decay of conductivity with Δr.

For the randomly aligned conducting sticks, theory predicts the value of the critical density, N_c, as:

$$l\sqrt{\pi N_C} = 4.236 \quad (4.3)$$

where *l* is the length of the sticks (SWNTs), which can be obtained through SEM and AFM analysis. Assuming an average SWNT length of 2 μm, a theoretical critical density of 1.43 conducting sticks (CS)/μm² can be calculated. A slight deviation in the parameters (i.e., bundling) can, however, have dramatic effect on the critical density.[131]

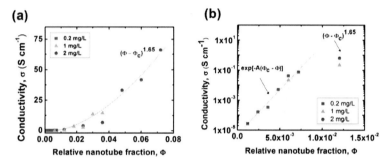

Figure 4.6 Variation of conductivity with respect to relative SWNT concentration in the thin films. The line in panel (a) corresponds to a fit using Eq. 4.2. (b) Semilog magnification of panel (a) at low Φ, showing the trend for the residual conductance below metallic percolation.

A figure of merit for evaluating the performance of SWNT thin films has been proposed on the basis of the ratio of optical conductivity to DC conductivities, σ_{OP}/σ_{DC}. Assuming the thickness of a metallic film is small compared to the wavelength of light, the relationship between its transmittance and sheet resistance in air can be modeled by:[56,132]

$$\frac{1}{T} = \left(1 + \frac{1}{2R_S}\sqrt{\frac{\mu_0}{\varepsilon_0}}\frac{\sigma_{OP}}{\sigma_{DC}}\right)^2 = \left(1 + \frac{188}{R_S}\frac{\sigma_{OP}}{\sigma_{DC}}\right)^2 \quad (4.4)$$

where μ_0 and ε_0 are the free space permeability and permittivity, with values of $4\pi \times 10^{-7}$ NA^{-2} and 8.854×10^{-12} C² N^{-1} m^{-2}, respectively. It has been shown that the optical conductivity in the visible range depends only on the overall network density, while the DC conductivity strongly depends on factors such as SWNT–SWNT connectivity, SWNT length, metallic-to-semiconducting SWNT ratio, and alignment. Optical properties of SWNT thin films have been studied mainly at normal incidence and interpreted using the standard model for metallic thin films.[56] Such models assume that the ratio between the static (direct current) and optical conductivities remains constant at any given photon energy. Hu et al. argued that optical properties of the SWNT thin films scale with concentration of the SWNTs only. However, the dielectric properties of the individual isolated SWNTs or bundles, where confinement effects are expected to dominate, are different than in SWNT thin films, where the electronic confinement is released due to variable connectivity of the network.[56] The effects of confinement using spectroscopic ellipsometry at low grazing angles have been elucidated by Fanchini et al. on vacuum-filtrated SWNT thin films.[133] The optical response of individual SWNT thin films of different densities, getting rid of the effects of the environment using effective medium theory (EMT) was determined. It was shown that the dielectric function of the nanotubes (ε_{NT}) is sample dependent and the absence of or reduction in electronic confinement plays a non-negligible role. The imaginary part (ε_2) of the dielectric function is an optical constant that is strongly related to the electronic absorption processes in solids. Figures 4.7a and 4.7b show the dielectric responses of the SWNTs in thin films after the effects of voids and bundle–bundle interactions have been eliminated. SWNT networks deposited below a critical filtrated volume of 50 mL, $\varepsilon_{NT,2}(E)$ typically exhibit a broad peak shifting from E_π = 5.0 eV (at 10 mL) to E_π = 4.25 eV (at 40 mL) (Fig. 4.7a). Such a downshift suggests that even at low filtration volumes where the bundles are physically isolated, their dielectric function depends on the amount of confinement. The maximum intensities of the π–π* transition peaks in individual SWNTs (E_π = 5.74 eV) was found to occur at higher energies. In graphite, the maximum intensity of the π–π* interband transitions occurs at 4.5 eV and it is superimposed on the metallic Drude background. It is therefore interesting to notice that once reaching a minimum, E_π increases again in more connected SWNT networks (Fig. 4.7b), approaching the value of graphite. Several reasons such as strong SWNT–SWNT interactions that cannot be described at the dipole–dipole level[134], the reduction of the π-bond strength due to disorder, and different levels of anisotropy in different samples or excitonic effects[135] could lead to the observed trend of E_π. These results clearly show that the dielectric function of the SWNTs (ε_{NT}) is sample dependent and the absence of or reduction in electronic confinement plays a non-negligible role.

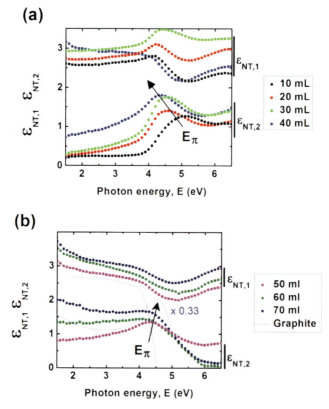

Figure 4.7 Calculated complex dielectric functions $\varepsilon_{NT,1}(E)$ and $\varepsilon_{NT,2}(E)$ of a single SWNT bundle as a function of the photon energy for (a) low and (b) high filtration volumes. The appearance of the Drude background at 50 mL indicates the onset of metallic optical behavior.

In addition to the different π–π^* peak values, other important changes in the electronic structure of the thin films occur at high filtration volumes, as can be observed by comparing Figs. 4.7a and 4.7b. The immediately noticeable difference is that $\varepsilon_{NT,2}(E)$ clearly diverges at low photon energies in networks deposited from filtration volumes of 40 mL, making the optical spectra similar to those of graphite[136] Such a divergence of $\varepsilon_2(E)$ at low energy is increasingly important at higher filtration volumes where it can be fitted in the framework of the Drude model for metals, a definite indication of metallic conductivity.[135] It should also be noticed that the Drude background is absent in Fig. 4.7a. Thus, the presence of metallic SWNTs and bundles is not a sufficient condition for metallic conductivity as long as bundles do not form an interconnected network. The metallic behavior arises when percolation amongst metallic SWNTs in the network occurs. The percolation thresholds obtained by electrical and optical measurements are found to be different. This can be attributed to the fact that the electrical percolation of the samples is not genuinely metallic percolation but rather a tunneling percolation, as theoretically predicted by Balberg et al.[137]

The level of transparency in the SWNT thin films is comparable to some of the best transparent, semiconducting and conducting oxides. Furthermore, the transmittance is nearly independent of the wavelength, in contrast to ITO. This is especially true in the ultraviolet (UV) region where ITO is opaque. The bandgap of the SWNTs (0.5–0.7 eV for 1–1.5 nm diameter semiconducting SWNTs[2,3]) is much smaller than those of semiconducting oxides (>3eV[138]). The transparency of the SWNT networks cannot be attributed to their bandgap but instead to their low-polarization-dependent optical absorption cross section that results from their small size and highly anisotropic shape. Moreover, absorption and reflection of unpolarized incident light is suppressed for polarization components perpendicular to the SWNT axis, which reduces the optical density. Transparency of the SWNT thin films can also be attributed to the low carrier density that leads to low plasma frequency. Thin films are highly transparent in the visible and infrared (IR) spectral ranges. Transmittance curves for SWNT thin films with different densities are shown in Fig. 4.8a–c at different filtration volumes. For comparison, transmittance of ITO is also provided. An optical image of SWNT thin films revealing their transparency is given in Fig. 4.8d. MWNTs have also been considered for transparent conductive coatings, but their significantly higher diameter increases the light absorption.[116]

Transparent conductors require sheet resistances of 50 Ω/square or less at 85% transparency (i.e., ITO). Touch-screen applications are less demanding, where a sheet resistance of 500–1000 Ω/sq at 85% transparency is sufficient for practical applications. Current state-of-the-art SWNT thin films are well above the figure of merit for touch screens; however, improvements are needed for ITO replacement applications. The performance gap is small though. The list of reported sheet resistance and transmittance values of the SWNT thin films includes but is not limited to 40 Ω/square at 70% and 70 Ω/square at 80% following nitric acid treatment,[115] 120 Ω/square at 80% for undoped films,[125] 160 Ω/square at 87% following SOCl$_2$ doping,[139] 170 Ω/square at 81%, and 80 Ω/square at 72% by polymer-assisted dispersion.[102] It can be generalized that SWNT thin films from arc- or laser-grown SWNTs give lower sheet resistance values at the same transmittance values compared to CVD-grown counterparts.[116,139–141]

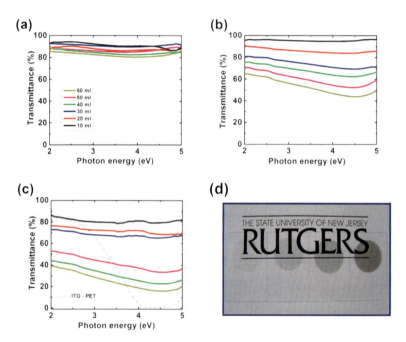

Figure 4.8 Transmittance of SWNT thin films prepared via vacuum filtration at different densities. SWNT thin-film densities are adjusted through solution concentration and filtration volumes. Three different solution concentrations have been elaborated in panels (a–c), each with six filtration volumes given on the curves. A transmission spectrum for ITO has been provided for comparison. (d) Digital photograph of films in (c) at four different filtrate volumes, revealing the degree of transparency.

Lastly, SWNT thin films, resembling a spider web, show higher mechanical flexibility than that of crystalline inorganic films and are structurally well suited for flexible devices and applications.

4.5 SWNT Functionalization Treatments

Significant effort has been devoted to separation of SWNTs according to their electronic properties or enrichment of either metallic or semiconducting species. Various approaches such as dielectrophoresis[142,143], DNA-assisted separation[80,144–146], polymer-assisted extraction[147,148], cycloaddition[149], selective oxidation[150], octadecylamine-assisted separation,[151] and density gradient ultracentrifugation (DGU)[47–49] have been employed.

Of these, DGU offers a scalable method for sorting SWNTs according to their diameter, bandgap, and electronic type, as shown in Fig. 4.9a–d. Figure 4.9a shows a digital image of ultracentrifuge tubes following DGU treatment, and coinciding optical absorption spectra from various locations for small-diameter SWNTs (0.7–1.1 nm) are shown in Fig. 4.9b. Figures 4.9c and 4.9d show a digital image of ultracentrifuge tubes following DGU and coinciding optical absorption spectra from various locations for large-diameter SWNTs (1.1–1.6 nm), respectively. This method works on the principle of differences in buoyant densities of SWNTs. SWNTs encapsulated with bile salts and mixtures containing anionicalkyl surfactants are placed in the centrifuge. The resulting centripetal force sediments SWNTs with respect to their buoyant densities and spatially separates them in the gradient. Sorting of SWNTs with diameters ranging from 0.7–1.6 nm has been demonstrated. The observation of different colors in the topmost bands in the DGU tubes after centrifugation verifies sorting of SWNTs by diameter. Optical absorbance spectra for different fractions further confirm sorting by electronic structure. Additional DGU cycles improve the isolation of individual chiralities and achieve a narrow distribution (>97% of the SWNTs are found to be within 0.02 nm diameter range) of the SWNT types. Results from thin-film transistor (TFT) devices employing separated SWNTs confirm the efficiency of separation according to the electronic type.[47]

An alternative route to improve enrichment of either semiconducting or metallic SWNTs is doping. The state-of-the-art SWNT thin-film transistors exposed to air exhibit unipolar p-type behavior from the physisorbed oxygen. On the other hand, chemically bonded dopants (functional groups) are more desirable and stable. A few doping schemes for SWNTs include charge transfer from hexachloroantimonate ions[152], thiol radicals[153], diazonium salts[154–158], and bromine functionalization.[159,160] Diazonium salt doping minimizes the contribution of metallic SWNTs, allowing semiconducting ones to dominate the transport. Postdeposition functionalization with diazonium salts is a simple and wafer-scalable approach.[154] The mechanism for improved semiconducting behavior is attributed to the fact that metallic tubes are, on average, more reactive to diazonium reagents. Briefly, availability of electrons at the Fermi level stabilizes the charge transfer complex formed by the diazonium reagents on the SWNTs, facilitating the reaction. On the contrary, the absence of electrons near the Fermi level makes semiconducting SWNTs less likely to react with diazonium, although reactions can also occur in some very small-diameter tubes.

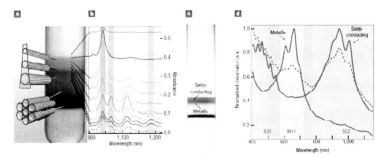

Figure 4.9 (a) Sorting of small-diameter (0.7–1.1 nm) SWNTs encapsulated with SC by DGU. (b) Optical absorbance spectra for different fractions, confirming separation by diameter. (c) Sorting of large-diameter (1.1–1.6 nm) SWNTs encapsulated with a cosurfactant mixture of SC and SDS according to electronic type by DGU. (d) Optical absorbance spectra confirming sorting by electronic type. (Reproduced with permission from Ref. 46. Copyright 2008 Nature Publishing Group.) *Abbreviation*: SC, sodium cholate.

Bromine attachment has also been investigated for the separation of metallic SWNTs[159] and doping of semiconducting ones.[161,162] Rao et al. first reported a reduction in resistivity due to charge transfer, observed by shifts in the SWNT Raman peaks.[162] In the experiments, bromine was physisorbed onto SWNTs, leading to strong instabilities in conductivity after exposure to air. Chemical attachment of bromine can be done by nucleophilic substitution in the form of acyl bromide (–COBr) groups on carboxylated SWNTs. The attachment of Br not only occurs on metallic but also on the semiconducting ones. Density of states calculations and SWNT TFT results revealed strong depletion of the amount of extended electronic states in the metallic SWNTs. Furthermore, the downshift of the Fermi level and the appearance of acceptor states facilitate the formation of holes in the valence band at finite temperature. Thus in Br-doped thin films, the electrons in metallic SWNTs are not only confined but also reactivated by bromine, behaving as *p*-type semimetallic elements. Depletion of midgap electron states also allows the fabrication of transistors at higher SWNT densities, which in turn increases the mobility without lowering the ON/OFF ratio. The stability of covalently Br-functionalized SWNT networks is demonstrated by the fact that the performance of TFTs is independent of the measurement atmosphere.[160]

In addition to Br doping, chlorine (Cl) doping by simply dipping the as-deposited SWNT networks in acid followed by thionyl chloride ($SOCl_2$)[140,141,163] has also been demonstrated. The results show improved conductance without deteriorating the transparency, as shown in Fig. 4.10a. The reason for conductivity improvement in only acid-treated SWNT thin films is the removal of the residual surfactant at the nanotube junctions and creation of oxygen-doping sites. For the $SOCl_2$-treated SWNTs, the chlorine in the form of Cl or COCl provides higher electron density, which imparts higher conductivity. Chlorine functionalization has also been found to be stable in various solvents like methanol and chloroform, solvents that are routinely used in organic electronics. The functional groups have been found to be stable well above the annealing temperatures of organic devices. The flexibility of Cl-doped SWNT networks, in contrast to ITO, is demonstrated in Fig. 4.10b.

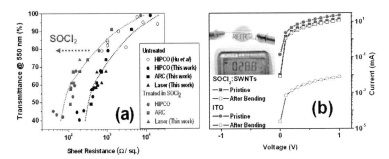

Figure 4.10 (a) Film transmittance vs. sheet resistance for the various films investigated (laser, arc discharge, and HiPCO synthesized) and measured after and before the HNO_3-$SOCl_2$ treatment, and (b) current vs. voltage characteristics of ITO and SWNT thin films after bending at 45° ten times. (Reproduced with permission from Ref. 141. Copyright 2007 American Institute of Physics.)

4.6 Applications and Devices

4.6.1 Photovoltaic Devices

The intriguing optoelectronic properties of the SWNT thin films have motivated researchers to investigate their use in various electronic devices and applications. Applications from a variety of technologies/fields have been presented. A list of applications includes but is not limited to energy (photovoltaic devices, light-emitting diodes [LEDs], batteries, supercapacitors, smart windows), electronics (TFTs, logic circuits, radio, speakers), and security (sensors, electromagnetic shielding, optical and IR sensors), some of which will be discussed here.

Organic photovoltaic (OPV) devices can be manufactured using solution-based methods, enabling mass production at low cost. Commonly used flexible OPV devices utilize ITO-coated glass/PET substrates as the cathode, a blend of (bulk heterojunction) a semiconducting polymer (donor) and a fullerene derivative (acceptor) as the active layer, and an evaporated aluminum thin film as the anode. The most promising combination of materials for the donor and acceptor materials are P3HT and [6,6]-phenyl-C61-butyric acid methyl (PCBM) ester, respectively. Photovoltaic conversion efficiencies exceeding 5% have been achieved by these devices.[164,165]

ITO is the standard material used in transparent and conducting applications such as photovoltaic devices and displays. Driven by the high demand for liquid crystal (LC) and plasma displays, the price of indium has increased enormously in the last few years. This is also due to the scarcity of the material. Further, commercially available ITO is deposited by sputtering, which adds to its cost and necessitates temperatures in excess of 250°C. This raises some compatibility issues with polymeric substrates. In addition to cost, a major technological problem associated with ITO on cheap, flexible substrates is its reliability. Straining ITO leads to a change in its electrical properties (Fig. 4.10b). In addition, it has been shown that the conductivity of ITO is a source of series resistance that decreases the fill factor and hence the power conversion efficiency, especially in large-area OPV devices. Lastly, chemical stability of ITO against acids and bases is relatively poor. Therefore, SWNT thin films have been investigated as a potential alternative. ITO and SWNT thin films have comparable work functions (~5 eV), as obtained by techniques such as field emission[166], ultraviolet photoelectron spectroscopy (UPS)[167] measurements, and simulations.[168] Work function is particularly important for the construction of devices like OPVs and organic light-emitting diodes (OLEDs).

Nanotubes in the form of MWNTs were first integrated in OPVs as the hole collector by Ago et al.[169] Other photovoltaic devices with SWNTs as electron recipients and transporters[170-172] have also been demonstrated. Kymakis et al. used SWNTs as electron acceptors via creation of nano heterojunctions with P3OT for exciton dissociation. SWNT thin-film electrodes in OPVs were first demonstrated in 2005.[173] The efficiency of SWNT devices on glass substrates was found to be about 1% in the initial report but have been improved to >3.0%. Although the sheet resistance of SWNT networks is higher than that of ITO, the excellent efficiency values of SWNT electrodes can be attributed to the three-dimensional nature of the thin films. That is, the porous network of SWNT thin films is impregnated by the photoactive P3HT:PCBM blend so that the SWNTs are in intimate contact with the bulk heterojunction matrix. This structure facilitates hole transfer from the blend to the SWNTs. More recently, SWNT thin films with lower roughness, higher transmittance, and similar sheet resistance values deposited on PET substrates have been utilized to achieve efficiency values around 3.1%[174], as shown in Fig. 4.11a,b. A major limitation of SWNT thin films could be the fact that they exhibit optical anisotropy, as demonstrated by transmission and ellipsometric measurements at several incidence angles.[175]

In addition to OPVs, SWNT thin films have also been incorporated into other types of solar cells. For instance, flexible organic-inorganic hybrid solar cells (zinc oxide nanowires (ZnO): P3HT) have been demonstrated through the use of SWNT thin-film electrodes.[176] Thin-film solar cells with heterojunctions between n-type silicon (Si) substrates and nanotubes (both single and double walled) have been reported.[177-180] In such devices, electron– hole pairs generated at heterojunctions split and are transported through SWNTs (holes) and n-Si (electrons). SWNT thin films in this application work as a photogeneration site, a charge carrier transport layer, and a collector.[177,178] Recently, n-Si/p-SWNT heterojunction solar cells fabricated with HNO$_3$-treated SWNTs were found to decrease internal resistance and improve the fill factor, leading to enhanced charge separation and transport. Photovoltaic conversion efficiency for such a device was reported as high as 13.8%.[181] ITO is also the standard for the transparent electrodes in smart windows. Electrochemically driven devices necessitate the use of corrosive electrolytes that are harmful to ITO. SWNT thin films have been demonstrated to be effective as electrodes in electrochromic devices, where polyaniline and sulphuric acids were used as the active material and the electrolyte, respectively.[182] Excellent chemical stability was observed with SWNT thin films.

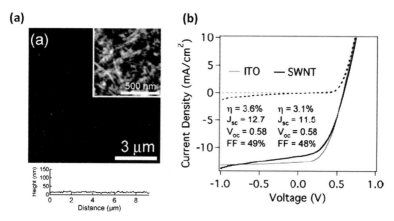

Figure 4.11 (a) AFM image of an ultrasonically sprayed SWNT film that is treated with nitric acid. Inset shows a higher magnification showing individual and small-bundle SWNTs. Below the image is a line scan of the films, showing excellent uniformity and low roughness. (b) Current density vs. voltage characteristics of OPV devices constructed on ITO (dark curves) and SWNT thin films (light curves) shown in (a). Dashed and solid curves indicate devices in dark and light, respectively. (Reproduced with permission from Ref. 174. Copyright 2009 Wiley-VCH.)

The ability to functionalize SWNT networks with small molecules makes them attractive for light detection. Porphyrin (a light-sensitive small molecule)-decorated SWNT thin films have been used to detect photo-induced electron transfer by Hecht et al.[183] This process mimics part of the natural photosynthesis procedure and could be promising for future photovoltaic technologies. Similar effects have been observed through deposition of other moieties such as a light-sensitive polymer, poly{(m-phenylene-vinylene)-co-[(2,5-dioctyloxy-p-phenylene)vinylene]} (PmPv), by Star et al.[184]

and cadmium selenide (CdSe) nanoparticles by Hu et al.[185] on SWNT networks.

4.6.2 Light-Emitting Diodes

SWNT thin films have also been demonstrated as electrodes in organic and polymeric LEDs (OLEDs). In OLEDs, electrons and holes are injected from both electrodes into an active molecular/macromolecular medium, where radiative recombination leads to light emission. Injection efficiency influences not only the brightness and efficiency but also the lifetime and stability of these devices. ITO is the standard hole injection anode material for OLEDs. The OLED device structure is similar to that of OPVs. ITO replacement with SWNT thin films has been demonstrated in OLEDs with remarkable performance.[87,139,186,187] Transmittance values of the SWNT thin films used in OLED devices are shown in Fig. 4.12a. SWNT roughness values of 1–2 nm are required for OLEDs to ensure uniform charge injection. Initial SWNT electrodes for OLEDs were rough, but progress toward planarization through a conducting polymer top coating has decreased the roughness to 4–5 nm, which led to a performance increase of up to 450 times.[188] The results of optimized OLED devices are shown in Fig. 4.12b. Planarization has also been achieved by deposition of a thin ITO layer on top of SWNT networks.[189] SWNT thin films have also been utilized for both anode and cathode electrodes in blue LEDs fabricated by lamination. These devices revealed significant flexibility without failure, even after bending to a radius of 2.5 mm.[190]

4.6.3 Supercapacitors and Batteries

Printed flexible electronics also require flexible power sources. Two of the most important electrochemical energy sources are batteries and capacitors. Electrochemical storage systems vary greatly in performance in terms of energy and power they can store per weight. Batteries have high energy densities, while capacitors have high power densities. They both consist of two electrodes separated by a porous, electrically insulating separator soaked in an ionically conductive electrolyte. Supercapacitors (or electrochemical double-layer capacitors [EDLCs]) have typical capacitance values that are four to five orders of magnitude larger than ordinary dielectric

88 | Single-Walled Carbon Nanotube Thin-Film Electronics

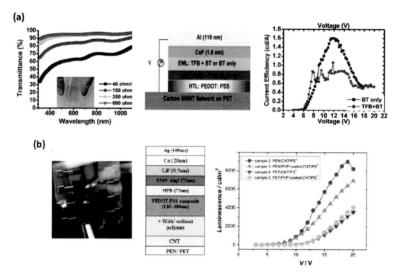

Figure 4.12 (a) Transmittance of the SWNT thin films used for anodes, structure of the OLED device with an SWNT thin film anode, and current efficiency vs. voltage curves for these devices. (Reproduced with permission from Ref. 187. Copyright 2006 American Chemical Society.) (b) Digital picture, schematic of the OLED device with planarized SWNT thin film electrodes, and luminescence vs. voltage response characteristics of OLEDs. (Reproduced with permission from Ref. 188. Copyright 2009 American Chemical Society.)

capacitors per centimeter cube area. This is due to i) separation of charges in electrochemical double layers on the order of fractions of nanometers as compared to hundreds of nanometers and ii) utilization of very high specific-area porous electrodes. Supercapacitors provide current boosts on high load demands. CNTs are natural candidates for supercapacitor electrodes. They are lightweight and electrochemically stable and have a high surface area and electronic conductivity. In the last few years, many examples have appeared in the literature where different forms of CNTs have been used in a multitude of ways to produce supercapacitor devices. Simon et al. present a brief review of materials for electrochemical capacitors.[191] Relatively thick SWNT films were first used as electrodes with a liquid electrolyte as a simple approach for the fabrication of supercapacitors.[192] The use of SWNT thin films eliminated the need for separate metallic current collectors. Two years after, the same group represented supercapacitors with SWNT thin-film electrodes and a printable polymer electrolyte.[193] The schematic and results of these supercapacitor devices utilizing polymer electrolytes with SWNT thin-film electrodes are shown in Fig. 4.13a,b. All components of the presented supercapacitor can be printed or deposited using solution-based, room-temperature processes. Stretchable supercapacitors with SWNT thin-film electrodes have been also presented with almost no change in electrochemical performance during stretching.[194] Stretchability is associated with buckled SWNT films. Buckling is simply achieved through solution deposition of SWNT thin films on top of a prestrained polydimethylsiloxane (PDMS) film, followed by their release. SWNT thin films are also used as charge collectors in primary zinc-carbon batteries that are fabricated through solution-based processes.[195]

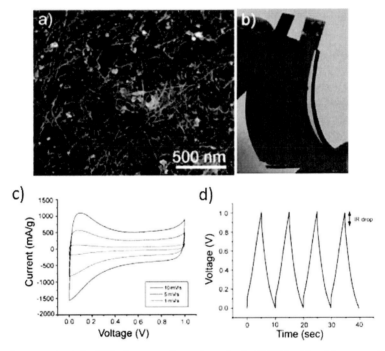

Figure 4.13 (a) SEM image of as-deposited SWNT thin films used for supercapacitor electrodes, (b) digital image of the supercapacitor with SWNT electrodes and a polymer electrolyte fabricated on PET, (c) cyclovoltammogramm of a SWNT thin-film supercapacitor with a polymer electrolyte, and (d) corresponding charge/discharge curves for that device. (Reproduced with permission from Ref. 193. Copyright 2009 American Chemical Society.)

4.6.4 Sensors

CNTs offer a wide range of advantages as a sensor material. The most important one is their miniature size. Moreover, they have very high sensitivity to various types of gases/molecules. In addition, they can be operated at room temperature, minimizing power consumption.[196] Slow molecular desorption may lead to long recovery times; however, solutions like heating in air to 200°C[196], UV exposure[197], and gating have been provided.[198] SWNT thin films can also be used as sensors.[197,198] The use of SWNT thin films has several advantages compared to individual nanotube sensor assembly. First, there is no precision required for assembling the SWNT network sensor. Second, multiple SWNTs are in contact with analytes, improving the sensor performance and detection limit. Thin films are also tolerant to individual SWNT channel failure. Novak et al. employed SWNT thin-film sensors to detect dimethyl methylphosphonate (DMMP), a simulant for the nerve agent sarin. Sensors were simply formed by metallization. DMMP is a strong electron donor, and it causes a reduction of the hole density in SWNTs, increasing their resistance. These sensors were shown to be reversible and capable of detecting sub-ppm levels along with being intrinsically selective against interfering signals from hydrocarbon vapors and humidity. SEM images of SWNTs casted on interdigitated electrodes and the sensor response in terms of conductance change for nitrotoluene are shown in Fig. 4.14a. Step function sensor response with concentration is observed. Wang et al. demonstrated that a receptor containing polymers can be used to disperse SWNTs and those nanotubes upon deposition onto a substrate in the form of thin films can be used as chemiresistive sensors. This receptor containing polymers increases the sensitivity due to strong hydrogen-bonding interactions with the analyte, as well as improving the SWNT dispersion.[199] Highly flexible chemiresistive sensors with SWNT thin films decorated with palladium (Pd) nanoparticles have been fabricated for hydrogen detection. Detection levels as low as 30 ppm were achieved in air at room temperature. Sensitivities much higher than traditional sensors have been obtained through this low-cost alternative.[200,201]

Conductance response in SWNT sensors is due to charge transfer from analyte molecules. Adsorbates also form polarizable layers that increase the SWNT capacitance.[202] An ptical micrograph of a SWNT thin film chemicapacitor is shown in Fig. 4.14b. SWNT thin

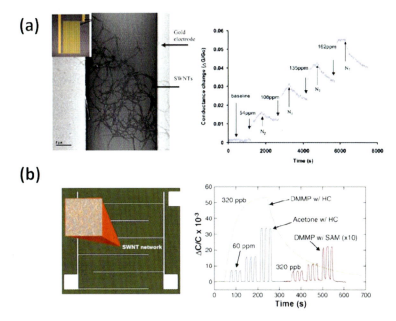

Figure 4.14 (a) SEM images of the SWNTs across interdigitated gold electrodes and corresponding sensor response for nitrotoluene (reproduced with permission from Ref. 197. Copyright 2003 American Chemical Society). (b) Optical micrograph of an SWNT thin-film chemicapacitor with the corresponding sensor response to acetone and DMMP. Response to DMMP evaluated both with and without polymer coating (reproduced with permission from Ref. 202. Copyright 2005 American Association for the Advancement of Science [AAAS]).

films can be seen between the electrodes. The corresponding sensor response is shown for acetone and DMMP. The response to acetone was evaluated for three different concentrations, as indicated by the blue curve for a chemoselective polymer-coated sensor. The response time of the same polymer-coated sensor to DMMP was found to be longer due to slow diffusion of DMMP, as indicated by the green curve. The response time was lowered by deposition of self-assembled monolayers, as described by the red curve. This sort of chemical selectivity suggests promise of using SWNT thin films for sensing applications. Snow et al. conducted simultaneous capacitance and conductance measurements and found that the ratio of the conductance to capacitance is an intrinsic signature of a chemical vapor, independent of its concentration, which facilitates identification.[203] For most vapors, the capacitance response is found to be more sensitive, recovering much faster over a broad range of analytes.

4.6.5 Electromagnetic Interference Shielding

Electromagnetic interference (EMI) shielding of radio and microwave frequencies and radiation from cell phones and computers is important for military and health applications. Commonly used EMI-shielding materials include thin metal foils and grids. Conducting polymers and composites are attractive alternatives but face degradation problems. Polymers containing SWNTs are promising for EMI shielding, with results indicating that shielding up to 1 GHz is feasible.[204] EMI shielding in SWNTs is due to reflection.[205] Microwave conductivity measurements on SWNT thin films yield promising results as transparent materials for EMI shielding.[206] EMI shielding with SWNTs has also been demonstrated up to terahertz frequencies.[207] Flexible films of SWNTs deposited on PET substrates, while optically transparent, exhibited good shielding efficiency that can be engineered through the thickness of the film and postdeposition acid treatments.

4.6.6 IR Properties and Applications

Optoelectronic materials that have IR spectral properties can be utilized in military and industrial applications. Telecommunications, thermal imaging, sensors, and solar cells are some examples. IR photoresponse of thick, suspended SWNT films has been demonstrated in vacuum at low temperatures, and a resistance drop of 0.7% upon illumination has been obtained. This change is attributed to a thermal effect, where excitons decay into heat and warming of SWNTs reduces their resistance.[208] The scenario is a little different when SWNT thin films are embedded into electrically and thermally insulating matrix. Pradhan et al. demonstrated SWNT thin films coated with a polycarbonate give a significant photoresponse upon IR illumination in air at room temperature. This change is attributed to a photoeffect, where excitons dissociate into free electrons and holes thermally or by a large electric field, increasing the conductivity of the SWNTs. It has also been shown that the semiconducting SWNTs are responsible for the observed photoresponse.[209] IR properties of SWNT thin films have been measured by Hu et al., who found them to be more than 90% transparent up to 22 μm. IR transmittance of SWNT thin films has been found to be the highest due to their low carrier density compared to other transparent conductors such as ITO, graphene, silver nanowires, and PEDOT:PSS. SWNT thin films, therefore, can be used as transparent conductors in IR solar cells and also in applications where heat dissipation is required.[210]

4.6.7 Thin-Film Transistors

The field effect in SWNT TFTs was initially reported by Snow et al.[50], who proposed it to be a combination of the field dependence of the carrier concentration in the semiconducting SWNTs and the gating of the Schottky barriers present at the metal–nanotube contacts[211,212] and at the semiconducting SWNT/metallic SWNT internanotube contacts.[213] Figure 4.15a shows transfer characteristics of SWNT networks with different nanotube densities. A large ON/OFF ratio (>10^4) can be achieved for networks with low nanotube densities (1 tube/μm^2), while significant leakage current can be observed for those at higher nanotube densities (>3 tubes/μm^2). Low leakage current can also be achieved by increasing the channel length to reduce the probability of percolation between metallic SWNTs. Randomly oriented networks of SWNTs consist of a large number of junctions so that charge carriers traveling along a nanotube must tunnel into other nanotubes to drive current across the network. The total resistance of SWNT networks therefore arises from junction resistance as well as intrinsic resistance of individual SWNTs. Fuhrer et al. investigated the transport properties of individual SWNT cross junctions and demonstrated that the tunneling transmission coefficients depend on the SWNT species that form the junctions. Specifically, transmission coefficients at the junction formed by two metallic SWNTs (MM), two semiconducting SWNTs (SS), and metallic and semiconducting SWNTs (MS) were found to be on the order of 0.1, 0.06, and 0.0002, respectively.[213] That is, the resistance of MS junctions is ~2–3 orders of magnitude higher than that of MM or SS junctions due to the presence of Schottky barriers. The carrier transport along the semiconducting SWNT at the MS junction is even more difficult than tunneling into the metallic SWNT, because the tunnel barrier is twice the width of the Schottky barrier.

The field effect at low SWNT densities allows the fabrication of transparent and flexible devices. TFTs can be fabricated from randomly oriented or highly aligned SWNTs, the latter showing significantly improved performance but requiring transfer onto desired substrates. Cao et al. fabricated all CNT TFTs using high-density, random SWNT thin films for the gate, source, and drain electrodes and low-density SWNT thin films for the active channel layer.[58] Transfer characteristics of all-SWNT transparent and flexible TFTs for different channel lengths (30, 55, 75, 112, 150, 225, and 300 μm from top to bottom) and a channel width of 750 μm are shown in Fig. 4.15b. The inset shows the schematic of the devices with the SEM image of the interface between a source/drain electrode (left) and a semiconducting channel (right).

For pristine SWNT TFTs, results from several studies in the literature clearly show an inverse relation between mobility and the ON/OFF ratio.[54,55,57,58,107,108,111,214] This is attributed to the SWNT density and the concentration of metallic SWNTs in the active layer. Mobility values in excess of 25 cm^2/Vs have been reported for pristine, random SWNT TFTs.[58] These mobility values are significantly better than organic semiconductors and amorphous silicon. As discussed before, the ratio of semiconducting SWNTs is twice that of metallic counterparts. Only a single semiconducting SWNT transport path is needed between source and drain electrodes for gating the TFT. Possibility of metallic pathways is reduced significantly through striping of the SWNT thin films in the channel, as shown in Fig. 4.15c. This eliminates the cross talk between the SWNTs and improves the ON/OFF ratio (significantly decreasing the OFF current), with a small sacrifice in transconductance.[57,60,108] Placing stripes also improves the mobility values of the random networks up to 80 cm^2/Vs. Transfer characteristics of such devices are shown in Fig. 4.15c with a channel width and length of 200 μm and 100 μm, respectively. An SEM image of SWNT thin-film stripes are shown in the inset.

In high-density networks, percolation amongst metallic SWNTs is more likely. At moderate densities, only the semiconducting SWNTs form such a percolating network and the thin film exhibits semiconducting properties.[56,132,215] Thus, best field-effect devices consist of networks where nanotube density allows for percolation among the semiconducting SWNTs only.[109] Recently, simulations and electric force microscopy (EFM) measurements provided new insights

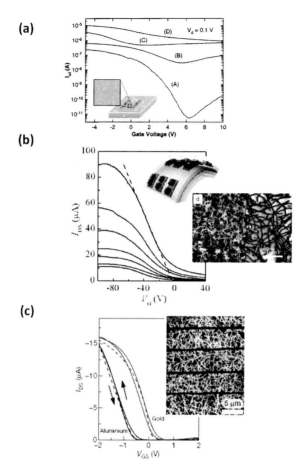

Figure 4.15 (a) Transfer characteristics of SWNT network TFTs with different nanotube densities. (a) 1 tube/μm² and (b–d) >3 tubes/μm². Schematic of the device shown in the inset. (Reproduced with permission from Ref. 50. Copyright 2003 American Institute of Physics.) (b) Transfer characteristics of all-SWNT, transparent, and flexible TFTs with different channel lengths. Inset shows the schematic of the devices and the SEM image of high- and low-density SWNT thin films used for the contacts and the channel, respectively. Channel width is 750 μm, and source drain voltage is −1.0 V. (Reproduced with permission from Ref. 58. Copyright 2006 Wiley-VCH.) (c) Measured and simulated transfer characteristics for striped SWNT TFTs with a channel width and length of 200 μm and 100 μm. Inset shows the SEM image of the stripes. (Reproduced with permission from Ref. 60. Copyright 2008 Nature Publishing Group.)

into the transport mechanism of SWNT TFTs. Results obtained using the setup shown in Fig. 4.16a indicate that large resistance values of MS nanotube junctions severely impede carrier transport for low-density networks below the metallic SWNT percolation threshold. The potential across a biased network is nonuniform, and a sudden voltage drop was observed at nanotube junctions, leading to different electrical performance among identical SWNT networks (Fig. 4.16b). An electronic-phase diagram describing three different gating mechanisms for the SWNT TFTs was proposed on the basis of these results. The regions on the phase diagram can be described as i) a binary network, ii) a blocked semiconductor network, and iii) an unblocked semiconductor network (Fig. 4.16c). In binary networks, neither semiconducting nor metallic SWNTs percolate; however, the current flows between the two subnetworks. In blocked semiconductor networks, percolation is achieved among semiconducting SWNTs; however, they are blocked by crossings with metallic SWNTs. Therefore, in these two cases the channel conductance is modulated by Schottky barrier widths. In the third case, pure, unblocked, semiconducting SWNTs connect source and drain electrodes. These results explain the relatively large variations among reports in the literature on the ON/OFF ratios and ON currents for SWNT TFTs. Other phases include subpercolation (shaded dark-gray region in Fig. 4.16c), where films do not form a percolating network of either metallic or semiconducting SWNTs. The light-gray area in Fig. 4.16c, labeled as *metallic network phase*, on the other hand, represents SWNT films that are shorted out by percolating metallic SWNTs. This work also points out advantages of postdeposition treatments on improving the device characteristics.[216]

Dramatic improvements in TFT characteristics have been obtained from aligned arrays of SWNTs where SWNT–SWNT junctions are absent.[59] A schematic of the arrayed SWNT TFT devices fabricated on quartz substrates in a top-gated design is shown in Fig. 4.17a. An SEM image of the SWNTs in the channel region is shown in Fig. 4.17b. Mobility values for large-channel-length devices (>25 µm) approaching 1,000 cm^2/Vs have been reported, as shown in Fig. 4.17c. The aligned SWNTs can also be transferred onto flexible substrates. Figures 4.17d and 4.17e represent transfer and output characteristics of arrayed SWNT TFTs fabricated on silicon substrates with a bilayer dielectric (epoxy/SiO$_2$) in a bottom-gate configuration, respectively. Channel resistances in arrayed devices

Applications and Devices | 97

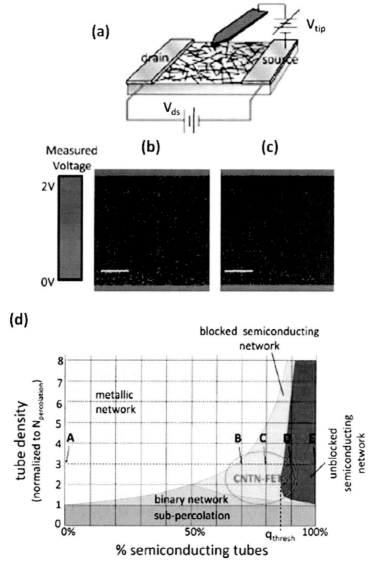

Figure 4.16 (a) Schematic showing EFM measurements. (b, c) Voltage maps for identical SWNT thin films showing drastic differences in electrical performance. (d) Calculated phase diagram for the behavior of SWNT TFTs for a mixture of metallic and semiconducting SWNTs. Tube density is normalized to the percolation threshold. (Reproduced with permission from Ref. 216. Copyright 2009 American Chemical Society.)

have been found to be low compared to random thin films. ON/OFF ratios of the as-fabricated arrays are extremely small due to metallic SWNTs connecting source and drain electrodes but can be improved by electrical breakdown and burning of metallic SWNTs through Joule heating. The effect of the burning process is shown in Fig. 4.17d, where triangles and circles represent transfer characteristics before and after burning, respectively. Corresponding output characteristics following burning are provided in Fig. 4.17e.

The channel length and width of the devices are 12 µm and 200 µm, respectively. The gate voltage varies from −5 V to 5 V (top to bottom). Devices fabricated from SWNTs transferred from the quartz growth substrate to a doped silicon substrate. (Reproduced with permission from Ref. 59. Copyright 2007 Nature Publishing Group.)

4.6.8 Other Devices

More sophisticated devices such as inverters and logic circuits have also been demonstrated with SWNT thin films. However, for these hierarchical devices, *n*-type doping of SWNT networks is required. Polyethyleneimine (PEI) deposition onto thin films has been the most widely utilized route for achieving unipolar *n*-type SWNT TFTs and diodes[57] and complementary inverter devices.[59,108,217] Figure 4.18a shows the transfer characteristics of a *p-n* diode at zero gate bias.

Channel length and width are 8 µm and 200 µm, respectively. Upon the application of forward bias, the current was found to increase rapidly with the voltage. However, when reverse biased, virtually no current flow was observed up to a breakdown voltage of 9 V. Schematics of these devices are also provided in Fig. 4.18a. Transfer characteristics of *p*- and *n*-type SWNT TFTs are shown in Fig. 4.18b, together with the output characteristics of a complementary inverter. Individual SWNT TFTs reported to have mobility values of around 10^{-2} cm^2/Vs and an ON/OFF ratio of 10^4–10^5. A voltage gain of approximately 20 has been reported for SWNT thin-film inverters.[217] More complex logic circuits like NAND and NOR gates using SWNT TFTs have also been demonstrated.[60,217,218]

An alternative approach to using *n*-type SWNT TFTs is to utilize an intrinsically *n*-type material to complement *p*-type SWNT TFTs. For this purpose, individual[219] and networks[220] of ZnO nanowires have been used. Circuit- and system-level integration of SWNT

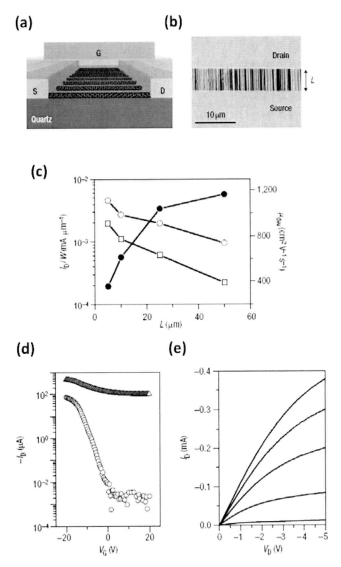

Figure 4.17 (a) Schematic of the top-gated, arrayed SWNT TFT devices fabricated on quartz substrates. (b) SEM image of the channel region of such a device. (c) Width-normalized ON and OFF currents, open circles and squares, respectively, and mobility values of arrayed SWNT TFT devices with respect to channel length. (d) Transfer characteristics for the devices before and after the burning process with the corresponding (e) output characteristics following burning.

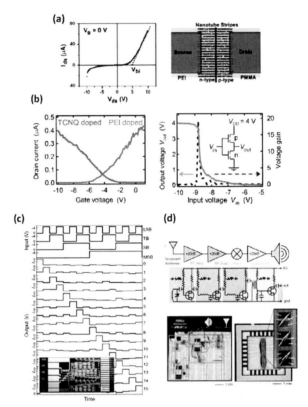

Figure 4.18 (a) Transfer characteristics of a *p-n* diode at zero gate voltage. Schematic of the *p*- and *n*-type channels are shown on the right-hand side. (Reproduced with permission from Ref. 57. Copyright 2004 American Chemical Society.) (b) Transfer characteristics of the PEI and tetracyanoquinodimethane (TCNQ)-doped SWNT TFTs with the corresponding output characteristics of the inverter. Inset shows the schematic of the device. (Reproduced with permission from Ref. 217. Copyright 2008 American Institute of Physics.) (c) Input and output characteristics of the 4-bit decoder composed of 88 SWNT TFTs. First four traces and remaining traces are inputs and outputs, respectively. Inset shows the optical image of these devices. (Reproduced with permission from Ref. 60. Copyright 2008 Nature Publishing Group.) (d) Block and circuit diagrams of a radio system that utilize SWNT TFTs for all of its active components. Optical image of this system with magnified view of SWNT TFTs wire bonded into packages. (Reproduced with permission from Ref. 221. Copyright 2008 American Academy of Sciences.)

thin films has been demonstrated. Cao et al. fabricated a circuit incorporating 88 SWNT TFTs in 4 inverters and a NOR array, with the output of the inverter serving as one of the inputs for the NOR gate. The input and output characteristics of such devices are shown in Fig. 4.18c, with an optical image provided in the inset. The devices have been demonstrated to decode a binary-encoded input of 4 data bits into 16 individual data output lines at kilohertz frequencies.[60] SWNT thin films have also been implemented in analog circuits. Kocabas et al. demonstrated SWNT transistor radios, where all key functional devices, including resonant antennas, fixed radio frequency (RF) amplifiers, RF mixers, and audio amplifiers were composed of SWNT thin films.[221] Block and circuit diagrams of such a system are shown in Fig. 4.18d, together with the constructed circuit and packaged devices. This radio, as demonstrated, was able to receive signals broadcast by a commercial radio station. These results suggest that SWNT thin-film devices can be used for various interesting applications with great success. New device architectures with SWNT thin films are also underway. One of them is the vertical FET, where the source, the active layer, and the drain are stacked vertically relative to the gate.[222] This structure allows transistors in which gate-controlled light emission is possible.

4.7 Conclusions and Outlook

SWNT thin films offer interesting electrical, optical, mechanical, and chemical properties and provide a practical approach to electronic devices compared to individual SWNTs. Films are transparent and conducting, both of which can be tuned, and can be deposited over large areas, at room temperature, from a solution using inexpensive and scalable processes. Their transparency and conductivity presently allow them to be used commercially for some applications. They are highly flexible and provide chemical stability. Progress toward large-scale synthesis, separation, and deposition of SWNTs along with continuous improvement of properties should allow them to find use in high-end transparent conductor applications.

In addition to the transparent conducting applications, TFTs from random networks as well as aligned SWNTs provide pathways for high-performance flexible and transparent active electronic devices capable of driving digital and analog circuitry. The transport in a random network follows the percolation model in two dimensions. Substantial progress in theoretical understanding of the transport in SWNT networks has been made, and this has helped to improve the properties of devices utilizing thin films. The mobility values of random SWNT thin films already exceed state-of-the-art organic and amorphous inorganic materials, while highly aligned SWNT networks offer mobility values in excess of polycrystalline silicon. The primary concern with SWNT TFTs is the ON/OFF ratio, which must be improved from the present values of $\sim 10^5$ to $>10^6$.

References

1. M. S. Dresselhaus, G. Dresselhaus, P. C. Eklund, eds., *Science of Fullerenes and Carbon Nanotubes*, California: Academic Press, 1996.
2. M. S. Dresselhaus, G. Dresselhaus, P. Avouris, eds., *Carbon Nanotubes: Synthesis, Structure, Properties and Applications*, New York: Springer-Verlag, 2001.
3. A. Jorio, M. S. Dresselhaus, G. Dresselhaus, eds., *Carbon Nanotubes: Advanced Topics in Synthesis, Structure, Properties and Applications*, Berlin, Heidelberg: Springer-Verlag, 2008.
4. S. Bandow, M. Takizawa, K. Hirahara, M. Yudasaka, S. Iijima, *Chem. Phys. Lett.*, **337**, 48–54 (2001).
5. S. C. Lyu, B. C. Liu, C. J. Lee, H. K. Kang, C.-W. Yang, C. Y. Park, *Chem. Mater.*, **15**, 3951–3954 (2003).
6. D. S. Bethune, C. H. Klang, M. S. de Vries, G. Gorman, R. Savoy, J. Vazquez, R. Beyers, *Nature*, **363**, 605–607 (1993).
7. C. Journet, M. L. de La Chapelle, W. K. Maser, P. Bernier, A. Loiseau, S. Lefrant, P. Deniard, R. Lee, J. E. Fischer, *Nature*, **388**, 756–758 (1997).
8. T. W. Ebbesen, P. M. Ajayan, *Nature*, **358**, 220–222 (1992).
9. A. Thess, R. Lee, P. Nikolaev, H. Dai, P. Petit, J. Robert, C. Xu, Y. H. Lee, S. G. Kim, A. G. Rinzler, D. T. Colbert, G. E. Scuseria, D. Tomanek, J. E. Fischer, R. E. Smalley, *Science*, **273**, 483–487 (1996).
10. M. Yudasaka, T. Komatsu, T. Ichihashi, S. Iijima, *Chem. Phys. Lett.*, **278**, 102–106 (1997).
11. T. Guo, P. Nikolaev, A. Thess, D. T. Colbert, R. E. Smalley, *Chem. Phys. Lett.*, **243**, 49–54 (1995).
12. M. Endo, K. Takeuchi, K. Kobori, K. Takahashi, H. W. Kroto, A. Sarkar, *Carbon*, **33**, 873–881 (1995).

13. J. Kong, *Chem. Phys. Lett.*, **292**, 567–574 (1998).
14. A. M. Cassell, J. a Raymakers, J. Kong, H. Dai, *J. Phys. Chem. B*, **103**, 6484–6492 (1999).
15. J. Hafner, *Chem. Phys. Lett.*, **296**, 195–202 (1998).
16. J. Kong, H. T. Soh, A. M. Cassell, C. F. Quate, H. Dai, *Nature*, **395**, 878–881 (1998).
17. N. R. Franklin, Y. Li, R. J. Chen, A. Javey, H. Dai, *Appl. Phys. Lett.*, **79**, 4571–4573 (2001).
18. Y. Homma, Y. Kobayashi, T. Ogino, T. Yamashita, *Appl. Phys. Lett.*, **81**, 2261–2263 (2002).
19. R. Seidel, M. Liebau, G. S. Duesberg, F. Kreupl, E. Unger, A. P. Graham, W. Hoenlein, W. Pompe, *Nano Lett.*, **3**, 965–968 (2003).
20. R. G. Lacerda, A. S. Teh, M. H. Yang, K. B. K. Teo, N. L. Rupesinghe, S. H. Dalal, K. K. K. Koziol, D. Roy, G. A. J. Amaratunga, W. I. Milne, M. Chhowalla, D. G. Hasko, F. Wyczisk, P. Legagneux, *Appl. Phys. Lett.*, **84**, 269–271 (Jan. 2004).
21. H. Dai, A. G. Rinzler, P. Nikolaev, A. Thess, D. T. Colbert, R. E. Smalley, *Chem. Phys. Lett.*, **260**, 471–475 (1996).
22. Y. Li, W. Kim, Y. Zhang, M. Rolandi, D. Wang, H. Dai, *J. Phys. Chem. B*, **105**, 11424–11431 (2001).
23. S. M. Bachilo, L. Balzano, J. E. Herrera, F. Pompeo, D. E. Resasco, R. B. Weisman, *J. Am. Chem. Soc.*, **125**, 11186–11187 (2003).
24. Y. Zhang, A. Chang, J. Cao, Q. Wang, W. Kim, Y. Li, N. Morris, E. Yenilmez, J. Kong, H. Dai, *Appl. Phys. Lett.*, **79**, 3155–3157 (2001).
25. S. Huang, B. Maynor, X. Cai, J. Liu, *Adv. Mater.*, **15**, 1651–1655 (2003).
26. Y. Li, D. Mann, M. Rolandi, W. Kim, A. Ural, S. Hung, A. Javey, J. Cao, D. Wang, E. Yenilmez, Q. Wang, J. F. Gibbons, Y. Nishi, H. Dai, *Nano Lett.*, **4**, 317–321 (2004).
27. G. Hong, B. Zhang, B. Peng, J. Zhang, W. M. Choi, J.-Y. Choi, J. M. Kim, Z. Liu, *J. Am. Chem. Soc.*, **131**, 14642–14643 (2009).
28. L. Ding, A. Tselev, J. Wang, D. Yuan, H. Chu, T. P. McNicholas, Y. Li, J. Liu, *Nano Lett.*, **9**, 800–805 (2009).
29. A. R. Harutyunyan, G. Chen, T. M. Paronyan, E. M. Pigos, O. a Kuznetsov, K. Hewaparakrama, S. M. Kim, D. Zakharov, E. a Stach, G. U. Sumanasekera, *Science*, **326**, 116–120 (2009).
30. S. Iijima, *Nature*, **354**, 56–58 (1991).
31. S. Iijima, T. Ichihashi, *Nature*, **363**, 603–605 (1993).

32. G. Moore, *Electron. Mag.*, **38**, 114 (1965).
33. S. J. Tans, M. H. Devoret, H. Dai, A. Thess, R. E. Smalley, L. J. Geerligs, C. Dekker, *Nature*, **386**, 474–477 (1997).
34. A. Javey, J. Guo, Q. Wang, M. Lundstrom, H. Dai, *Nature*, **424**, 654–657 (2003).
35. K. Tsukagoshi, B. W. Alphenaar, H. Ago, *Nature*, **401**, 572–574 (1999).
36. L. E. Hueso, J. M. Pruneda, V. Ferrari, G. Burnell, J. P. Valdes-Herrera, B. D. Simons, P. B. Littlewood, E. Artacho, A. Fert, N. D. Mathur, *Nature*, **445**, 410–413 (2007).
37. W. Liang, M. Bockrath, D. Bozovic, J. H. Hafner, M. Tinkham, H. Park, *Nature*, **411**, 665–669 (2001).
38. Z. Yao, C. L. Kane, C. Dekker, *Phys. Rev. Lett.*, **84**, 2941–2944 (2000).
39. S. J. Tans, A. R. M. Verschueren, C. Dekker, *Nature*, **393**, 49–52 (1998).
40. R. Martel, T. Schmidt, H. R. Shea, T. Hertel, P. Avouris, *Appl. Phys. Lett.*, **73**, 2447–2449 (1998).
41. A. Bachtold, P. Hadley, T. Nakanishi, C. Dekker, *Science*, **294**, 1317–1320 (2001).
42. A. Javey, Q. Wang, A. Ural, Y. Li, H. Dai, *Nano Lett.*, **2**, 929–932 (2002).
43. A. Javey, H. Kim, M. Brink, Q. Wang, A. Ural, J. Guo, P. McIntyre, P. McEuen, M. Lundstrom, H. Dai, *Nat. Mater.*, **1**, 241–246 (2002).
44. A. Javey, J. Guo, D. B. Farmer, Q. Wang, D. Wang, R. G. Gordon, M. Lundstrom, H. Dai, *Nano Lett.*, **4**, 447–450 (2004).
45. L. Hu, D. S. Hecht, G. Gruner, *Chem. Rev.*, **110**, 5790–5844 (2010).
46. M. C. Hersam, *Nat. Nanotechnol.*, **3**, 387–394 (2008).
47. M. S. Arnold, A. A. Green, J. F. Hulvat, S. I. Stupp, M. C. Hersam, *Nat. Nanotechnol.*, **1**, 60–65 (2006).
48. A. A. Green, M. C. Hersam, *Nano Lett.*, **8**, 1417–1422 (2008).
49. M. S. Arnold, S. I. Stupp, M. C. Hersam, *Nano Lett.*, **5**, 713–718 (2005).
50. E. S. Snow, J. P. Novak, P. M. Campbell, D. Park, *Appl. Phys. Lett.*, **82**, 2145–2147 (2003).
51. Z. Wu, Z. Chen, X. Du, J. M. Logan, J. Sippel, M. Nikolou, K. Kamaras, J. R. Reynolds, D. B. Tanner, A. F. Hebard, A. G. Rinzler, *Science*, **305**, 1273–1276 (2004).
52. A. Behnam, L. Noriega, Y. Choi, Z. Wu, A. G. Rinzler, A. Ural, *Appl. Phys. Lett.*, **89**, 093107 (2006).
53. R. K. Das, B. Liu, J. R. Reynolds, A. G. Rinzler, *Nano Lett.*, **9**, 677–683 (2009).

54. K. Bradley, J.-C. P. Gabriel, G. Gruner, *Nano Lett.,* **3**, 1353–1355 (2003).
55. E. Artukovic, M. Kaempgen, D. S. Hecht, S. Roth, G. Gruner, *Nano Lett.,* **5**, 757–760 (2005).
56. L. Hu, D. Hecht, G. Gruner, *Nano Lett.,* **4**, 2513–2517 (2004).
57. Y. Zhou, A. Gaur, S.-H. Hur, C. Kocabas, M. A. Meitl, M. Shim, J. A. Rogers, *Nano Lett.,* **4**, 2031–2035 (2004).
58. Q. Cao, S.-H. Hur, Z.-T. Zhu, Y. G. Sun, C.-J. Wang, M. A. Meitl, M. Shim, J. A. Rogers, *Adv. Mater.,* **18**, 304–309 (2006).
59. S. J. Kang, C. Kocabas, T. Ozel, M. Shim, N. Pimparkar, M. A. Alam, S. V. Rotkin, J. A. Rogers, *Nat. Nanotechnol.,* **2**, 230–236 (2007).
60. Q. Cao, M. A. Alam, J. A. Rogers, N. Pimparkar, J. P. Kulkarni, H.-S. Kim, C. Wang, M. Shim, K. Roy, *Nature*, **454**, 495–500 (2008).
61. K. Tohji, H. Takahashi, Y. Shinoda, N. Shimizu, B. Jeyadevan, I. Matsuoka, Y. Saito, A. Kasuya, S. Ito, Y. Nishina, *J. Phys. Chem. B*, **101**, 1974–1978 (1997).
62. Z. Shi, Y. Lian, F. Liao, X. Zhou, Z. Gu, Y. Zhang, S. Iijima, *Solid State Commun.,* **112**, 35–37 (1999).
63. E. Mizoguti, F. Nihey, M. Yudasaka, S. Iijima, T. Ichihashi, K. Nakamura, *Chem. Phys. Lett.,* **321**, 297–301 (2000).
64. E. Dujardin, T. W. Ebbesen, A. Krishnan, M. M. J. Treacy, *Adv. Mater.,* **10**, 1472–1475 (1998).
65. A. G. Rinzler, J. Liu, H. Dai, P. Nikolaev, C. B. Huffman, F. J. Rodríguez-Macías, P. J. Boul, A. H. Lu, D. Heymann, D. T. Colbert, R. S. Lee, J. E. Fischer, A. M. Rao, P. C. Eklund, R. E. Smalley, *Appl. Phys. A: Mater. Sci. Proc.,* **67**, 29–37 (1998).
66. A. C. Dillon, T. Gennett, K. M. Jones, J. L. Alleman, P. A. Parilla, M. J. Heben, *Adv. Mater.,* **11**, 1354–1358 (1999).
67. S. Bandow, A. M. Rao, K. A. Williams, A. Thess, R. E. Smalley, P. C. Eklund, *J. Phys. Chem. B*, **101**, 8839–8842 (1997).
68. K. B. Shelimov, R. O. Esenaliev, A. G. Rinzler, C. B. Huffman, R. E. Smalley, *Chem. Phys. Lett.,* **282**, 429–434 (1998).
69. A. R. Harutyunyan, B. K. Pradhan, J. Chang, G. Chen, P. C. Eklund, *J. Phys. Chem. B*, **106**, 8671–8675 (2002).
70. G. Sánchez-Pomales, L. Santiago-Rodríguez, N. E. Rivera-Vélez, C. R. Cabrera, *Phys. Status Solidi A*, **204**, 1791–1796 (2007).
71. I. W. Chiang, B. E. Brinson, A. Y. Huang, P. A. Willis, M. J. Bronikowski, J. L. Margrave, R. E. Smalley, R. H. Hauge, *J. Phys. Chem. B*, **105**, 8297–8301 (2001).

72. B. Kitiyanan, W. E. Alvarez, J. H. Harwell, D. E. Resasco, *Chem. Phys. Lett.*, **317**, 497–503 (2000).
73. J. Chen, M. A. Hamon, H. Hu, Y. Chen, A. M. Rao, P. C. Eklund, R. C. Haddon, *Science*, **282**, 95–98 (1998).
74. E. T. Mickelson, C. B. Huffman, A. G. Rinzler, R. E. Smalley, R. H. Hauge, J. L. Margrave, *Chem. Phys. Lett.*, **296**, 188–194 (1998).
75. Y. Wang, Z. Iqbal, S. Mitra, *Carbon*, **43**, 1015–1020 (2005).
76. T. Ramanathan, H. Liu, L. C. Brinson, *J. Polym. Sci. Part B: Polym. Phys.*, **43**, 2269–2279 (2005).
77. P. W. Chiu, G. S. Duesberg, U. Dettlaff-Weglikowska, S. Roth, *Appl. Phys. Lett.*, **80**, 3811–3813 (2002).
78. L. A. Girifalco, M. Hodak, R. S. Lee, *Phys. Rev. B*, **62**, 13104–13110 (2000).
79. M. J. OConnell, P. Boul, L. M. Ericson, C. B. Huffman, Y. Wang, E. H. Haroz, C. Kuper, J. Tour, K. D. Ausman, R. E. Smalley, *Chem. Phys. Lett.*, **342**, 265–271 (2001).
80. M. Zheng, A. Jagota, E. D. Semke, B. A. Diner, R. S. Mclean, S. R. Lustig, R. E. Richardson, N. G. Tassi, *Nat. Mater.*, **2**, 338–342 (2003).
81. M. J. OConnell, S. M. Bachilo, C. B. Huffman, V. C. Moore, M. S. Strano, E. H. Haroz, K. L. Rialon, P. J. Boul, W. H. Noon, C. Kittrell, J. Ma, R. H. Hauge, R. B. Weisman, R. E. Smalley, *Science*, **297**, 593–596 (2002).
82. M. F. Islam, E. Rojas, D. M. Bergey, A. T. Johnson, A. G. Yodh, *Nano Lett.*, **3**, 269–273 (2003).
83. C. Richard, F. Balavoine, P. Schultz, T. W. Ebbesen, C. Mioskowski, *Science*, **300**, 775–778 (2003).
84. M. D. Lay, J. P. Novak, E. S. Snow, *Nano Lett.*, **4**, 603–606 (2004).
85. J. I. Paredes, M. Burghard, *Langmuir*, **20**, 5149–5152 (2004).
86. V. C. Moore, M. S. Strano, E. H. Haroz, R. H. Hauge, R. E. Smalley, J. Schmidt, Y. Talmon, *Nano Lett.*, **3**, 1379–1382 (2003).
87. Y. Wang, C.-an Di, Y. Liu, H. Kajiura, S. Ye, L. Cao, D. Wei, H. Zhang, Y. Li, K. Noda, *Adv. Mater.*, **20**, 4442–4449 (2008).
88. G. Eda, G. Fanchini, A. Kanwal, M. Chhowalla, *J. Appl. Phys.*, **103**, 093118 (2008).
89. J. L. Hudson, M. J. Casavant, J. M. Tour, *J. Am. Chem. Soc.*, **126**, 11158–11159 (2004).
90. C. A. Dyke, J. M. Tour, *Nano Lett.*, **3**, 1215–1218 (2003).
91. J. Chen, A. M. Rao, S. Lyuksyutov, M. E. Itkis, M. A. Hamon, H. Hu, R. W.

Cohn, P. C. Eklund, D. T. Colbert, R. E. Smalley, R. C. Haddon, *J. Phys. Chem. B*, **105**, 2525–2528 (2001).

92. F. Pompeo, D. E. Resasco, *Nano Lett.*, **2**, 369–373 (2002).
93. V. Georgakilas, K. Kordatos, M. Prato, D. M. Guldi, M. Holzinger, A. Hirsch, *J. Am. Chem. Soc.*, **124**, 760–761 (2002).
94. H. Peng, L. B. Alemany, J. L. Margrave, V. N. Khabashesku, *J. Am. Chem. Soc.*, **125**, 15174–15182 (2003).
95. J. Liu, A. G. Rinzler, H. Dai, J. H. Hafner, R. K. Bradley, P. J. Boul, A. Lu, T. Iverson, K. Shelimov, C. B. Huffman, F. Rodriguez-Macias, Y.-S. Shon, T. R. Lee, D. T. Colbert, R. E. Smalley, *Science*, **280**, 1253–1256 (1998).
96. B. J. Landi, H. J. Ruf, J. J. Worman, R. P. Raffaelle, *J. Phys. Chem. B*, **108**, 17089–17095 (2004).
97. S. Giordani, S. D. Bergin, V. Nicolosi, S. Lebedkin, M. M. Kappes, W. J. Blau, J. N. Coleman, *J. Phys. Chem. B*, **110**, 15708–15718 (2006).
98. S. Giordani, S. Bergin, V. Nicolosi, S. Lebedkin, W. J. Blau, J. N. Coleman, *Phys. Status Solidi B*, **243**, 3058–3062 (2006).
99. J. L. Bahr, E. T. Mickelson, M. J. Bronikowski, R. E. Smalley, J. M. Tour, *Chem. Commun.*, 193–194 (2001).
100. S. De, P. E. Lyons, S. Sorel, E. M. Doherty, P. J. King, W. J. Blau, P. N. Nirmalraj, J. J. Boland, V. Scardaci, J. Joimel, J. N. Coleman, *ACS Nano*, **3**, 714–720 (2009).
101. W. Wang, K. A. S. Fernando, Y. Lin, M. J. Meziani, L. M. Veca, L. Cao, P. Zhang, M. M. Kimani, Y.-P. Sun, *J. Am. Chem. Soc.*, **130**, 1415–1419 (2008).
102. S. L. Hellstrom, H. W. Lee, Z. Bao, *ACS Nano*, **3**, 1423–1430 (2009).
103. J. Geng, B.-S. Kong, S. B. Yang, S. C. Youn, S. Park, T. Joo, H.-T. Jung, *Adv. Funct. Mater.*, **18**, 2659–2665 (2008).
104. H. Gu, T. M. Swager, *Adv. Mater.*, **20**, 4433–4437 (2008).
105. E. Kymakis, G. A. J. Amaratunga, *Appl. Phys. Lett.*, **80**, 112–114 (2002).
106. E. S. Snow, P. M. Campbell, M. G. Ancona, J. P. Novak, *Appl. Phys. Lett.*, **86**, 033105 (2005).
107. S.-H. Hur, J. A. Rogers, O. O. Park, *Appl. Phys. Lett*, **86**, 243502 (2005).
108. S. H. Hur, C. Kocabas, A. Gaur, O. O. Park, M. Shim, J. A. Rogers, *J. Appl. Phys.*, **98**, 114302 (2005).
109. H. E. Unalan, G. Fanchini, A. Kanwal, A. Du Pasquier, M. Chhowalla, *Nano Lett.*, **6**, 677–682 (2006).
110. Q. Cao, J. A. Rogers, *Nano Res.*, **1**, 259–272 (2008).

111. T. Ozel, A. Gaur, J. A. Rogers, M. Shim, *Nano Lett.*, **5**, 905–911 (2005).
112. N. Saran, K. Parikh, D. S. Suh, E. Munoz, H. Kolla, S. K. Manohar, *J. Am. Chem. Soc.*, **126**, 4462–4463 (2004).
113. M. A. Meitl, Y. X. Zhou, A. Gaur, S. Jeon, M. L. Usrey, M. S. Strano, J. A. Rogers, *Nano Lett.*, **4**, 1643–1647 (2004).
114. R. C. Tenent, T. M. Barnes, J. D. Bergeson, A. J. Ferguson, B. To, L. M. Gedvilas, M. J. Heben, J. L. Blackburn, *Adv. Mater.*, **21**, 3210–3216 (2009).
115. H.-Z. Geng, K. K. Kim, K. P. So, Y. S. Lee, Y. Chang, Y. H. Lee, *J. Am. Chem. Soc.*, **129**, 7758–7759 (2007).
116. M. Kaempgen, G. S. Duesberg, S. Roth, *Appl. Surf. Sci.*, **252**, 425–429 (2005).
117. N. P. Armitage, J.-C. P. Gabriel, G. Gruner, *J. Appl. Phys.*, **95**, 3228–3230 (2004).
118. B. Dan, G. C. Irvin, M. Pasquali, *ACS Nano*, **3**, 835–843 (2009).
119. T. Mustonen, K. Kordás, S. Saukko, G. Tóth, J. S. Penttilä, P. Helistö, H. Seppä, H. Jantunen, *Phys. Status Solidi B*, **244**, 4336–4340 (2007).
120. W. R. Small, M. in het Panhuis, *Small*, **3**, 1500–1503 (2007).
121. H. Xin, A. T. Woolley, *Nano Lett.*, **4**, 1481–1484 (2004).
122. J. Hone, M. C. Llaguno, N. M. Nemes, A. T. Johnson, J. E. Fischer, D. A. Walters, M. J. Casavant, J. Schmidt, R. E. Smalley, *Appl. Phys. Lett.*, **77**, 666–668 (2000).
123. J. E. Fischer, W. Zhou, J. Vavro, M. C. Llaguno, C. Guthy, R. Haggenmueller, M. J. Casavant, D. E. Walters, R. E. Smalley, *J. Appl. Phys.*, **93**, 2157–2163 (2003).
124. C. Lim, D.-H. Min, S.-B. Lee, *Appl. Phys. Lett.*, **91**, 243117 (2007).
125. Y. Zhou, L. Hu, G. Gruner, *Appl. Phys. Lett.*, **88**, 123109 (2006).
126. M. H. Kim, J.-Y. Choi, H. K. Choi, S.-M. Yoon, O. O. Park, D. K. Yi, S. J. Choi, H.-J. Shin, *Adv. Mater.*, **20**, 457–461 (2008).
127. D. Stauffer, A. Aharony, *Introduction to Percolation Theory*, London: Taylor and Francis, 1992.
128. M. Sahimi, *Applications of Percolation Theory*, London: Taylor and Francis, 1994.
129. I. Balberg, C. H. Anderson, S. Alexander, N. Wagner, *Phys. Rev. B*, **30**, 3933–3943 (1984).
130. M. Stadermann, S. J. Papadakis, M. R. Falvo, J. Novak, E. Snow, Q. Fu, J. Liu, Y. Fridman, J. J. Boland, R. Superfine, S. Washburn, *Phys. Rev. B*, **69**, 201402 (2004).

131. B. Vigolo, C. Coulon, M. Maugey, C. Zakri, P. Poulin, *Science*, **309**, 920–923 (2005).
132. D. Hecht, L. Hu, G. Gruner, *Appl. Phys. Lett.*, **89**, 133112 (2006).
133. G. Fanchini, H. E. Unalan, M. Chhowalla, *Appl. Phys. Lett.*, **88**, 191919 (2006).
134. R. M. A. Azzam, N. M. Bashara, *Ellipsometry and Polarized Light*, Amsterdam: North Holland, 1987.
135. F. Wang, G. Dukovic, L. E. Brus, T. F. Heinz, *Science*, **308**, 838–841 (2005).
136. E. A. Palik, ed., *Handbook of Optical Constants of Solids*, New York: Academic Press, 1975.
137. I. Balberg, *Phys. Rev. Lett.*, **59**, 1305–1308 (1987).
138. G. Thomas, *Nature*, **389**, 907–908 (1997).
139. D. H. Zhang, K. Ryu, X. L. Liu, E. Polikarpov, J. Ly, M. E. Tompson, C. W. Zhou, *Nano Lett.*, **6**, 1880–1886 (2006).
140. Z. Li, H. R. Kandel, E. Dervishi, V. Saini, Y. Xu, A. R. Biris, D. Lupu, G. J. Salamo, A. S. Biris, *Langmuir*, **24**, 2655–2662 (2008).
141. B. B. Parekh, G. Fanchini, G. Eda, M. Chhowalla, *Appl. Phys. Lett.*, **90**, 121913 (2007).
142. R. Krupke, F. Hennrich, H. V. Löhneysen, M. M. Kappes, *Science*, **301**, 344–347 (2003).
143. R. Krupke, S. Linden, M. Rapp, F. Hennrich, *Adv. Mater.*, **18**, 1468–1470 (2006).
144. M. Zheng, A. Jagota, M. S. Strano, A. P. Santos, P. W. Barone, S. G. Chou, B. A. Diner, M. S. Dresselhaus, R. S. McLean, G. B. Onoa, G. G. Samsonidze, E. D. Semke, M. L. Usrey, D. J. Walls, *Science*, **302**, 1545–1548 (2003).
145. X. Tu, S. Manohar, A. Jagota, M. Zheng, *Nature*, **460**, 250–253 (2009).
146. M. S. Strano, Z. Zheng, A. Jagota, G. B. Onoa, D. A. Heller, P. W. Barone, M. L. Usrey, *Nano Lett.*, **4**, 543–550 (2004).
147. F. Chen, B. Wang, Y. Chen, L.-J. Li, *Nano Lett.*, **7**, 3013–3017 (2007).
148. A. Nish, J.-Y. Hwang, J. Doig, R. J. Nicholas, *Nat. Nanotechnol.*, **2**, 640–646 (2007).
149. M. Kanungo, H. Lu, G. G. Malliaras, G. B. Blanchet, *Science*, **323**, 234–237 (2009).
150. Y. Miyata, Y. Maniwa, H. Kataura, *J. Phys. Chem. B*, **110**, 25–29 (2006).
151. D. Chattopadhyay, L. Galeska, F. A. Papadimitrakopoulos, *J. Am. Chem. Soc.*, **125**, 3370–3375 (2003).

152. J. Chen, C. Klinke, A. Afzali, P. Avouris, *Appl. Phys. Lett.*, **86**, 123108 (2005).
153. X. Li, Y. Liu, D. Shi, Y. Sun, G. Yu, D. Zhu, H. Liu, X. Liu, D. Wu, *Appl. Phys. Lett.*, **87**, 243102–243103 (2005).
154. L. An, Q. Fu, C. Lu, J. Liu, *J. Am. Chem. Soc.*, **126**, 10520–10521 (2004).
155. M. S. Strano, C. A. Dyke, M. L. Usrey, P. W. Barone, M. J. Allen, H. Shan, C. Kittrell, R. H. Hauge, J. M. Tour, R. E. Smalley, *Science*, **301**, 1519–1522 (2003).
156. C. Wang, Q. Cao, T. Ozel, A. Gaur, J. A. Rogers, M. Shim, *J. Am. Chem. Soc.*, **127**, 11460–11468 (2005).
157. M. S. Strano, *J. Am. Chem. Soc.*, **125**, 16148–16153 (2003).
158. C. W. Lee, X. Han, F. Chen, J. Wei, Y. Chen, M. B. Chan-Park, L.-J. Li, *Adv. Mater.*, (2009).
159. Z. Chen, X. Du, M.-H. Du, C. D. Rancken, H.-P. Cheng, A. G. Rinzler, *Nano Lett.*, **3**, 1245–1249 (2003).
160. G. Fanchini, H. E. Unalan, M. Chhowalla, *Appl. Phys. Lett.*, **90**, 092114 (2007).
161. R. Lee, H. Kim, J. Fischer, A. Thess, R. Smalley, *Nature*, **388**, 255–256 (1997).
162. A. Rao, P. Eklund, S. Bandow, A. Thess, R. Smalley, *Nature*, **388**, 257–259 (1997).
163. T. M. Barnes, J. L. Blackburn, T. J. Coutts, M. J. Heben, *ACS Nano*, **2** (2008).
164. G. Li, V. Shrotriya, J. Huang, Y. Yao, T. Moriarty, K. Emery, Y. Yang, *Nat. Mater.*, **4**, 864–868 (2005).
165. M. Reyes-Reyes, K. Kim, D. L. Carroll, *Appl. Phys. Lett.*, **87**, 083506 (2005).
166. N. D. Jonge, M. Allioux, M. Doytcheva, M. Kaiser, K. B. K. Teo, R. G. Lacerda, W. I. Milne, *Appl. Phys. Lett.*, **85**, 1607 (2004).
167. J. van de Lagemaat, T. M. Barnes, G. Rumbles, S. E. Shaheen, T. J. Coutts, C. Weeks, I. Levitsky, J. Peltola, P. Glatkowski, *Appl. Phys. Lett.*, **88**, 233503 (2006).
168. B. Shan, K. Cho, *Phys. Rev. Lett.*, **94**, 236602 (2005).
169. H. Ago, K. Petritsch, M. S. P. Shaffer, A. H. Windle, R. H. Friend, *Adv. Mater.*, **11**, 1281–1285 (1999).
170. S. Bhattacharyya, E. Kymakis, G. A. J. Amaratunga, *Chem. Mater.*, **16**, 4819–4823 (2004).

171. E. Kymakis, *Sol. Energy Mater. Sol. Cells*, **80**, 465–472 (2003).
172. E. Kymakis, I. Alexandrou, G. A. J. Amaratunga, *J. Appl. Phys.*, **93**, 1764–1768 (2003).
173. A. D. Pasquier, H. E. Unalan, A. Kanwal, S. Miller, M. Chhowalla, *Appl. Phys. Lett.*, **87**, 203511 (2005).
174. R. C. Tenent, T. M. Barnes, J. D. Bergeson, A. J. Ferguson, B. To, L. M. Gedvilas, M. J. Heben, J. L. Blackburn, *Adv. Mater.*, **21**, 3210–3216 (2009).
175. G. Fanchini, S. Miller, B. B. Parekh, M. Chhowalla, *Nano Lett.*, **8**, 2176–2179 (2008).
176. H. E. Unalan, P. Hiralal, D. Kuo, B. Parekh, G. Amaratunga, M. Chhowalla, *J. Mater. Chem.*, **18**, 5909 (2008).
177. Z. Li, V. P. Kunets, V. Saini, Y. Xu, E. Dervishi, G. J. Salamo, A. R. Biris, A. S. Biris, *Appl. Phys. Lett.*, **93**, 243117 (2008).
178. Z. Li, V. P. Kunets, V. Saini, Y. Xu, E. Dervishi, G. J. Salamo, A. R. Biris, A. S. Biris, *ACS Nano*, **3**, 1407–1414 (2009).
179. J. Wei, Y. Jia, Q. Shu, Z. Gu, K. Wang, D. Zhuang, G. Zhang, Z. Wang, J. Luo, A. Cao, D. Wu, *Nano Lett.*, **7**, 2317–2321 (2007).
180. Y. Jia, J. Wei, K. Wang, A. Cao, Q. Shu, X. Gui, Y. Zhu, D. Zhuang, G. Zhang, B. Ma, L. Wang, W. Liu, Z. Wang, J. Luo, D. Wu, *Adv. Mater.*, **20**, 4594–4598 (2008).
181. Y. Jia, A. Cao, X. Bai, Z. Li, L. Zhang, N. Guo, J. Wei, K. Wang, H. Zhu, D. Wu, P. M. Ajayan, *Nano Lett.*, dx.doi.org/10.1021/nl2002632 (2011).
182. L. Hu, G. Gruner, D. Li, R. B. Kaner, J. Cech, *J. Appl. Phys.*, **101**, 016102 (2007).
183. D. S. Hecht, R. J. A. Ramirez, M. Briman, E. Artukovic, K. S. Chichak, J. F. Stoddart, G. Gruner, *Nano Lett.*, **6**, 2031–2036 (2006).
184. A. Star, Y. Lu, K. Bradley, G. Gruner, *Nano Lett.*, **4**, 1587–1591 (2004).
185. L. Hu, Y. L. Zhao, K. Ryu, C. Zhou, J. F. Stoddart, G. Gruner, *Adv. Mater.*, **20**, 939–946 (2008).
186. C. M. Aguirre, S. Auvray, S. Pigeon, R. Izquierdo, P. Desjardins, R. Martel, *Appl. Phys. Lett.*, **88**, 183104 (2006).
187. J. Li, L. Hu, L. Wang, Y. Zhou, G. Gruner, T. J. Marks, *Nano Lett.*, **6**, 2472–2477 (2006).
188. E. C.-W. Ou, L. Hu, G. C. R. Raymond, O. K. Soo, J. Pan, Z. Zheng, Y. Park, D. Hecht, G. Irvin, P. Drzaic, G. Gruner, *ACS Nano*, **3**, 2258–2264 (2009).
189. J. Li, G. Ner, L. Hu, J. Liu, L. Wang, T. J. Marks, G, *Appl. Phys. Lett*, **93**, 083306 (2008).

190. Z. Yu, L. Hu, Z. Liu, M. Sun, M. Wang, G. Gruner, Q. Pei, *Appl. Phys. Lett.*, **95**, 203304 (2009).
191. P. Simon, Y. Gogotsi, *Nat. Mater.*, **7**, 845–854 (2008).
192. M. Kaempgen, J. Ma, G. Gruner, G. Wee, S. G. Mhaisalkar, *Appl. Phys. Lett.*, **90**, 264104 (2007).
193. M. Kaempgen, C. K. Chan, J. Ma, Y. Cui, G. Gruner, *Nano Lett.*, **9**, 1872–1876 (2009).
194. C. Yu, C. Masarapu, J. Rong, B. Wei, H. Jiang, *Adv. Mater.*, **21**, 4793–4797 (2009).
195. A. Kiebele, G. Gruner, *Appl. Phys. Lett.*, **91**, 144104 (2007).
196. J. Kong, N. R. Franklin, C. Zhou, M. G. Chapline, S. Peng, K. Cho, H. Dai, *Science*, **287**, 622 (2000).
197. J. Li, Y. Lu, Q. Ye, M. Cinke, J. Han, M. Meyyappan, *Nano Lett.*, **3**, 929–933 (2003).
198. J. P. Novak, E. S. Snow, E. J. Houser, D. Park, J. L. Stepnowski, R. A. McGill, *Appl. Phys. Lett.*, **83**, 4026 (2003).
199. F. Wang, H. Gu, T. M. Swager, *J. Am. Chem. Soc.*, **130**, 5392–5393 (2008).
200. Y. Sun, H. H. Wang, *Adv. Mater.*, **19**, 2818–2823 (2007).
201. Y. Sun, H. H. Wang, *Appl. Phys. Lett.*, **90**, 213107 (2007).
202. E. S. Snow, F. K. Perkins, E. J. Houser, S. C. Badescu, T. L. Reinecke, *Science*, **307**, 1942–1945 (2005).
203. E. S. Snow, F. K. Perkins, *Nano Lett.* **5**, 2414–2417 (2005).
204. J. Sandler, *Polymer*, **44**, 5893–5899 (2003).
205. N. Li, Y. Huang, F. Du, X. He, X. Lin, H. Gao, Y. Ma, F. Li, Y. Chen, P. C. Eklund, *Nano Lett.*, **6**, 1141–1145 (2006).
206. H. Xu, S. M. Anlage, L. Hu, G. Gruner, *Appl. Phys. Lett.*, **90**, 183119 (2007).
207. M. A. Seo, J. H. Yim, Y. H. Ahn, F. Rotermund, D. S. Kim, S. Lee, H. Lim, *Appl. Phys. Lett.*, **93**, 231905 (2008).
208. M. E. Itkis, F. Borondics, A. Yu, R. C. Haddon, *Science*, **312**, 413–416 (2006).
209. B. Pradhan, K. Setyowati, H. Liu, D. H. Waldeck, J. Chen, *Nano Lett.*, **8**, 1142–1146 (2008).
210. L. Hu, D. S. Hecht, G. Gruner, *Appl. Phys. Lett.*, **94**, 081103 (2009).
211. S. Heinze, J. Tersoff, R. Martel, V. Derycke, J. Appenzeller, P. Avouris, *Phys. Rev. Lett.*, **89**, 106801 (2002).

212. T. W. Tombler, C. Zhou, J. Kong, H. Dai, *Appl. Phys. Lett.,* **76**, 2412–2414 (2000).
213. M. S. Fuhrer, J. Nygård, L. Shih, M. Forero, Y.-G. Yoon, M. S. C. Mazzoni, H. J. Choi, J. Ihm, S. G. Louie, A. Zettl, P. L. McEuen, *Science*, **288**, 494–497 (2000).
214. T. Takenobu, T. Takahashi, T. Kanbara, K. Tsukagoshi, Y. Aoyagi, Y. Iwasa, *Appl. Phys. Lett.,* **88**, 033511 (2006).
215. S. Kumar, J. Murthy, M. Alam, *Phys. Rev. Lett.,* **95**, 3–6 (2005).
216. M. A. Topinka, M. W. Rowell, D. Goldhaber-Gordon, M. D. McGehee, D. S. Hecht, G. Gruner, *Nano Lett.,* **9**, 1866–1871 (2009).
217. R. Nouchi, H. Tomita, A. Ogura, H. Kataura, M. Shiraishi, *Appl. Phys. Lett.,* **92**, 253507 (2008).
218. P.-W. Chiu, C.-H. Chen, *Appl. Phys. Lett.,* **92**, 063511 (2008).
219. G. Jo, W.-K. Hong, J. I. Sohn, M. Jo, J. Shin, M. E. Welland, H. Hwang, K. E. Geckeler, T. Lee, *Adv. Mater.,* **21**, 2156–2160 (2009).
220. H. E. Unalan, Y. Zhang, P. Hiralal, S. Dalal, D. Chu, G. Eda, K. B. K. Teo, M. Chhowalla, W. I. Milne, G. Amaratunga, *Appl. Phys. Lett.,* **94**, 163501 (2009).
221. C. Kocabas, H. Kim, T. Banks, J. A. Rogers, A. A. Pesetski, J. E. Baumgardner, S. Krishnaswamy, H. Zhang, *Proc. Natl. Acad. Sci. U S A,* **105**, 1405 (2008).
222. B. Liu, M. A. McCarthy, Y. Yoon, D. Y. Kim, Z. Wu, F. So, P. H. Holloway, J. R. Reynolds, J. Guo, A. G. Rinzler, *Adv. Mater.,* **20**, 3605–3609 (2008).

Chapter 5

Single-Walled Carbon Nanotube-Based Solution-Processed Organic Optoelectronic Devices

Ming Shao[a] and Bin Hu[b]

[a]Oak Ridge National Laboratory, Center for Nanophase Materials Sciences,
PO Box 2008, MS-6488 Oak Ridge, TN 37831-6493, USA
[b]403 Ferris Hall, Department of Materials Science and Engineering,
University of Tennessee, Knoxville, TN 37996-2100, USA
shaom@ornl.gov, bhu@utk.edu

5.1 Introduction

Solution-processed optoelectronic devices have attracted great interest due to the advantages of low cost and large-area flexible processing. Recently, much attention has been paid to combine nanomaterial single-walled carbon nanotubes (SWCNTs) for improving the performance of organic optoelectronic devices. One potential application of SWCNTs in organic optoelectronics could be as a new class of transparent and conductive electrodes for displays, solar cells, and solid-state lighting devices.[1–3] Compared with the standard electrode material indium tin oxide (ITO), SWCNT-based electrodes would have the same conductive properties with high transparencies, would be flexible, and could be deposited at room temperature on a wide variety of substrates. In parallel with the effort

---

Luminescence: The Instrumental Key to the Future of Nanotechnology
Edited by Adam M. Gilmore
Copyright © 2014 Pan Stanford Publishing Pte. Ltd.
ISBN 978-981-4241-95-3 (Hardcover), 978-981-4267-72-4 (eBook)
www.panstanford.com

on the SWCNTs as the electrode, people are interested in doping SWCNTs into a conjugated polymer. The blend of SWCNTs with a conjugated polymer can not only offer the attractive properties of both components but also introduce new properties by morphologic modification and electronic interaction of each component.

In this chapter, we demonstrate the approach of doping SWCNTs into a conjugated polymer for organic optoelectronic devices: organic light-emitting diode (OLED), organic photovoltaic (OPV) cell, etc. The electroluminescence efficiency of an OLED can be enhanced by optimizing the doping concentration of SWCNTs into a hole-transporting polymer and a light-emitting material. We found the dispersed SWCNTs play different roles when used as polymer:SWCNT composites in OLEDs. In hole-transporting material, the SWCNTs improve the conductivity of the composite film and they also act as the exciton-quenching centers when their concentration is high enough. When used in a light-emitting layer, SWCNTs act as an n-type dopant to improve the electron transport. Thus, the efficiency of the OLED is enhanced by the balanced bipolar injection. Our results elucidate the role of SWCNTs in OLEDs and are helpful for improving OLED performance by using SWCNTs.

Besides the electroluminescence response, the PV responses upon doping SWCNTs into a conjugated polymer are investigated. The dispersed SWCNTs can influence the charge transport and exciton dissociation at the tube–chain interface in the SWCNT:polymer composites. In the low SWCNT doping concentration, the SWCNT mainly improves the bipolar injection. As the SWCNT doping concentration continues to increase, the interfacial exciton dissociation becomes dominated, giving rise to an increased PV response.

5.2 Effects of SWCNTs on the Electroluminescent Performance of Organic Light-Emitting Diodes

Recently, much attention has been paid to using nanomaterials for improving the electroluminescent (EL) performance of OLEDs.[4–7] Among those nanomaterials, SWCNTs are expected to be used in OLEDs due to their facile synthesis process, excellent mechanical properties, and good carrier-transporting ability. Although SWCNTs have been reported to improve the EL efficiency when used as polymer:SWCNTs composites, their role in OLEDs is not clear yet. Several possible explanations have been proposed, including a) improving hole injection and transportation[8], b) improving the conductivity of polymer films[9], c) blocking holes in the polymer composite[10], and d) trapping holes of SWCNTs in a hole-conducting polymer.[11] However, these explanations are not suitable for an explanation of all the phenomena occurring in the reported literature. We investigated the roles of SWCNTs on the EL performance of OLEDs when mixing them with a hole-conducting material and a light-emitting material, respectively. The effects of SWCNTs in OLEDs have been elucidated. Our findings provide a clear recognition of the role of SWCNTs in OLEDs and will be helpful for improving OLED performance by using SWCNTs.

The SWCNTs were synthesized by a laser ablation method[12] and were purified by hydrothermal and chemical treatments. The hole injection material poly(3,4-ethylenedioxythiophene) (PEDOT) doped with poly(styrene sulfonate) (PSS) (Baytron P 4083) (PEDOT) was acquired from the Bayer Company. The hole-transporting material poly(9-vinylcarbazole) (PVK) and the electron-transporting material 2-(4-biphenylyl)-5-(4-tert-butylphenyl)-1,3,4-oxadiazole (PBD) were purchased from the Sigma Aldrich Company and used as received. The light-emitting conjugated polymer polyfluorence (PFO; purchased from the H. W. Sand Company) and Super-Yellow (purchased from the Merck Company) were used as received. The electrophosphorescent material iridium (III) tris(2-(4-totyl) pyridinato-N,C^2) [Ir(mppy)$_3$] was purchased from American Dye Source Inc. To investigate the role of SWCNTs in the performance of the OLEDs, three types of device architecture were fabricated. The first type of devices has the structure of ITO/PEDOT:SWCNT (x wt%, 40 nm)/PFO (80 nm)/Ca (20 nm)/Al (80 nm), where x is 0, 0.001, 0.01, and 0.02, respectively relative to solid PEDOT content. The second one is ITO/PEDOT:SWCNT (0.01 wt% relative to solid PEDOT content, 40 nm)/PVK:PBD:Ir(mppy)$_3$ (69 wt% : 30 wt% : 1 wt%, 60 nm)/Ca (20 nm)/Al (80 nm), which use Ir(mppy)$_3$ as a phosphorescent emissive material. The third one has a structure of ITO/PEDOT (40 0m)/Super-Yellow:SWCNT (x wt%, 80 nm)/Ca (20 nm)/Al (80 nm), where x is 0 and 0.005, respectively.

To get the uniform dispersion of SWCNTs in PEDOT, we suspended the SWCNTs in deionized water and ultrasonicated the

SWCNT solution for 30 minutes, and then we mixed the SWCNT solution with the PEDOT solution and deionized water to form the composites at the designed doping concentration. For the Super-Yellow:SWCNT composite preparation, SWCNTs dispersed in chloroform were added to the Super-Yellow chloroform solution to give the desired doping concentration. The PEDOT:SWCNTs composites were spin-coated onto an ITO-coated glass substrate, and the thickness of the film was 40 nm. The light-emitting layers for the three types of devices were all formed by spin coating from their chloroform solutions. Finally, the calcium (Ca) and aluminum (Al) electrodes were deposited sequentially by thermal evaporation under the vacuum of 2×10^{-6} Torr, and the deposition rate was typically about 1 Å/s. The thicknesses and morphology of the films were measured by Vecco atomic force microscopy (AFM) using a tapping mode. The resistivity and conductivity of PEDOT:SWCNT composite films on glass substrates were measured by a four-probe method with a Keithley source meter. The current and voltage characteristics of the OLEDs were measured with a Keithley 2400 source meter. The EL spectra of OLEDs were recorded by a HORIBA Jobin Yvon spectrometer. The luminance and efficiency of OLEDs were measured by an OLED testing system calibrated by the National Institute of Standards and Technology.

We investigated the roles of SWCNTs in a hole-conducting layer on the performance of OLEDs at first. Figure 5.1a shows the current density–voltage (*J*–*V*) characteristics of OLEDs employing PEDOT:SWCNT composites with various doping concentrations under the forward bias. By increasing the SWCNT concentration in PEDOT, the current density of devices becomes higher under the same voltage, which means the hole injection and/or transport is improved by introducing SWCNTs in the PEDOT layer of devices. However, the EL intensity didn't show a monotonous increase when increasing the SWCNT concentration in PEDOT under the same current density. From Fig. 5.1b we can see, when the SWCNT concentration in PEDOT is no higher than 0.01 wt%, the higher the doping concentration, the stronger the EL intensity. But when the doping concentration of SWCNT reaches 0.02 wt%, it will deteriorate the EL performance of devices and the EL intensity becomes lower than that of the device without SWCNT doping. Our results have some difference from those reported by Woo et al.[11] In their case, the doping concentration of SWCNTs in PEDOT is no less than 0.05

wt%. Although they observed the enhancement of current density when increasing the SWCNT doping concentration in PEDOT, which is similar with our case, they only found reduced EL brightness in devices employing PEDOT:SWCNT composites relative to the device without using SWCNTs. As a result, their explanation of SWCNTs as hole traps in a PEDOT film should be reconsidered, since there will be no EL intensity enhancement in our case if the injected holes were initially trapped by SWCNTs in a PEDOT film.

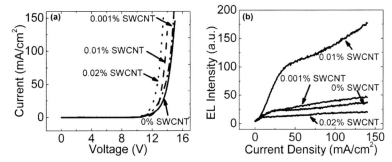

Figure 5.1 The *J–V* curves (a) and the EL intensity–current density curves (b) of devices with the structure of ITO/PEDOT:SWCNT (*x* wt%, 40 nm)/PFO (80 nm)/Ca (20 nm)/Al (80 nm).

We investigated the conductivity of PEDOT:SWCNT composite films with different doping concentrations of SWCNTs. It is shown in Table 5.1 that when the SWCNT concentration changed from 0 wt% to a high value of 0.66 wt%, the resistivity of the composite film decrease gradually and the conductivity increased monotonously. This can be ascribed to the increase in partial conductive paths when introducing SWCNTs into the polymer film[13], since the metallic components in SWCNTs have better conductivity than PEDOT. As a result, the enhancement of EL intensity shown in Fig. 5.1b when increasing the SWCNT concentration in PEDOT from 0.001 wt% to 0.01 wt% can be ascribed to the increase of the conductivity in the PEDOT film with SWCNT doping, thus leading to the improvement of hole transport in devices. It should be noted that although SWCNTs can be beneficial in improving the carrier transport, they also can act as trap sites to quench the excitons.[14–15] It is well known that when excitons are formed in the emissive layer, they can migrate during their lifetime and the singlet excitons usually have a migration distance of ~20 nm.[16] Thus if the excitons are formed near the PEDOT:SWCNT/

PFO interface, they can easily migrate to the SWCNT-trapping sites and thus be quenched, provided that the SWCNT concentration is high enough near the interface.

Table 5.1 The resistivity (ρ) and conductivity (C) of PEDOT:SWCNT composite films with various SWCNT concentrations

SWCNT concentration (%)	ρ ($\Omega \cdot$cm)	C (S/cm)
0	44902	2.23×10^{-5}
0.0004	40778	2.45×10^{-5}
0.0042	31709	3.15×10^{-5}
0.0010	29775	3.36×10^{-5}
0.6622	23666	4.23×10^{-5}

We notice that PEDOT has the highest occupied molecular orbital (HOMO) energy level of −5.2 eV, while PFO has a HOMO energy level of −6.0 eV[16], so there is an energy barrier of 0.8 eV for holes to overcome. In our devices with the structure of ITO/PEDOT:SWCNT/PFO/Ca/Al, the lowest unoccupied molecular orbital (LUMO) energy level of the light-emitting material PFO aligns with the Fermi level of the Ca cathode. Consequently, the electron transport from the cathode into the light-emitting layer is not hindered. However, since an energy barrier between the HOMO levels prevents holes to enter the light-emitting layer, the exciton recombination zone in the PFO layer must be close to the interface of the PEDOT:SWCNT/PFO interface. As a result, the excitons formed in the PFO layer can migrate to the PEDOT:SWCNT layer and be quenched by the SWCNTs if the SWCNT concentration is high enough in the PEDOT:SWCNT film. That is the reason why we see reduced EL intensity in OLEDs when the SWCNT concentration is as high as 0.02 wt% in the PEDOT:SWCNT film compared with a pure PEDOT film.

Figure 5.2a shows the luminance-voltage (*L–V*) curves of OLEDs using pure PEDOT and PEDOT:SWCNT (0.01 wt%) composites. The maximum luminance of the device with SWCNTs in PEDOT is 2,150 cd/m^2, which is 1.4 times as high as the one without SWCNTs. Figure 5.2b,c illustrates the EL efficiency versus the current density curves of the device with and without SWCNT doping. We can see the maximum luminous efficiency and the maximum external quantum efficiency (EQE) of the device with SWCNTs in PEDOT are 0.32 cd/A (1.2 times than that without SWCNTs) and 0.22% (1.4 times than that without SWCNTs), respectively. The enhancement of EL efficiency in devices employing SWCNTs in PEDOT can be ascribed to the improvement of hole transport, thus leading to more balanced holes and electrons in the light-emitting layer.

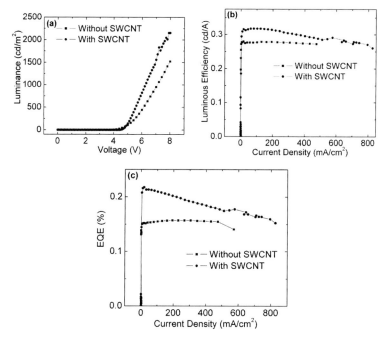

Figure 5.2 The L–V curves (a), the luminous efficiency (b), and the EQE (c) characters of devices without and with SWCNT doping. The device structure is ITO/PEDOT (40 nm)/PFO (80 nm)/Ca (20 nm)/Al (80 nm) and ITO/PEDOT:SWCNT (0.01 wt%, 40 nm)/PFO (80 nm)/Ca (20 nm)/Al (80 nm), respectively.

As we mentioned above, the improvement of EL performance in devices with the SWCNT concentration no higher than 0.01 wt% lies in their role in increasing hole transport in the conducting PEDOT film. Consequently, for OLEDs with holes as the minor carriers in the light-emitting layer (such as our PFO case) the EL performance of devices can be improved by doping a very low amount of SWCNTs into the hole-conducting polymer. To investigate the impact of the mixture of the hole-conducting polymer and SWCNTs on the EL performance of OLEDs with balanced electrons and holes in the light-emitting layer, we have fabricated devices with the structure of ITO/PEDOT:SWCNT

(0.01 wt%)/PVK:PBD:Ir(mppy)$_3$/Ca/Al. It has been reported that when the mixture of PVK, PBD, and Ir(mppy)$_3$ is used as a light-emitting layer in OLEDs, highly efficient electrophosphorescence has been achieved due to balanced electrons and holes in the light-emitting layer.[17]

Figure 5.3a shows the *J–V* curves of the electrophosphorescent devices with and without SWCNTs doped in the PEDOT films. As can be seen from the figure, the *J–V* curve of the device with a PEDOT:SWCNT composite shifts to a lower voltage compared with the one without SWCNTs, which is similar to the PFO case. However, the maximum EL efficiency in the device with SWCNTs is reduced relative to the one without SWCNTs (shown in Fig. 5.3a,b). As we expected, a low concentration of SWCNTs in a PEDOT film will

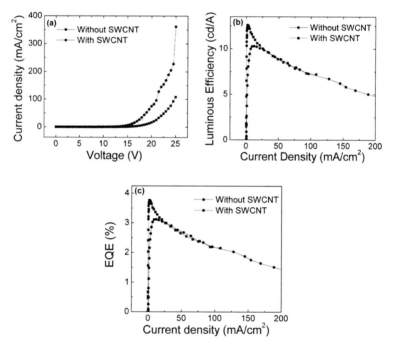

Figure 5.3 The *J–V* curves (a), the luminous efficiency (b), and the EQE (c) characters of the electrophosphorescent devices without and with SWCNT doping. The device structure is ITO/PEDOT (40 nm)/PVK:PBD:Ir(mppy)$_3$ (69 wt% : 30 wt% : 1 wt%, 60 nm)/Ca (20 nm)/Al (80 nm) and ITO/PEDOT:SWCNT (0.01 wt%, 40 nm)/PVK:PBD:Ir(mppy)$_3$ (69 wt% : 30 wt% : 1 wt%, 60 nm)/Ca (20 nm)/Al (80 nm), respectively.

enhance the hole transport in devices and shift the J–V curve to a lower voltage. Nevertheless, in OLEDs with balanced holes and electrons, introducing more holes to the light-emitting layer will cause unbalanced electrons and holes and thus reduce the EL efficiency.

Figure 5.4 shows the AFM images of the PEDOT film with and without SWCNT dopants. In the case of a pure PEDOT film, the islands have big sizes. But in the PEDOT:SWCNT (0.01 wt%) composite film, the islands have small sizes, which means the contact between the interface of PEDOT and the light-emitting layer will be improved.

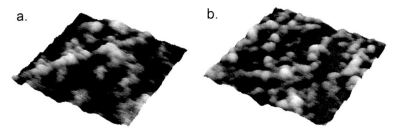

Figure 5.4 The AFM images of the pure PEDOT film (a) and the PEDOT:SWCNT (0.01 wt%) composite film (b) measured by the tapping mode.

Besides acting as dopants in a hole-conducting polymer in OLEDs, SWCNTs can also be used in the light-emitting polymer layer. To investigate an SWCNT's role in the EL performance when it is in the light-emitting layer, we have fabricated devices with the structure of ITO/PEDOT/Supper-Yellow:SWCNT/Ca/Al. Figure 5.5a shows the J–V characters of devices with and without SWCNTs in the light-emitting layer.

Introducing SWCNTs into the Supper-Yellow layer shifts the J–V curve to the lower voltage compared with a pure Supper-Yellow layer. Since Supper-Yellow is a poly(p-phenylene vinylene) (PPV) derivate and it can transport holes more easily than electrons[18], OLEDs based on Super-Yellow are hole-dominated devices and electrons are minor carriers. It is well known that SWCNTs have the electron-withdrawing ability[19], so they can act as an n-type material to improve the electron transport when doped into a polymer host. Consequently, the J–V curve of the device with the Supper-Yellow:SWCNT composite shifts to a lower voltage due to improved electron transport in the Supper-Yellow layer. As can be seen from

Fig. 5.5b,c, the EL efficiency has also been improved in devices with the Supper-Yellow:SWCNT composite. The maximum luminous efficiency and maximum EQE reach 9.8 cd/A and 2.9%, respectively, which are 1.6 times and 1.7 times than the one with a pure Super-Yellow layer, respectively. Such enhancement in EL efficiency is also the result of the improvement of electron transport in the Super-Yellow layer due to the SWCNTs acting as the n-type dopants, thus leading to more balanced electrons and holes in the light-emitting layer. Figure 5.5d shows the EL spectra of devices with and without SWCNTs in the light-emitting layer, and we can see that mixing Super-Yellow with SWCNTs does not change the EL emission spectra.

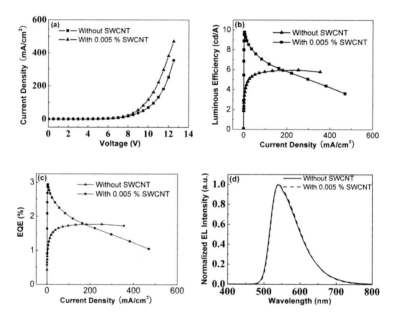

Figure 5.5 The *J–V* curves (a), the luminous efficiency characters (b), the EQE characters (c), and EL spectra (d) of devices without and with SWCNT doping. The device structure is ITO/PEDOT (40 nm)/Super-Yellow (80 nm)/Ca (20 nm)/Al (80 nm) and ITO/PEDOT (40 nm)/Super-Yellow:SWCNT (0.005 wt%, 80 nm)/Ca (20 nm)/Al (80 nm), respectively.

Here we elucidate that SWCNTs play different roles when used as polymer:SWCNT composites in OLEDs. When SWCNTs are mixed in a hole-conducting layer, they can have a positive or a negative influence on the EL performance of OLEDs, which depends on the SWCNT concentration. If the SWCNT concentration is low enough (in our case ≤0.01 wt%), they can play a role in improving the conductivity, thus enhancing the hole transport in the hole-conducting layer. Also, they improve the contact at the interfaces by inducing much smaller islands on the surface compared with the ones on the surface of a pure hole-conducting layer. But if the SWCNT concentration is too high (in our case higher than 0.02 wt%), they can act as exciton-quenching centers to quench the excitons migrating to the interface between the hole-conducting layer and the light-emitting layer. When SWCNTs are incorporated in an emitting polymer layer, they are beneficial to improve the electron transport since SWCNTs can act as n-type dopants in the light-emitting layer, profiting from the electron-withdrawing nature of SWCNTs.

5.3 CNT Effect on Photovoltaic Response in Conjugated Polymers

Conjugated polymers as soluble semiconductors have demonstrated facile material processing, ready formation of hetero interfaces, strong light absorption, and efficient emission in optoelectronic devices.[20] However, low charge conductivities often form an obstacle for obtaining advanced functionalities from such organic devices. As compared to conjugated polymers, carbon nanotubes (CNTs) have been found to be one-dimensional conduction materials with high charge mobilities.[21] Therefore, combining the optical properties of conjugated polymers with the long-range charge transport characteristics of CNTs may generate advanced optoelectronic functionalities.[22] It is because of this consideration that conjugated polymer:CNT composites have attracted much interest for their applications in optoelectronic devices such as LEDs[23–26], PV cells[27–31], and field-effect transistors (FETs).[32] However, dispersing nanotubes into a polymer matrix inevitably forms SWCNT–polymer interfaces with close physical contacts due to random entanglement of soft polymer chains around the rigid nanotubes.[33,34] We note that the SWCNT–polymer interface is essentially a critical factor that determines the optical and electronic properties of such polymer:CNT composites.[35] In principle, there are two processes associated with the SWCNT–polymer interface in a polymer:nanotube composite under electrical or photo- excitation.

First, charge transport may occur between the CNTs and the polymer chains due to the close interfacial contacts, which can function as a bridge to improve the bipolar charge injection into the conjugated polymer and thus enhance the polymer EL output.[6] Secondly, excitons generated in the conjugated polymer chains may be dissociated at the SWCNT–polymer interfaces due to the interfacial band offset, resulting in free electrons and holes in the nanotubes and the polymer chains, respectively, for PV action.[36] In general, the SWCNT-induced EL enhancement and exciton dissociation mutually conflict in the SWCNT:poly 2-methoxy-5-2-ethylhexyloxy-1, 4-phenylenevinylene (MEHPPV) composite devices. As a result, it is often very challenging to obtain both enhanced EL and PV actions from polymer:nanomaterial composites. In this chapter we report the studies of the dependences of EL and PV enhancements on the concentration of SWCNTs in the SNWT:MEHPPV composites. The SWCNTs were synthesized by the laser vaporization method and purified by a dilute HNO_3 reflux/air oxidation procedure, as described elsewhere.[37,38] The MEHPPV was purchased from Sigma-Aldrich with the molecular weight of 51,000 and the polydispersity of 1.1. To obtain uniform dispersion of the SWCNTs in the MEHPPV, two solvents, orthodichlorobenzene (ODCB) and chloroform ($CHCl_3$), were used first to suspend the SWCNTs and dissolve the MEHPPV, respectively, under ultrasonication. Then the SWCNT and MEHPPV solutions were mixed to formulate the SWCNT:MEHPPV composite solutions at designed doping concentrations. The 80 nm thick light-emitting films were spin-cast from the composite solutions on ITO substrates. Single-layer LEDs of SWCNT:MEHPPV composites were fabricated with an architecture of ITO/composite/Al by thermally evaporating the Al electrode on the spin-cast thin films under a vacuum of 2×10^{-6} Torr. The thickness of spin-cast thin films was measured by a Dektak IIA surface profilometer. The EL, PV, and PL characteristics were measured with a Keithley 2400 sourcemeter, an integrating sphere, and a Fluorolog 3–22 spectrometer in a nitrogen atmosphere.

It can be seen in Fig. 5.6a that the presence of SWCNTs increases the MEHPPV EL intensity and reduces the threshold voltage from 8 V for pure MEHPPV to 2 V for the composite containing 0.1 wt% SWCNTs at forward bias. Especially, the 0.1 wt% SWCNTs clearly result in a significant charge injection and thus EL from MEHPPV at reverse bias. Figure 5.6b yields an alternating EL output. The similar

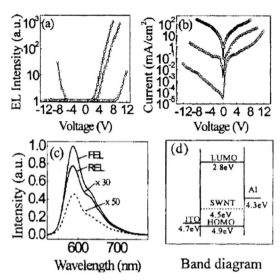

Figure 5.6 (a) EL–voltage and (b) current–voltage characteristics of the single-layer LEDs of ITO/ SWCNT x + MEHPPV/ Al composites at both forward and reverse biases. $x = 0$ circles, $x = 0.01$ wt % triangles, and $x = 0.10$ wt% squares. (c) EL spectra of the SWCNT:MEHPPV composite LED $x = 0.10$ wt% at both forward and reverse biases of 8 V. FEL and REL represent the forward and the reverse EL, respectively. The dashed line is the FEL spectrum of the undoped MEHPPV at a bias of 12 V. (d) A schematic band diagram of the SWCNT:MEHPPV composite LED. The work function of the SWCNTs is 4.5 eV.[39] The electron affinity and ionization potential are 2.8 and 4.9 eV, respectively[21], for the MEHPPV. The electrode work functions were assumed to be 4.7 eV for ITO and 4.3 eV for Al.[40]

EL spectra between the SWCNT 0.1 wt%:MEHPPV composite and the pristine MEHPPV (Fig. 5.6c) confirm that the light emission comes from the MEHPPV component in the composite LEDs at forward or reverse biases. Therefore, it is clear that the dispersed SWCNTs improve the charge injection and subsequently enhance the EL in the SWCNT:MEHPPV composite LEDs at both forward and reverse biases. We now consider why the dispersed SWCNTs can boost both charge injection and EL in the SWCNT:MEHPPV composites. The band diagram (Fig. 5.6d) indicates that the electrons and holes can be easily injected into the SWCNTs due to the absence of substantial potential barriers at the ITO/SWCNT and SWCNT/Al contacts in the SWCNT:MEHPPV composite LEDs at forward bias. At reverse bias, although small potential barriers exist at the ITO/SWCNT and SWCNT/Al contacts, electrons and holes can still be more significantly injected into the SWCNTs as compared to the MEHPPV component in the SWCNT:MEHPPV composites. Under the influence of an applied electric field, the injected electrons and holes in the SWCNTs can be further transferred into the LUMO and HOMO of the MEHPPV component at the SWCNT–polymer interface due to the large surface contact areas of SWCNTs and the field-enhanced effect at nanotube tips.[25] Therefore, the dispersed SWCNTs can improve the electron and hole injection into conjugated polymers through the tube–chain interfaces at both forward and reverse biases, leading to an enhanced reverse and forward organic EL, as observed in Fig. 5.6a,b.

It is noted that the band offset at the SWCNT–polymer interface usually causes exciton dissociation and thus increases the PV response, as shown in Fig. 5.7a. To study this SWCNT-induced exciton dissociation, we measured the photocurrent of the single-layer SWCNT:MEHPPV composite LEDs under the illumination of 0.2 mW/cm^2 at 450 nm. It can be seen in Fig. 5.7b that the photocurrent from the SWCNT:MEHPPV LEDs continuously increases upon SWCNT doping. For instance, the 0.1 wt% of SWCNTs increases the photocurrent by a factor of 6 at a reverse bias of 0.6 V. Therefore, it can be concluded that SWCNTs at a low doping concentration of 0.1 wt% enhance both the EL and the PV response in the SWCNT:MEHPPV composite LEDs. We note that the EL light emission and PV response are usually competing in an optoelectronic device. To further understand the competition between the SWCNT-enhanced EL and PV response, we measured both EL efficiency and photocurrent as a function of the SWCNT doping concentration based on the single-layer SWCNT:MEHPPV composite LEDs.

Figure 5.8 shows that the external quantum EL efficiency increases from 0.04% photon/electron with the SWCNT doping at low concentrations and rapidly reaches a maximum of about 0.25% photon/electron when the SWCNTs concentration was increased to about 0.02 wt%. A further increase of the SWCNTs concentration results in a drastic decrease of the EL efficiency. However, the photocurrent shows a continuous increase with the SWCNT doping from low to relatively high concentrations up to 0.2 wt%. Therefore, we can conclude that the dispersed SWCNTs at low concentrations mainly improve the bipolar charge injection through the SWCNT–polymer interface and thus enhance the polymer EL at both forward and reverse biases. When the SWCNTs concentration further increases 0.02 wt%, the EL efficiency is largely decreased and the enhancement of the PV response becomes a dominant effect in the SWCNT:MEHPPV composite. In summary, these SWCNT concentration-dependent EL and PV actions indicate that adjusting the dispersion of the SWCNTs in conjugated polymer materials can lead to individual control of dual EL and PV functions in CNT-doped conjugated polymer optoelectronic devices.

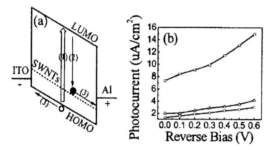

Figure 5.7 (a) Schematic PV process at reverse bias in the SWCNT:MEHPPV composite. 1: photoabsorption; 2: exciton dissociation; 3: charge conduction. (b) Photocurrent density shown as a function of applied reverse bias for different SWCNT doping concentrations in the SWCNT:MEHPPV composite LEDs: $x = 0$ circles, $x = 0.01$ wt% triangles, and $x = 0.10$ wt% squares.

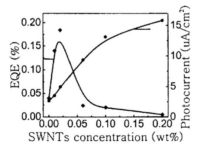

Figure 5.8 EL efficiency and photocurrent are shown as a function of the SWCNT concentration in SWCNT:MEHPPV composite LEDs. The EQEs of EL were measured at the current density of 10 mA/cm^2, while the photocurrent was recorded at a 0.5 V reverse bias and 0.2 mW/cm^2 illumination of 450 nm. The photocurrent density is defined as the difference between the currents measured with and without the photoillumination.

References

1. Z. C. Wu, Z. H. Chen, X. Du, J. M. Logan, J. Sippel, M. Nikolou, K. Kamaras, J. R. Reynolds, D. B. Tanner, A. F. Hebard, A. G. Rinzler, *Science*, **305**, 1273 (2004).
2. M. Zhang, S. L. Fang, A. A. Zakhidov, S. B. Lee, A. E. Aliev, C. D. Williams, K. R. Atkinson, R. H. Baughman, *Science*, **309**, 1215 (2005).
3. D. H. Zhang, K. M. Ryu, X. L. Liu, E. Polikarpov, J. Ly, M. E. Tompson, C. W. Zhou, *Nano Lett.*, **6**, 1880 (2006).
4. F. Massuyeau, J. L. Duvail, H. Athalin, J. M. Lorcy, S. Lefrant, J. Wéry, E. Faulques, *Nanotechnology*, **20**, 155701 (2009).
5. L. Qian, T. Feng, X. Zheng, S. Y. Quan, D. A. Liu, Y. M. Wang, Y. S. Wang, X. R. Xu, *Spectrosc. Spect. Anal.*, **26**, 601 (2006).
6. B. Hu, Y. Wu, Z. T. Zhang, S. Dai, J. Shen, *Appl. Phys. Lett.*, **88**, 022114 (2006).
7. Z. Tan, F. Zhang, T. Zhu, J. Xu, A. Y. Wang, J. D. Dixon, L. Li, Q. Zhang, S. E. Mohney, J. Ruzyllo, *Nano Lett.*, **7**, 3803 (2007).
8. Y. G. Ha, E. A. You, B. J. Kim, J. H. Choi, *Synth. Met.*, **153**, 205 (2005).
9. S. A. Curran, P. M. Ajayan, W. J. Blau, D. L. Carroll, J. N. Coleman, A. B. Dalton, A. P. Davey, A. Drury, B. McCarthy, S. Maier, A. Strevens, *Adv. Mater.*, **10**, 1091 (1998).
10. H. S. Woo, R. Czerw, S. Webster, D. L. Carroll, J. Ballato, A. E. Strevens, D. O'Brien, W. J. Blau, *Appl. Phys. Lett.*, **77**, 1393 (2000).
11. H. S. Woo, R. Czerw, S. Webster, D. L. Carroll, J. W. Park, J. H. Lee, *Synth. Met.*, **116**, 369 (2001).
12. J. Y. Kim, E. S. Kim, J. H. Choi, *J. Appl. Phys.*, **91**, 1944 (2002).
13. J. Y. Kim, M. Kim, J.-H. Choi, *Synth. Met.*, **139**, 565 (2003).
14. C. Yang, M. Wohlgenannt, Z. V. Vardeny, W. J. Blau, A. B. Dalton, R. Baughman, A. A. Zakhidov, *Phys. B*, **338**, 366 (2003).
15. E. Kymakis, G. A. J. Amaratung, *Appl. Phys. Lett.*, **80**, 112 (2002).
16. A. L. Burin, M. A. Ratner, *J. Phys. Chem. A*, **104**, 4704 (2000).
17. X. Yang, D. Neher, D. Hertel, T. K. Daubler, *Adv. Mater.*, **16**, 161 (2004).
18. H. Spreitzer, H. Becker, E. Kluge, W. Kreuder, H. Schenk, R. Demandt, H. Schoo, *Adv. Mater.*, **10**, 1340 (1998).
19. A. Kongkanand, P. V. Kamat, *ACS Nano*, **1**, 13 (2007).
20. R. H. Friend, R. W. Gymer, A. B. Holmes, J. H. Burroughes, R. N. Marks, C. Taliani, D. D. C. Bradley, D. A. Dos Santos, J. L. Brédas, M. Lögdlund, W. R. Salaneck, *Nature (London)*, **397**, 121 (1999).

21. T. W. Ebbesen, H. J. Lezec, H. Hiura, J. W. Bennett, H. F. Ghaemi, T. Thio, *Nature (London)*, **382**, 54 (1996).
22. P. J. F. Harris, *Int. Mater. Rev.*, **49**, 31 (2004).
23. S. A. Curran, P. M. Ajayan, W. J. Blau, D. L. Carroll, J. N. Coleman, A. B. Dalton, A. P. Davey, A. Drury, B. McCarthy, S. Maier, A. Strevens, *Adv. Mater. Weinheim, Ger.*, **10**, 1091 (1998).
24. H. S. Woo, R. Czerw, S. Webster, D. L. Carroll, J. Ballato, A. E. Strevens, D. O'Brien, W. J. Blau, *Appl. Phys. Lett.*, **77**, 1393 (2000).
25. P. Fournet, J. N. Coleman, B. Lahr, A. Drury, W. J. Blau, D. F. O'Brien, H. H. Hörhold, *J. Appl. Phys.*, **90**, 969 (2001).
26. J. Y. Kim, M. Kim, H. M. Kim, J. Joo, J. H. Choi, *Opt. Mater. Amsterdam, Neth.*, **21**, 147 (2002).
27. D. B. Romero, M. Carrard, W. de Heer, L. Zuppiroli, *Adv. Mater. Weinheim, Ger.*, **8**, 899 (1996).
28. H. Ago, K. Petritsch, M. S. P. Shaffer, A. H. Windle, R. H. Friend, *Adv. Mater. Weinheim, Ger.*, **11**, 1281 (1999).
29. E. Kymakis, G. A. J. Amaratunga, *Appl. Phys. Lett.*, **80**, 112 (2002).
30. S. Bhattacharyya, E. Kymakis, G. A. J. Amaratunga, *Chem. Mater.*, **16**, 4819 (2004).
31. B. J. Landi, S. L. Castro, H. J. Ruf, C. M. Evans, S. G. Bailey, R. P. Raffaelle, *Sol. Energy Mater. Sol. Cells*, **87**, 733 (2005).
32. A. Star, Y. Lu, K. Bradley, G. Grüner, *Nano Lett.*, **4**, 1587 (2004).
33. M. Yang, V. Koutsos, M. Zaiser, *J. Phys. Chem. B*, **109**, 10009 (2005).
34. K. P. Ryan, S. M. Lipson, A. Drury, M. Cadek, M. Ruether, S. M. O'Flaherty, V. Barron, B. McCarthy, H. J. Byrne, W. J. Blau, J. N. Coleman, *Chem. Phys. Lett.*, **391**, 329 (2004).
35. H. Ago, M. S. P. Shaffer, D. S. Ginger, A. H. Windle, R. H. Friend, *Phys. Rev. B*, **61**(2), 286 (2000).
36. C. Yang, M. Wohlgenannt, Z. V. Vardeny, W. J. Blau, A. B. Dalton, R. Baughman, A. A. Zakhidov, *Phys. B*, **338**, 366 (2003).
37. A. A. Puretzky, D. B. Geohegan, X. Fan, S. J. Pennycook, *Appl. Phys. A: Mater. Sci. Process*, **70**, 153 (2000).
38. A. C. Dillon, T. Gennett, K. M. Jones, J. L. Alleman, P. A. Parilla, M. J. Heben, *Adv. Mater. Weinheim, Ger.*, **11**, 1354 (1999).
39. S. J. Tans, A. R. M. Vershueren, C. Dekker, *Nature (London)*, **393**, 49 (1998).
40. I. D. Parker, *J. Appl. Phys.*, **75**, 1656 (1994).

Chapter 6

Exciton Energy Transfer in Carbon Nanotubes Probed by Photoluminescence

Ping Heng Tan,[a,b] Tawfique Hasan,[b] Francesco Bonaccorso,[b] and Andrea C. Ferrari[b]

[a]*State Key Laboratory for Superlattices and Microstructures, Institute of Semiconductors, Beijing 100083, China*
[b]*Department of Engineering, University of Cambridge, Cambridge CB3 0FA, UK*
phtan@semi.ac.cn, acf26@hermes.cam.ac.uk

6.1 Introduction

Photoluminescence (PL) spectroscopy is a common technique used for the characterization of semiconductors. The detection of PL from isolated semiconducting single-walled nanotubes (s-SWNTs)[1] has made the optical properties of SWNTs a subject of intense interest.[2–16] Debundling of SWNTs was believed to be an essential prerequisite to observe PL emissions from s-SWNTs. However, Refs. 12 and 17 observed optical signatures from SWNT bundles in aqueous solutions and assigned them to Förster resonance energy transfer (FRET).[12,17]

The exciton binding energies of s-SWNTs are very large, from tens of meVs to 1 eV, depending on diameter, chirality, and dielectric

---

Luminescence: The Instrumental Key to the Future of Nanotechnology
Edited by Adam M. Gilmore
Copyright © 2014 Pan Stanford Publishing Pte. Ltd.
ISBN 978-981-4241-95-3 (Hardcover), 978-981-4267-72-4 (eBook)
www.panstanford.com

screening.[3,4,18] Photoexcited electrons and holes in s-SWNTs can create a bound state of an electron and a hole via Coulomb interaction to form excitons. Excitons dominate the optical properties of SWNTs, even at room temperature, because of their large binding energy.[3,4]

Many works focused on dispersed SWNTs, assuming that sedimentation-based ultracentrifugation (SUF) would completely remove bundles.[1,2] However, small bundles with about 3–10 tubes still exist in most surfactant-assisted SWNT dispersions, even after SUF.[4,12] The formation of bundles in nanotube samples modifies their excitonic behavior, revealed by a change of their optical properties.[19-22] To study the intrinsic optical and electronic properties, completely isolated nanotubes must be used. On the other hand, bundles are necessary in SWNT-based devices, such as saturable absorbers in passively mode-locked lasers and noise suppression filters.[23-34] However, in such applications, the bundle size must be smaller than the device operation wavelength to avoid nonsaturable losses due to scattering.[35]

Thus, quantification of bundling in aqueous dispersions is a key requirement for basic research and applications.

In this chapter, we discuss the spectral features of exciton energy transfer (EET)-induced PL emission from acceptor s-SWNTs in bundles after the resonance excitation of donor s-SWNTs. Moreover, we explain how to distinguish such features from other spectral features in the PL and photoluminescence excitation (PLE) spectra. Then, we discuss the transfer efficiency of excitons and exciton relaxation pathways in bundles. Finally, we discuss how FRET may be utilized for nanotube-based photonic devices.

6.2 The Photoluminescence Spectrum of Nanotube Bundles

To confirm the presence of bundles and to study their optical properties, we compare as-prepared SWNT dispersions in D_2O with sodium dodecylbenzene sulfonate (SDBS) as a surfactant and the same sample after two months (Fig. 6.1a,b). The (6,5) and (7,5) emission peaks observed in Ref. 2 are indicated with two dash-dotted lines. These show a ~1–5 nm blue shift compared to our measurements but on samples prepared with sodium dodecyl sulfate (SDS) as a surfactant.

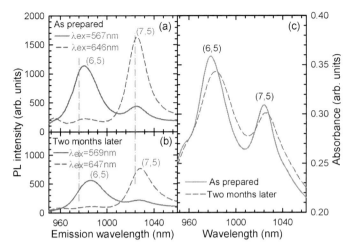

Figure 6.1 PL emission spectra of (6,5) and (7,5) for (a) an as-prepared dispersion and (b) after two months. (c) Absorption spectra of (6,5) and (7,5) for an as-prepared dispersion and after two months. The dash-dotted lines indicate the PL-peak positions observed in Ref. 2.

Since it was shown that dispersions prepared with similar methods using SDS and SDBS do not have any spectral shift[36], the effect of dielectric screening on individual SWNTs wrapped by SDS and SDBS is nearly identical in the 850–1,100 nm emission range. We thus assign the red shift in our dispersions compared to Ref. 2 to bundling. The emission peaks of (6,5) and (7,5) from the same dispersion after two months (Fig. 6.1b) red shift by ~3–5 nm relative to the as-prepared dispersion, Fig. 6.1a. This suggests that the bundle size further increases after two months.

The corresponding absorption spectrum in Fig. 6.1c further confirms this, since the absorption peaks are red-shifted and broadened.

Figure 6.2 is a contour plot (known as a "PLE map"), showing the PL emission of the dispersion as-prepared and after two months, as a function of emission and excitation wavelength. The main features in the map are the exciton-exciton resonances. Each resonance spot can be labeled as (λ_{ex}, λ_{em}), where λ_{ex} and λ_{em} are the excitation and emission wavelengths. λ_{ex} corresponds to the energy of the excitonic states eh_{ii} associated with the ith electronic interband transitions E_{ii} (i = 1, 2, 3, 4) in the single-particle picture, while λ_{em} represents the emission energy of the lowest optical-active excitonic transition eh_{11}. All the observed resonances are indicated with crosses. Some spots, observed in previous results[2] but not in this work, are marked with ×. The (n,m) assignments for all resonances are shown in Fig. 6.2.

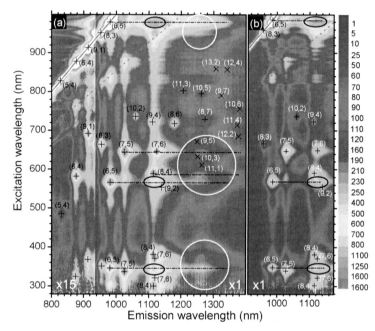

Figure 6.2 PLE map for (a) an as-prepared dispersion and (b) after two months. Crosses represent eh_{11} emission of SWNTs for excitation matching their eh_{11}, eh_{22}, eh_{33}, and eh_{44}. Resonances related to (eh_{22}, eh_{11}) observed in Ref. 2 are indicated as ×. Each resonance is labeled with the chiral index of the corresponding SWNT. The solid lines in the upper-left corners represent resonances with same excitation and recombination energies. The dotted lines are phonon sidebands. Open ellipses mark emission from (8,4), (7,6), and (9,4) with excitation matching the (6,5) eh_{11}, eh_{22}, and eh_{33}. Open circles mark emission in the 1,200–1,350 nm range, with excitation matching eh_{11}, eh_{22}, and eh_{33} of (6,5), (7,5), and (8,4).

In addition to the well-known exciton–exciton resonances, Fig. 6.2a,b shows new features indicated by open ellipses and circles. The exciton–exciton resonances decrease in intensity after two months, again suggesting SWNTs aggregate with time. However, the relative intensity of the (980 nm, 1,118 nm) band becomes stronger after two months. The two peaks near (568 nm, 1,118 nm), (346 nm, 1,118 nm) (open ellipses, Fig. 6.2b) are better resolved due to the lower intensity of the (8,4), (7,6), and (9,2) exciton-exciton resonances compared to the pristine dispersion. In the 1,200–1,350 nm emission range, exciton-exciton resonances of SWNTs such as (13,2), (12,4), (10,5), (9,7), (10,6), (8,7), (11,4), (12,2), (9,5), (10,3), and (11,1), indicated by crosses and ×, have weak emission intensities due to their lower concentration in CoMoCAT samples.[65] However, the new features marked by open circles show much stronger intensity than the above exciton-exciton resonances. The excitation energies of all the new features marked by open circles match eh_{11}, eh_{22}, or eh_{33} of (6,5), (7,5), and (8,4), as indicated by horizontal dash-dotted lines in Fig. 6.2a,b. These features are induced by the resonant excitation of nanotubes with a large bandgap, causing emission from smaller-bandgap SWNTs. This implies EET between SWNTs in bundles, where the larger-gap tubes act as exciton donors (D) and the smaller-gap tubes act as exciton acceptors (A). The EET features can be denoted as (eh^D_i, eh^A_i), with i = 1, 2, 3, ...

The exciton-phonon interaction is very strong in nanotubes.[37] This can mix an exciton with phonons and form a sideband above the main absorption peak due to strong exciton-phonon coupling.[37] Such sidebands of eh_{ii} excitons (i = 1, 2, . . .) have been resolved in PLE maps of SWNT dispersions[8,9,12] ∼0.215 eV above the eh_{ii} excitons. We denote such sidebands as eh_{11} + K (dotted lines in Fig. 6.2).

Due to the strong exciton-phonon coupling, the main excitonic absorption peak transfers part of its spectral weight to the phonon sideband.[37] Phonon sidebands could be involved in the energy transfer process between adjacent tubes in bundles. However, the observed sideband feature is much weaker than corresponding exciton-exciton resonances.[8,9,12] Thus, the associated EET-features from sidebands should be much weaker and overshadowed by strong exciton-exciton resonances of other nanotube species. Ideally, to unambiguously identify and appropriately assign the weak features related to energy transfer from excitonic phonon sidebands, dispersions enriched with few chiralities and small bundles are required. For this purpose, we enrich SWNTs in a sodium cholate (SC) dispersion via density gradient utracentrifugation (DGU).[38,39] In DGU, aqueous dispersions of surfactant-encapsulated tubes are ultracentrifuged in a preformed density gradient medium (DGM).

During the process, they move along an ultracentrifuge tube, dragged by the centrifugal force, until they reach the point where their buoyant density equals that of the surrounding DGM, that is, their isopycnic point.[38,39] After DGU, using an upward displacement fraction technique, we extract small aliquots of sorted SWNTs layer by layer.[38,39] In particular we focus on one of the middle fractions. Spectroscopic characterization shows that this dispersion is enriched with small-diameter tubes such as (6,5), (9,1), and (6,4).[39] These tubes are detected in the PLE contour plot, as shown in Fig. 6.3. The PL peaks of the (6,5) eh_{11} excitonic transitions are ~7 nm red-shifted with respect to the topmost fraction of a DGU-enriched sample in Refs. 38 and 39. This indicates the presence of bundles in our dispersion. A further evidence of bundling is the EET-induced (eh_{22}^D, eh_{11}^A) features from (6,5) to (7,5) and (8,4), indicated by the horizontal solid line in Fig. 6.3.

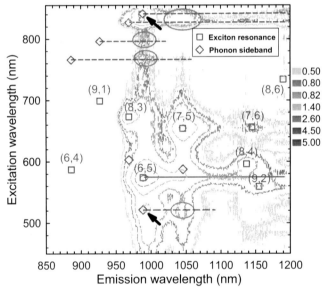

Figure 6.3 PLE contour plot for the middle faction of an SC-encapsulated dispersion. Open squares are (eh_{22}, eh_{11}) exciton resonances, each labeled with the corresponding SWNT chiral index. Open diamonds indicate phonon sidebands for each nanotube species. Arrows indicate phonon sidebands for the (6,5) eh_{11} and eh_{22} excitons. Open ellipses show EET-induced features from the resonant excitation of phonon sidebands of SWNTs.

The features marked by open diamonds in Fig. 6.3 are related to the phonon sidebands, whose excitation energy is ~0.215 eV higher than the eh_{11} and eh_{22} energies. Moreover, the PLE map in Fig. 6.3 shows some intense features marked with open ellipses. Their excitation energy corresponds to the phonon sidebands of (6,5), (6,4), (9,1), and (8,3), as indicated by the dashed lines. Therefore, we attribute these features to PL emission from a smaller-gap nanotube induced by the resonant excitation of phonon sidebands eh_{11} + K, or eh_{22} + K of a larger-gap one. For example, the feature at (521 nm, 1,045 nm) corresponds to the resonant excitation of the (6,5) eh_{22} + K and emission from the (7,5) eh_{11}. This is a new type of EET feature, resulting from resonant excitation of eh_{ii} + K (i = 1, 2, 3, . . .) phonon sidebands of large-gap SWNTs. The EET-induced features from eh_{ii} (i = 1, 2, 3, . . .) excitons and eh_{ii} + K (i = 1, 2, 3, . . .) sidebands are very common in the PLE maps of nanotube dispersions. For example, Figs. 1b and 5b in Ref. 40 reported a PL peak below eh_{11} for (6,5) and (7,5) when the excitation matched the corresponding eh_{22}. McDonald et al.[40] assigned such features to "mid gap emission" (see Figs. 1b and 5b of Ref. 40). We note that this so-called "midgap emission" at (570 nm, 1,130 nm) in Fig. 5b of Ref. 40 matches the (7,6) and (9,4) emission wavelength. We thus assign this feature not to midgap emission but to EET from (6,5) to (7,6) and (9,4). The PL emission associated with EET was also observed in pairs of semiconducting s-SWNTs by high-resolution optical microscopy and spectroscopy[41], as well as in pairs of suspended SWNTs[42], and in other nanotube dispersions, freestanding bundles, or double-wall nanotubes.[43-55]

We find that the existence and/or the gradual formation of bundles is inevitable even after debundling procedures.[12] The presence of bundles can be probed by monitoring the EET features in PL and PLE. Figure 6.4 shows the PL and PLE spectra of as-prepared dispersions[12] in SDBS/D$_2$O and after two months. The PL spectra are collected resonantly exciting the (6,5) eh_{22} at ~568 nm. (9,4), (8,4), (7,6), (9,2) and (8,6) are nonresonant at this excitation wavelength. Thus, the PL intensity from (9,4), (8,4), (7,6), (9,2), and (8,6) is expected to be very weak compared to that of (6,5). We note that their intensity is comparable to that of (6,5), especially after two months. We also consider that (6,5) have larger PL cross sections[19] and have a higher concentration in the dispersion, as revealed in the absorption spectra (dotted line in Fig. 6.4a) with respect to other tubes. The strong PL emission in the 1,080–1,200 nm range

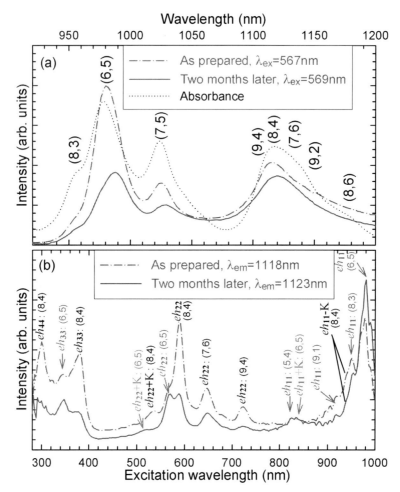

Figure 6.4 (a) PL (dash-dotted line) and absorption (dotted) spectra of an as-prepared dispersion in D_2O/SDBS and PL (solid line) of the same dispersion after two months. (b) PLE spectra of an as-prepared dispersion (dash-dotted line) and after two months (solid line). The PLE and PLE wavelengths are indicated.

is attributed to EET from (6,5) eh_{22} to eh_{11} of (9,4), (8,4), (7,6), (9,2), and (8,6). In the PLE map, each vertical line provides a corresponding PLE spectrum, revealing high-lying excitonic transitions of specific tubes at its emission wavelength. For the PLE spectra in Fig. 6.4b, we set the emission wavelength close to eh_{11} of (8,4), (7,6), and (9,4). The absorption peaks of the (8,4) eh_{44}, eh_{33}, and eh_{22} excitons, and eh_{22} + K and eh_{11} + K phonon sidebands, and of (7,6) and (9,4) eh_{22} can be identified for the as-prepared dispersion, as indicated in Fig. 6.4b. After two months, the PL intensity of these peaks decreases more than 50%. However, some absorption features are better resolved with respect to the pristine dispersion. After checking the eh_{ii} (i = 1, 2, 3) energy of (6,5), (5,4), (9,1), and (8,3), the EET from the (6,5) eh_{ii} (i = 1, 2, 3) and (5,4), (9,1), and (8,3) eh_{11} to the (8,4), (7,6), and (9,4) eh_{11} can be identified.

The broad features at (514 nm, 838 nm) cannot be assigned to any eh_{ii}; however, their energy matches the (6,5) eh_{22} + K and eh_{11} + K phonon sidebands. Thus, such features are induced by EET from (6,5) eh_{22} + K and eh_{11} + K phonon sidebands to (8,4), (7,6), and (9,4) eh_{ii}. All these EET features are labeled in Fig. 6.4b by arrows.

Thus, in the presence of bundles, the spectra can be explained by EET between donor and acceptor tubes. EET occurs at the eh_{11} of acceptors when the donor eh_{ii} (i = 1, 2, 3, 4) excitons or eh_{ii} + K (i = 1, 2, 3, 4) phonon sidebands are resonantly excited.

6.3 Mechanism and Efficiency of EET in Nanotube Bundles

In low-dimensional systems, exciton tunneling, photon exchange, and FRET are efficient EET mechanisms.[56-61] FRET occurs between a donor (D) molecule in the excited state and an acceptor (A) molecule in the ground state.[56] The donor molecules typically emit at shorter wavelengths that overlap with the absorption spectrum of the acceptor. FRET is a very efficient EET mechanism via resonant, near-field, dipole–dipole interactions.[56] It is commonly observed in biological systems, conjugated polymers, wires, and dots[56,58-61], where it dominates at short and intermediate distances (0.5–10 nm). Its efficiency is determined by the spectral overlap of donor emission and acceptor absorption, by the donor–acceptor distance, (R_{DA}), and by the relative orientation of emission and absorption dipoles.

The transfer rate is proportional to R_6^{DA}.[56] The FRET efficiency in bundles is expected to be high.[12] Indeed, the emission-absorption overlap between large- and small-gap tubes depends on the specific donor–acceptor couple. However, excitons can be cascadedly transferred from donor to acceptor, even when a small emission-absorption overlap is present, via intermediate gap tubes within a bundle. Moreover, phonon-assisted absorption[5] of the acceptor nanotube is likely to increase spectral overlap. SWNTs form bundles by aligning themselves parallel too each other, giving a maximum dipole orientation factor. Nanotubes in bundles also have a small wall-to-wall distance. This makes higher multipolar contributions possible.[56,58]

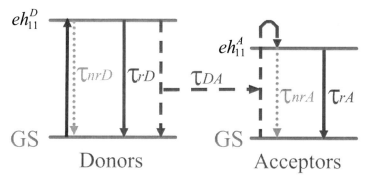

Figure 6.5   Scheme of exciton relaxation of two nanotubes in a bundle after resonant excitation of the donor eh_{11}. GS labels the ground states of the donor and the acceptor. In addition to radiative (r) and nonradiative (nr) relaxation of excitons, labeled τ_D and τ_{rD} for the donor and τ_A and τ_{rA} for the acceptor, EET with the rate τ_{DA} offers another pathway for de-excitation of the donor tube and excitation of the acceptor.

To estimate the energy transfer lifetime between donor and acceptor from experimental results, one can consider the exciton relaxation of two adjacent tubes with different gaps, following the resonant eh_{11} excitation of the larger-gap tube, as sketched in Fig. 6.5. The rate equations of the donor–acceptor system are:

$$\partial n_D/\partial t = G_{pe} - n_D(1/\tau_{nrD} + 1/\tau_{rD}) - n_D/\tau_{DA} \qquad (6.1)$$

$$\partial n_A/\partial t = n_D/\tau_{DA} - n_A(1/\tau_{nrA} + 1/\tau_{rA}) \qquad (6.2)$$

where τ_{DA} is the energy transfer lifetime between the donor and the acceptor, n_D is the population of excitons in the donor and n_A in the acceptor, τ_{nrD}, τ_{rD}, τ_{nrA}, and τ_{rA} are the radiative (r) and nonradiative (nr) lifetimes, and G_{pe} the exciton density in the donor created by photoexcitation. We get the ratio of acceptor emission intensity ($I_A =$

n_A/τ_{rA}) to the donor ($I_D = n_D/\tau_{rD}$) from Eqs. 6.1 and 6.2 at the steady state,

$$I_A/I_D = \frac{1/\tau_{DA}}{1/\tau_{rA} + 1/\tau_{nrA}} \frac{\tau_{rD}}{\tau_{rA}} \tag{6.3}$$

The eh_{11} radiative lifetime is ~20–180 ps, depending on temperature.[62] For tube diameters ≈ 0.75–0.95 nm, it is ~20–30 ps at room temperature.[63] This is much shorter than the theoretical radiative lifetime (~10 ns).[64] Thus, the observed lifetimes are determined by nonradiative recombinations. Eq. 6.3 can be simplified as,

$$I_A/I_D \approx \tau_{nrA}/\tau_{DA}. \tag{6.4}$$

We measured a very high I_A/I_D in bundles.[12] For example, under (5,4) and (6,5) eh_{11} excitation in Fig. 6.2a, the ratio of PL intensity of all acceptor tubes to that of the (5,4) and (6,5) donors is ~75 and 1.0, respectively. Some (5,4) and (6,5) tubes could be isolated. Thus, the above intensity ratio represents a lower limit. From Eq. 6.4, the energy transfer lifetime can be estimated as ~0.3 ps. This means that exciton relaxation in bundles by EET can be comparable or even faster than nonradiative recombination. This fast relaxation suppresses the PL emission from donors but, on the other hand, significantly increases the acceptors' luminescence.

In principle, the EET efficiency can be defined as the ratio between the excitons generated in the acceptors by EET and all the donor eh_{11} photoexcited excitons. If we assume $\tau_{nrA} \approx \tau_{nrD}$, the EET efficiency, ηDA, is given by:

$$\eta_{DA} = \frac{n_D}{\tau_{DA}}/G_{pe} \approx \frac{\tau_{nrA}/\tau_{DA}}{1 + \tau_{nrA}/\tau_{DA}} \approx \frac{I_A/I_D}{1 + I_A/I_D} \tag{6.5}$$

If we neglect the loss of excitons from excited excitonic states to eh_{11} for the donor nanotubes, the above results can also be used to estimate the efficiency of different EET processes, such as (eh^D, eh^A) and (eh^D + K, eh^A).

Considering the energy transfer from the (5,4) and (6,5) eh_{11} to the eh_{11} of their acceptors in Fig. 6.2a, and from the eh_{22} + K of (6,5) donors to the (7,5) acceptors eh_{11} in Fig. 6.3, we can estimate an efficiency of ~50–98% for eh_{11} resonant excitation and ~33% for eh_{22} + K resonant excitation. These estimates indicate that the EET efficiency is very high when the eh_{ii} or phonon sideband of donors are resonantly excited. Again, because some isolated tubes may exist in the dispersion, these estimates need be considered lower limits.

To quantify the EET efficiency and rate, Förster theory should be considered.[56] The energy transfer efficiency depends on the spectral overlap of emission and absorption bands. In the case of nanotubes in bundles, the donor and acceptor are two parallel tubes with a small wall-to-wall distance; thus the total transfer rate from a locally excited donor can be estimated by integrating the rates along the axis of the acceptor nanotube, as detailed by Qian et al.[41] The experimental results reproduce the low quantum yield (10^{-2}–10^{-3}) for PL emission from s-SWNTs.[1]

We also consider two other possible mechanisms for EET: tunneling and photon exchange. Tunneling requires coupling of exciton wave functions.[57] Its rate decays rapidly with R_{DA} and is very sensitive to the eh_{11} energy difference. The 16 species detected in our experiment have diameters ≈ 0.65–1.05 nm, $\Delta(eh_{11})$ ≈ 0.06–0.5 eV, and chiral angle variation ≈ 5–26°.[65] Therefore, the efficiency should strongly depend on the specific donor–acceptor couple. However, the experimental results show no (n,m) preference. Therefore, exciton tunneling is not expected to be the dominant EET mechanism.

On the other hand, photon exchange consists in exciton–photon coupling with no direct donor–acceptor interaction. It has a smaller dependence on R_{DA} than FRET; thus it can become significant for much longer distances than FRET. However, the lack of significant EET features in isolated tubes in solution[2,65], combined with the low quantum efficiency[1], suggest that even if photon exchange may weakly contribute to EET between nanotube bundles or between isolated SWNTs, it is not dominant between adjacent tubes in a given bundle.

6.4 How to Distinguish EET-Induced Features from Other Sidebands in the PL Spectrum?

In addition to the features associated with exciton–exciton resonances, phonon sidebands, and EET, other sidebands are also observed. In general, a sideband can be classified as an excitation or an emission sideband. The excitation sideband includes phonon-assisted absorptions, eh_{ii} + phonons[5,6,42], and phonon sidebands eh_{ii}

+ K,[8,9,12,14], while the emission sidebands could be due to EET[12,17,41], K–phonon sidebands eh_{ii}–K[13–16], triplet dark excitons[16], and emission of eh_{ii} – phonons.[5,13]

For the emission sideband, triplet dark excitons can only be observed in samples after pulsed-laser irradiation.[16] The intensity of phonon-assisted emissions, such as eh_{11} – G, is very weak.[16] Thus, we focus on how to distinguish the EET and eh_{ii} – K phonon sidebands. The eh_{ii} – K phonon sideband was initially interpreted as "deep dark excitonic (DE) states" by Kiowski et al.[13] In these, the energy separation between eh_{11} and "deep DE states" (Fig. 3 in Ref. 13) depends on the SWNT diameter and varies from ~95 meV for 0.75 nm tubes to ~140 meV for 1.1 nm tubes. However, recently, several groups demonstrated no diameter dependence for this sideband.[14–16] The explanation of the energy separation between the eh_{11} and eh_{ii}–K sideband for small-diameter tubes is a key issue to clarify the above contradictory results. One possible explanation is that the so-called "DE states" observed by Kiowski et al.[13] may be derived from triplet dark excitons. This assumption can be excluded by comparing the observed "DE states" (solid circles) and triplet dark excitons (solid diamonds), as shown in Fig. 6.7. In fact, considering the donor/acceptor pairs (6,5)/(10,2), (9,1)/(8,7), (7,5)/(8,4), and (10,2)/(8,6), the energy separation between the eh_{11} of those donor/acceptor pairs is in good agreement with the energy difference between eh_{11} and the so-called "DE states" for the corresponding donor tubes, as shown in Fig. 6.3 of Kiowski et al.[13] Therefore, the "DE states" features reported by Kiowski et al.[13] for small-diameter tubes, such as (6,5), (9,1), (7,5), and (10,2), can be in fact assigned as EET sidebands of donor/acceptor pairs, such as (6,5)/(10,2), (9,1)/(8,7), (7,5)/(8,4), and (10,2)/(8,6).

To distinguish between EET and eh_{ii}–K phonon sidebands, we compare the PLE maps of a dispersion in SDBS and the top fraction of a sorted dispersion in SC by DGU[39], as shown in Fig. 6.6. The former is enriched with bundles, while the latter is highly enriched in isolated tubes, where ~90% of s-SWNTs are (6,5).[39] As shown in Fig. 6.6a,b, for dispersions in SDBS containing bundles, the EET features can be very strong due to the bundling of (6,5) and (8,4), forming donor–acceptor pairs. In Fig. 6.6c, because of the removal of (8,4) acceptors in the topmost fraction, and the high enrichment of isolated (6,5), the EET from (6,5) to (8,4) is not seen. Instead, only the (6,5) eh_{11}–K sidebands can be seen, down-shifted ~21 meV from the EET from (6,5) to (8,4). Figure 6.6c shows that the PL intensity of the eh_{ii}–K sidebands for the top fraction is only ~7% of the main eh_{11} peaks, which have a much smaller intensity than EET features. EET is a major relaxation channel for exciton decay in bundles, which can quench the PL emission from donors. It is expected for the eh_{ii}–K sidebands to be weaker in bundles than in isolated tubes. Therefore, in the dispersions containing bundles, the eh_{ii}–K sidebands may be too weak to be clearly observed if they are partially shadowed by the tails of strong EET features, as shown in Fig. 6.6a,b.

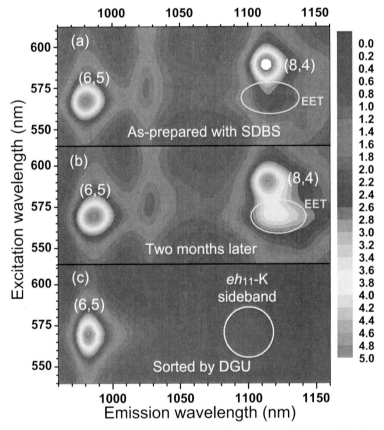

Figure 6.6 PLE maps of (a) an as-prepared dispersion in SDBS and (b) after two months. (c) PLE map of an as-prepared DGU dispersion in SC. All the maps are normalized by the intensity of (6,5) (eh_{22},eh_{11}) resonances. The ellipses show the EET-induced features, and the circle indicates the (6,5) eh_{11}–K emission sideband.

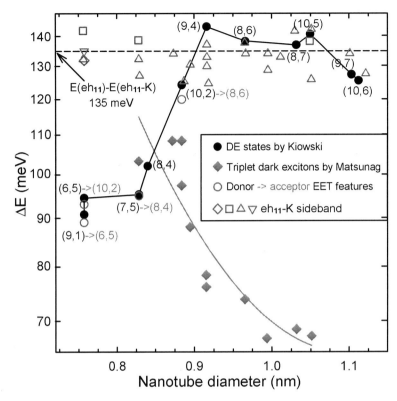

Figure 6.7 Energy separation (*E*) between eh_{11} and its lower energy sideband as a function of tube diameter: so-called DE states[13], EET features[12], triplet dark excitons[16], and eh_{11}-K sidebands (♦,[14,15,6] and ▽ in Fig. 6.6). The typical value (~135 meV) of energy separation between eh_{11} and eh_{11}-K phonon sideband is indicted.

Table 6.1 compares EET and eh_{11}-K sidebands. The eh_{11}-K sidebands can be observed in isolated nanotube dispersions if they are not shadowed by the intense exciton–exciton resonances of other nanotube species. The intensity and peak position of eh_{11}-K satellites of a given nanotube species are only determined by the main eh_{11} PL emission peaks of this nanotube. But, the EET peak position depends on both donors and acceptors, with donors determining the excitation position and acceptors the emission. The EET intensity is dominated by the bundling of donors and acceptors. EET features can be comparable to, even much stronger, than the PL emission of the corresponding donors or acceptors. Therefore, it is easy to distinguish the EET features from eh_{11}–K satellites in the PLE maps of a dispersion.

Table 6.1 Comparison between EET features and eh_{11}–K sidebands in PLE maps of nanotube dispersions

	EET	eh_{ii}–K
Satellites	emission bundles	emission isolated or bundles
Dispersions Intensity	strong, dependent on D-A pairs	weak, dependent on the main peak fixed, ~135 meV below eh_{11}–K
Position	dependent on D-A pairs	

6.5 Relaxation Pathways of Excitons in Nanotube Bundles

The dynamics of creation, relaxation, nonradiative decay, and radiative recombination of excitons and corresponding sidebands is a fundamental issue in nanotube research. The relaxation of excitons in bundles includes two parts. The first is the exciton relaxation in an individual tube (donor or acceptor). The second is the exciton relaxation between donor and acceptor tubes via EET. The total decay rate of an individual tube is $\tau r + \tau nr \approx 20$ps–1 at room temperature, dominated by τ_{nr}.[62,63,66] The intersubband exciton relaxation from eh_{22} to eh_{11} transitions is within ~150 fs, with a time constant ≈ 40 fs.[67] The maximum EET rate τ_{DA} is ~0.3 ps^{-1}, as estimated by Eq. 6.4. Therefore, the donor tubes transfer their exciton energy to acceptors via the lowest optical excitonic states of the donor tubes.

Figure 6.8 shows PLE spectra of isolated (6,5) eh_{11} and eh_{11}–K in the top fraction of an SC-encapsulated SWNT dispersion sorted by DGU, in an excitation range of eh_{22} + K to eh_{11} + K. The corresponding PLE map is shown in Fig. 6.6c. The excitonic transitions of eh_{22} + K, eh_{22}, and eh_{11} + K are identified. Additionally, there exists an EET-induced feature from the (8,3) eh_{11} + K phonon sideband to the (6,5) eh_{11} sideband. Multiplied by 13, the PL intensity of the eh_{11}–K sideband becomes similar to that of eh_{11} over the entire excitation range. This suggests that for PL emission at the eh_{11} + K sideband, the high-lying excitonic states, such as eh_{22} + K, eh_{22}, and eh_{11} + K, do not directly relax down to eh_{11}–K states but relax to eh_{11} states. Part recombine at eh_{11}; others relax to eh_{11}-K states and recombine there. Relaxation pathways down to eh_{11} and eh_{11}-K directly from specific high-lying excitonic states would result in a different intensity ratio of eh_{11}-K to eh_{11}.

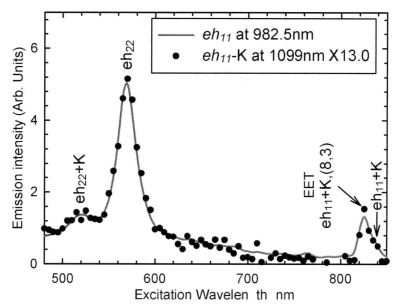

Figure 6.8 PLE spectra of eh_{11} and eh_{11}-K of isolated (6,5) tubes in the top fraction of an SC-encapsulated SWNT dispersion sorted by DGU. The detection wavelength is 982.5 nm for eh_{11} and 1,099 nm for eh_{11}-K. The PL intensity of eh_{11}-K is multiplied by a factor 13. The EET features from eh_{11} + K phonon sidebands of (8,3) to (6,5) eh_{11} are also identified.

The relaxation pathways of photoexcited excitons in bundles are illustrated in Fig. 6.9 in the case of the resonant excitation of eh_{22} and eh_{11} and/or their phonon sidebands. The weak relaxation pathway associated with phonon-assisted absorption and emission is not considered here. The EET recombination mechanism is as follows: photons are resonantly absorbed by the high-lying excitonic and exciton–phonon bound states of donors; the photoexcited excitons first quickly relax to the eh_{11}^D states; then some radiatively recombine at the eh^D of the donor nanotubes, or further relax down to eh_{11}^D –K and recombine there, while others resonantly transfer their energy to the excitons of acceptors.[12] Finally, the latter radiatively recombine at the corresponding eh_{11}^A or further relax down to eh_{11}^A −K and recombine.

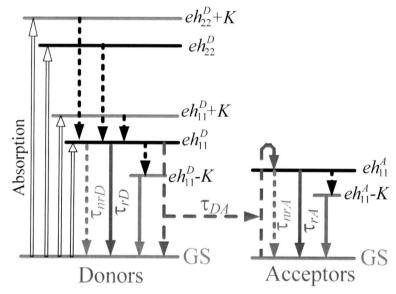

Figure 6.9 Schematic diagram of exciton absorption (open lines) and relaxation process in a donor–acceptor pair in s-SWNT bundles. Solid lines suggest possible radiative relaxation pathways. Dotted lines indicate possible nonradiative relaxation pathways. Dashed lines represent the EET pathways within a donor–acceptor pair.

6.6 How to Detect Bundles and Probe Their Concentration?

The quantification of bundling is important for both fundamental studies and applications. Bundling induces a red shift and broadening of excitonic transitions.[3,12,20] The red shift is attributed to the modification of Coulomb interactions by the dielectric screening induced by other adjacent tubes.[3] Thus, the red shift of the optical transitions in Raman, absorption, and PL spectra can be an indication of bundles.[3,12,20] The optical transitions of SWNTs are strongly modulated by different dielectric environments.[18,22,68–74] Therefore, care must be taken when using the peak shift of optical transitions to identify the presence of bundles, because this may also be induced by the dielectric environment rather than just bundling. Also, the red shift caused by bundling can be neutralized by the blue shift induced by a change in the dielectric environment. Therefore, detection of bundles through Raman, absorption, and PL measurements requires a reference sample with isolated nanotubes in the same solvent-surfactant environment to quantify the change of optical transitions due to bundling, which is not straightforward for all the pure solvent or solvent-surfactant systems.

The EET spectral features in one-photon PLE maps can be used as fingerprints of bundles. Two-photon excitation spectroscopy was used to estimate the exciton-binding energies in nanotubes.[3,4] Figure 6.10 represents the main features of the two-photon map in Fig. 1 of Ref. 4 and in Fig. 2 of Ref. 3, where HiPco SWNTs were used. This shows several peaks, indicated by solid squares and circles, with a larger excitation energy than the two-photon exciton resonance (open circles). There seems to be a Rydberg-like series of states for every SWNT; however, each of them matches the excitation energy of a larger-gap SWNT, as indicated by each horizonal dashed line in Fig. 6.10. These are analogous to the EET-induced peaks in one-photon excitation spectroscopy in Figs. 6.2 and 6.3. We attribute them to PL emission of small-gap tubes due to EET from larger-gap tubes in bundles. We assign the four features in Fig. 6.10a with ~1,390 nm excitation (solid squares) to EET from a (5,4) donor to (6,4), (9,1), (8,3), and (6,5) acceptors, although the concentration of (5,4) is very low in HiPco SWNTs and its two-photon exciton resonance is not observed in Fig. 6.10a. The EET-induced two-photon spectra show more distinct resonance peaks compared to Figs. 6.2 and 6.3. The EET from s-SWNTs to metallic ones is much faster than within s-SWNTs. Thus the PL emission from the latter in one-photon excitation can be heavily quenched due to the presence of m-SWNTs. Therefore, the EET signals should be very weak in the one-photon PLE maps for samples containing a high concentration of m-SWNTs, such as HiPco samples[2], where m-SWNTs constitute ~1/3 of the total. However, interestingly, two-photon excitation of HiPco dispersions[3,4] can exhibit strong EET features, as discussed above. The observation of significant EET features from (5,4) donors reveals that very high EET efficiency can occur in two-photon excitations. It also indicates that the EET rate from s- to m-SWNTs is not so fast to be totally quenched in two-photon excitation. Also, two-photon luminescence increases quadratically with the excitation power. Therefore, in comparison to one-photon excitation, two-photon excitation spectroscopy can be used as a more sensitive tool to probe bundling.

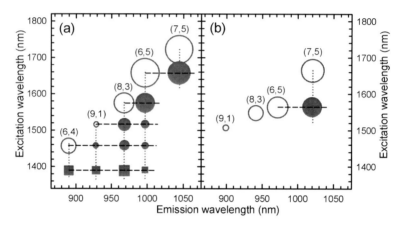

Figure 6.10 Main features in the two-photon PLE maps reported by (a) Maultzsch et al.[4] and (b) Wang et al.[3]

Open circles are two-photon PLE peaks, each labeled with the corresponding SWNT chiral index. Solid circles and squares are EET peaks.

Because the optical transitions of SWNTs are usually very broad due to bundling inhomogeneity, packing efficiency and dispersion of bundle size, absorption, and PL spectra are not sensitive probes of bundle size in dispersions and in composites. Moreover, if the concentration of some species of nanotubes is low, their PL and absorption features are expected to be shadowed by intense signal from nanotubes with a high concentration. However, the EET intensity can be much stronger than that of exciton–exciton resonances of donor or acceptor nanotubes. The position of EET features can also be far away from exciton–exciton resonances of donor or acceptor tubes. Thus, the EET optical signatures can be used to probe the presence of low-concentration donor or acceptor tubes and to quantify the bundling of a donor–acceptor pair. As shown in Fig. 6.10a, although the (5,4) concentration is very low in HiPco samples[2], the detection of an EET signal from (5,4) donors indicates their presence. The EET from eh_{ii}^D and $eh_{ii}^D + K$ sidebands ($i = 1, 2, \ldots$) of donors to eh_{11}^A excitons of acceptors is an intrinsic optical signature of bundles, independent of the influence of the surrounding dielectric environment.

Therefore, the EET-induced emissions from acceptors in the PL and PLE spectra are a direct, simple, and independent way to identify bundling.

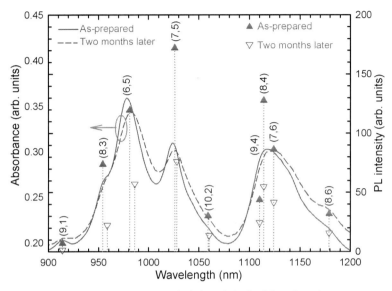

Figure 6.11 Absorption spectra (solid and dashed lines) and PL intensity of each exciton–exciton resonance (and) of an as-prepared dispersion and after two months.

The PL, PLE, and absorption spectra have been widely used to determine the concentration of nanotubes after correction by the theoretical cross-sections of different chiralities.[75–78] When bundles are present in dispersions or composites, the additional EET relaxation pathway decreases the PL intensity from donors and enhances that from acceptors. The presence of m-SWNTs also strongly quenches the luminescence from s-SWNTs. Figure 6.11 shows the absorption spectra and PL intensity of each exciton–exciton resonance of an as-prepared dispersion in SDBS/D_2O and after two months. It was reported that the theoretical PL intensity of an individual (6,5) tube is ~20% stronger than for a (7,5).[75] The (6,5) absorption intensity is also much stronger than that of (7,5), as shown in Fig. 6.3a. Thus, the (6,5) PL intensity is expected to be stronger than (7,5) if all nanotubes are individualized. However, Fig. 6.11a shows that in our case (7,5) tubes have much stronger PL emissions. The red shift in the absorption and PL spectra confirms that the SWNTs in the as-prepared dispersion contain small SWNT bundles. Note that the (6,5) absorption peak width (~45 meV) is ~31 meV broader than in (7,5). This implies that the (6,5) dielectric environment and the bundling status (e.g., size distribution of bundles) is more complex than for the (7,5). Compared to isolated tubes[2,65], additional EET relaxation channels exist in bundles. Therefore, for dispersions where the presence of bundles and their size distribution cannot be conclusively determined, the (eh_{22}, eh_{11}) intensity of each nanotube species does not necessarily reflect its abundance, contrary to what is sometimes suggested[76,79], even if the experimental PL is normalized with respect to concentrations and theoretical cross sections.[75] Considering that the absorption spectra of dispersions incubated for two months show a similar profile with respect to the as-prepared dispersions, we conclude that absorption spectroscopy is a more reliable technique to determine the concentration distribution of nanotubes in dispersions.

6.7 Exploiting EET for Photonic and Optoelectronic Applications

SWNTs show strong saturable absorption, that is, they become transparent under high-power irradiation.[30,67,80–82] SWNTs are thus suitable for application as mode lockers in ultrafast lasers and noise suppression filters.[23–33] SWNT-based saturable absorbers must be carefully designed to keep a balance between high modulation depth, depending on SWNT concentration, and unwanted nonsaturable losses, due to scattering from large bundles or entanglements, as well as from the polymer matrix, surfactant, residual catalyst particles, and carbon impurities.[30] To ensure fast device operation, the relaxation time of the saturable absorber must be as short as possible. The relaxation time is longer in isolated SWNTs than in bundles.[66,83,84] Hence, bundles are necessary in SWNT-based saturable absorbers, but their size must be smaller than the device operation wavelength to avoid nonsaturable losses due to scattering.[35] EET can thus be used to estimate bundling during the fabrication of such devices.

Dang et al.[53] used EET to monitor the effect of bundling on the efficiency of SWNT-based photovoltaic devices. They reported that bundling of large-gap s-SWNTs with smaller-gap s-SWNTs and m-SWNTs leads to a 1% reduction in power conversion efficiency in a dye-sensitized solar cell.[53]

Considering the application of bundles as optically pumped light-emitting devices, one must choose a pump laser whose wavelength or photon energy resonantly matches that of the nanotubes to achieve a high excitation efficiency for light emission. One could exploit EET, by matching the pump laser energy with the eh_{ii} or eh_{ii} + K of donor nanotubes, to induce an higher emission efficiency from acceptor tubes. This can also extend the wavelength or photon energy of the pump laser for the application of semiconducting bundles as optically pumped light emitters. Since m-SWNTs quench the luminescence from s-SWNTs, small bundles entirely made of s-SWNTs can be ideal for this application.

EET can be also used as a means to concentrate excitons, for example, for photovoltaic applications.[39,48] Han et al.[48] demonstrated a core-shell SWNT filament to concentrate excitons, demonstrating the viability of this concept.[48] Enriched SWNTs were dielectrophoretically spun to fabricate the core-shell filament, with small-gap SWNTs forming the inner or collector core, consisting predominantly of (8,7), (7,6), (11,3), (7,5), and (6,5), and large-gap SWNTs the outer absorbing shell, consisting of (6,5) only. When this filament is irradiated by a broadband light ranging from ultraviolet to near infrared, it exhibits a broadband absorption consistent with the SWNT population in the filament. This also confirms EET can occur between two SWNTs with a small emission–absorption overlap in a cascaded way via intermediate-gap tubes within a bundle.[12] Therefore, the emission of a core-shell nanotube filament is dominated by the smallest-gap tube—(8,7) in the case of Ref. 48—as a result of exciton funneling via EET to the core from the collector, as well as within the core by cascaded EET, from other species to the (8,7).[48] Similar structures consisting of enriched SWNTs can also be used for other devices to achieve an higher absorption bandwidth over a desired window, which could find application in various optoelectronic devices, including photodiodes and bolometers.[48]

6.8 Conclusions

We reviewed the use of PL, absorption, and PLE spectroscopy to investigate the optical properties of nanotube bundles. The existence and gradual formation of bundles in aqueous dispersions is inevitable even after ultrasonication followed by ultracentrifugation. With time, the emission and absorption spectra show weaker intensities and broader spectral profiles. Emission sidebands are observed in the PLE map, assigned to EET between bundled tubes. The EET features correspond to the excitation of donor tubes and the emission of acceptor tubes, and their intensity depends on bundling. The EET intensity can be much stronger than of exciton–exciton resonances of individual donor and acceptor tubes.

EET can be used for 1) nanotube bundle engineering to enhance and improve PL emission from small-bandgap nanotube acceptors for optoelectronics; 2) an additional relaxation channel for excitons, useful for ultrafast photonic devices; 3) investigation of optoelectronic interactions of hybrid structures of SWNTs and other materials, such as quantum dots, conjugated polymers, and biomolecules; 3) explanation of the complex spectral features in the PLE map of nanotube dispersions and composites; 4) provision of a direct, simple, and independent way to identify the presence of bundles or to confirm the presence of only individual isolated nanotubes; and 5) fingerprintin of bundles with different tube concentrations.

Acknowledgments

We acknowledge funding from the ERC grant NANOPOTS, the EPSRC grant EP/G042357/1, the Royal Society Wolfson Research Merit Award, a Newton International Fellowship, EU grants RODIN and Marie Curie ITN-GENIUS (PITN-GA-2010-264694), the Nokia Research Centre Cambridge, Chinese special fund for Major State Basic Research Project No. G2009CB929301, and NSFC (Nos. 10874177 and 10934007), King's college, Cambridge.

References

1. M. J. O'Connell, S. M. Bachilo, C. B. Huffman, V. C. Moore, M. S. Strano, E. H. Haroz, K. L. Rialon, P. J. Boul, W. H. Noon, C. Kittrell, J. P. Ma, R. H. Hauge, R. B. Weisman, R. E. Smalley, *Science*, **297**, 593 (2002).

2. S. M. Bachilo, M. S. Strano, C. Kittrell, R. H. Hauge, R. E. Smalley, R. B. Weisman, *Science*, **298**, 2361 (2002).

3. F. Wang, G. Dukovic, L. E. Brus, T. F. Heinz, *Science*, **308**, 838 (2005).

4. J. Maultzsch, R. Pomraenke, S. Reich, E. Chang, D. Prezzi, A. Ruini, E. Molinari, M. S. Strano, C. Thomsen, C. Lienau, *Phys. Rev. B*, **72**, 241402(R) (2005).
5. S. G. Chou, F. Plentz, J. Jiang, R. Saito, D. Nezich, H. B. Ribeiro, A. Jorio, M. A. Pimenta, Ge. G. Samsonidze, A. P. Santos, M. Zheng, G. B. Onoa, E. D. Semke, G. Dresselhaus, M. S. Dresselhaus, *Phys. Rev. Lett.*, **94**, 127402 (2005).
6. H. Htoon, M. J. O'Connell, S. K. Doorn, V. I. Klimov, *Phys. Rev. Lett.*, **94**, 127403 (2005).
7. Y. Z. Ma, L. Valkunas, S. L. Dexheimer, S. M. Bachilo, G. R. Fleming, *Phys. Rev. Lett.*, **94**, 157402 (2005).
8. F. Plentz, H. B. Ribeiro, A. Jorio, M. S. Strano, M. A. Pimenta, *Phys. Rev. Lett.*, **95**, 247401 (2005).
9. Y. Miyauchi, S. Maruyama, *Phys. Rev. B*, **74**, 035415 (2006).
10. Y. Miyauchi, M. Oba, S. Maruyama, *Phys. Rev. B*, **74**, 205440 (2006).
11. J. Lefebvre, P. Finnie, *Phys. Rev. Lett.*, **98**, 167406 (2007).
12. P. H. Tan, A. G. Rozhin, T. Hasan, P. Hu, V. Scardaci, W. I. Milne, A. C. Ferrari, *Phys. Rev. Lett.*, **99**, 137402 (2007).
13. O. Kiowski, K. Arnold, S. Lebedkin, F. Hennrich, M. M. Kappes, *Phys. Rev. Lett.*, **99**, 237402 (2007).
14. O. N. Torrens, M. Zheng, J. M. Kikkawa, *Phys. Rev. Lett.*, **101**, 157401 (2008).
15. Y. Murakami, B. Lu, S. Kazaoui, N. Minami, T. Okubo, S. Maruyama, *Phys. Rev. B*, **79**, 195407 (2009).
16. R. Matsunaga, K. Matsuda, Y. Kanemitsu, *Phys. Rev. B*, **81**, 033401 (2010).
17. P. H. Tan, T. Hasan, F. Bonaccorso, V. Scardaci, A. G. Rozhin, W. I. Milne, A. C. Ferrari, *Phys. E*, **40**, 2352 (2008).
18. V. Perebeinos, J. Tersoff, P. Avouris, *Phys. Rev. Lett.*, **92**, 257402 (2004).
19. S. Reich, C. Thomsen, P. Ordejon, *Phys. Rev. B*, **65**, 155411 (2002).
20. M. J. O'Connell, S. Sivaram, S. K. Doorn, *Phys. Rev. B*, **69**, 235415 (2004).
21. F. Wang, M. Y. Sfeir, L. M. Huang, X. M. H. Huang, Y. Wu, J. Kim, J. Hone, S. O'Brien, L. E. Brus, T. F. Heinz, *Phys. Rev. Lett.*, **96**, 167401 (2006).
22. A. G. Walsh, A. N. Vamivakas, Y. Yin, M. S. Ünlü, B. B. Goldberg, A. K. Swan, S. B. Cronin, *Nano Lett.* **7**, 1485 (2007).
23. S. Yamashita, S. Maruyama, Y. Murakami, Y. Inoue, H. Yaguchi, M. Jablonski, S. Y. Set, *Opt. Lett.*, **29**, 1581 (2004).

24. Y. Sakakibara, A. G. Rozhin, H. Kataura, Y. Achiba, M. Tokumoto, *Jpn. J. Appl. Phys.*, **44**, 1621 (2005).
25. G. Della Valle, R. Osellame, G. Galzerano, N. Chiodo, G. Cerullo, P. Laporta, O.Svelto, U. Morgner, A. G. Rozhin, V. Scardaci, A. C. Ferrari, *Appl. Phys. Lett.*, **89**, 231115 (2006).
26. V. Scardaci, A. G. Rozhin. F. Hennrich, W. I. Milne, A. C. Ferrari, *Phys. E*, **37**, 115 (2007).
27. V. Scardaci, Z. P. Sun, F. Wang, A. G. Rozhin, T. Hasan, F. Hennrich, I. H. White, W. I. Milne, A. C. Ferrari, *Adv. Mater.*, **20**, 4040 (2008).
28. Z. Sun, A. G. Rozhin, F. Wang, V. Scardaci, W. I. Milne, I. H. White, F. Hennrich, A. C. Ferrari, *Appl. Phys. Lett.*, **93**, 061114 (2008).
29. F. Wang, A. G. Rozhin, V. Scardaci, Z. Sun, F. Hennrich, I. H. White, W. I. Milne, A. C. Ferrari, *Nat. Nano*, **3**, 738 (2008).
30. T. Hasan, Z. Sun, F. Wang, F. Bonaccorso, P. H. Tan, A. G. Rozhin, A. C. Ferrari, *Adv. Mater.*, **21**, 3874 (2009).
31. Z. Sun, A. G. Rozhin, F. Wang, T. Hasan, D. Popa, W. ONeill, A. C. Ferrari, *Appl. Phys. Lett.*, **95**, 253102 (2009).
32. Z. Sun, T. Hasan, F. Wang, A. Rozhin, I. White, A. Ferrari, *Nano Res.*, **3**, 404 (2010).
33. S. J. Beecher, R. R. Thomson, N. D. Psaila, Z. Sun, T. Hasan, A. G. Rozhin, A. C. Ferrari, A. K. Kar, *Appl. Phys. Lett.*, **97**, 111114 (2010)
34. T. Hasan, V. Scardaci, P. H. Tan, F. Bonaccorso, A. G. Rozhin, Z. Sun, A. C. Ferrari, in O. Hayden, K. Nielsch, eds., *Molecular- and Nano-Tubes*, U.S.: Springer, 2011, 279.
35. C. F. Bohren, D. R. Huffman, *Absorption and Scattering of Light by Small Particles*, New York: Wiley-Interscience, 1998.
36. V. C. Moore, M. S. Strano, E. H. Haroz, R. H. Hauge, R. E. Smalley, *Nano Lett.*, **3**, 1379 (2003).
37. V. Perebeinos, J. Tersoff, P. Avouris, *Phys. Rev. Lett.*, **94**, 027402 (2005).
38. M. S. Arnold, A. A. Green, J. F. Hulvat, S. I. Stupp, M. C. Hersam, *Nat. Nanotechnol.*, **1**, 60 (2006).
39. F. Bonaccorso, T. Hasan, P. H. Tan, C. Sciascia, G. Privitera, G. Di Marco, P. G. Gucciardi, A. C. Ferrari, *J. Phys. Chem. C*, **114** (2010).
40. T. J. McDonald, J. L. Blackburn, W. K. Metzger, G. Rumbles, M. J. J. Heben, *Phys. Chem. C*, **111**, 17894 (2007).
41. H. Qian, C. Georgi, N. Anderson, A. A. Green, M. C. Hersam, L. Novotny, A. Hartschuh, *Nano Lett.*, **8**, 1363 (2008).

42. J. Lefebvre, P. Finnie, *J. Phys. Chem. C*, **113**, 7536 (2009).
43. T. Kato, R. Hatakeyama, *J. Am. Chem. Soc.*, **130**, 8101 (2008).
44. H. Hirori, K. Matsuda, Y. Kanemitsu, *Phys. Rev. B*, **78**, 113409 (2008).
45. F. M. Chen, W. J. Zhang, M. L. Jia, L. Wei, X. F. Fan, J. -L. Kuo, Y. Chen, M. B. Chan-Park, A. D. Xia, L. J. Li, *J. Phys. Chem. C*, **113**, 14946 (2009).
46. F. M. Chen, J. Ye, M. Y. Teo, Y. Zhao, L. P. Tan, Y. Chen, M. B. Chan-Park, L. J. Li, *J. Phys. Chem. C*, **113**, 20061 (2009).
47. R. S. Swathi, K. L. Sebastiana, *J. Chem. Phys.*, **132**, 104502 (2010).
48. J.-H. Han, G. L. C. Paulus, R. Maruyama, D. A. Heller, W.-J. Kim, P. W. Barone, C. Y. Lee, J. H. Choi, M.-H. Ham, C. Song, C. Fantini, M. S. Strano, *Nat. Mater.*, **9**, 833 (2010).
49. L. Luer, J. Crochet, T. Hertel, G. Cerullo, G. Lanzani, *ACS Nano*, **4**, 4265 (2010).
50. T. Koyama, Y. Miyata, Y. Asada, H. Shinohara, H. Kataura, A. Nakamura, *J. Phys. Chem. Lett.*, **1**, 3243 (2010).
51. T. Koyama, K. Asaka, N. Hikosaka, H. Kishida, Y. Saito, A. Nakamura, *J. Phys. Chem. Lett.*, **2**, 127 (2011).
52. T. Koyama, K. Asaka, N. Hikosaka, H. Kishida, Y. Saito, A. Nakamura, *J. Lumin.*, **131**, 494 (2011).
53. X. Dang, H. Yi, Ham, M.-H., J. Qi, D. S. Yun, R. Ladewski, M. S. Strano, P. T. Hammond, A. M. Belcher, *Nat Nano*, **6**, 377 (2011).
54. T. Ignatova, H. Najafov, A. Ryasnyanskiy, I. Biaggio, M. Zheng, S. V. Rotkin, *ACS Nano*, **5**, 6052 (2011).
55. T. Koyama, Y. Asada, N. Hikosaka, Y. Miyata, H. Shinohara, A. Nakamura, *ACS Nano*, **5**, 5881 (2011).
56. T. Förster, *Discuss. Faraday Soc.*, **27**, 7 (1959).
57. M. G. W. Alexander, M. Nido, W. W. Riihle, K. Koehler, *Phys. Rev. B*, **41**, 12295 (1990).
58. S. R. Adams, A. T. Haroontunian, Y. J. Buechler, S. S. Taylor, R. Y. Tsien, *Nature*, **349**, 694 (1991).
59. C. R. Kagan, C. B. Murray, M. Nirmal, M. G. Bawendi, *Phys. Rev. Lett.*, **76**, 1517 (1996).
60. K. Becker, J. Lupton, J. Mueller, A. L. Rogach, D. V. Talapin, H. Weller, J. Feldmann, *Nat. Mater.*, **5**, 777 (2006).
61. V. Biju, T. Itoh, Y. Baba, M. Ishikawa, *J. Phys. Chem. B*, **110**, 26068 (2006).

62. A. Hagen, M. Steiner, M. B. Raschke, C. Lienau, T. Hertel, H. Qian, A. J. Meixner, A. Hartschuh, *Phys. Rev. Lett.*, **95**, 197401 (2005).
63. H. Hirori, K. Matsuda, Y. Miyauchi, S. Maruyama, Y. Kanemitsu1, *Phys. Rev. Lett.*, **97**, 257401 (2006).
64. C. D. Spataru, S. Ismail-Beigi, L. X. Benedict, S. G. Louie, *Phys. Rev. Lett.*, **95**, 247402 (2005).
65. S. M. Bachilo, L. Balzano, J. E. Herrera, F. Pompeo, D. E. Resasco, R. B. Weisman, *J. Am. Chem. Soc.*, **125**, 11186 (2003).
66. S. Reich, M. Dworzak, A. Hoffmann, C. Thomsen, M. S. Strano, *Phys. Rev. B*, **71**, 033402 (2005).
67. C. Manzoni, A. Gambetta, E. Menna, M. Meneghetti, G. Lanzani, G. Cerullo, *Phys. Rev. Lett.*, **94**, 207401 (2005).
68. D. A. Heller, E. S. Jeng, T. K. Yeung, B. M. Martinez, A. E. Moll, J. B. Gastala, M. S. Strano, *Science*, **311**, 508 (2006).
69. Y. Ohno, S. Iwasaki, Y. Murakami, S. Kishimoto, S. Maruyama, T. Mizutani, *Phys. Rev. B*, **73**, 235427 (2006).
70. S. Giordani, S. D. Bergin, V. Nicolosi, S. Lebedkin, M. M. Kappes, W. J. Blau, J. N. Coleman, *J. Phys. Chem. B*, **110**, 15708 (2006).
71. Y. Miyauchi, R. Saito, K. Sato, Y. Ohno, S. Iwasaki, T. Mizutani, J. Jiang, S. Maruyama, *Chem. Phys. Lett.*, **442**, 394 (2007).
72. O. Kiowski, S. Lebedkin, F. Hennrich, S. Malik, H. Rosner, K. Arnold, C. Surgers, M. M. Kappes, *Phys. Rev. B*, **75**, 075421 (2007).
73. T. Hasan, V. Scardaci, P. H. Tan, A. G. Rozhin, W. I. Milne, A. C. Ferrari, *J. Phys. Chem. C*, **111**, 12594 (2007).
74. T. Hasan, P. H. Tan, F. Bonaccorso, A. Rozhin, V. Scardaci, W. Milne, A. C. Ferrari, *J. Phys. Chem. C*, **112**, 20227 (2008).
75. S. Reich, C. Thomsen, J. Robertson, *Phys. Rev. Lett.*, **95**, 077402 (2005).
76. T. Okazaki, T. Saito, K. Matsuura, S. Ohshima, M. Yumura, Y. Oyama, R. Saito, S. Iijima, *Chem. Phys. Lett.*, **420**, 286 (2006).
77. Z. T. Luo, L. D. Pfefferle, G. L. Haller, F. Papadimitrakopoulos, *J. Am. Chem. Soc.*, **128**, 15511 (2006).
78. C. Fantini, A. Jorio, A. P. Santos, V. S. T. Peressinotto, M. A. Pimenta, *Chem. Phys. Lett.*, **439**, 138 (2007).
79. A. Jorio, C. Fantini, M. A. Pimenta, D. A. Heller, M. S. Strano, M. S. Dresselhaus, Y. Oyama, J. Jiang, R. Saito, *Appl. Phys. Lett.*, **88**, 023109 (2006).
80. Y. C. Chen, N. R. Raravikar, L. S. Schadler, P. M. Ajayan, Y. P. Zhao, T. M. Lu, G. C. Wang, X. C. Zhang, *Appl. Phys. Lett.*, **81**(6), 975 (2002).

81. S. Tatsuura, M. Furuki, Y. Sato, I. Iwasa, M. Tian, H. Mitsu, *Adv. Mater.*, **15**, 534 (2003).
82. A. G. Rozhin, Y. Sakakibara, H. Kataura, S. Matsuzaki, K. Ishida, Y. Achiba, M. Tokumoto, *Chem. Phys. Lett.*, **405**, 288 (2005).
83. G. N. Ostojic, S. Zaric, J. Kono, M. S. Strano, V. C. Moore, R. H. Hauge, R. E. Smalley, *Phys. Rev. Lett.*, **92**, 117402 (2004).
84. J. Kono, G. N. Ostojic, S. Zaric, M. S. Strano, V. C. Moore, J. Shaver, R. H. Hauge, R. E. Smalley, *Appl. Phys. A*, **78**, 1093 (2004).

Chapter 7

Advances in Dispersal Agents and Methodology for SWNT Analysis

Tsuyohiko Fujigaya[a] and Naotoshi Nakashima[a,b]
[a]*Department of Applied Chemistry, Graduate School of Engineering, Kyushu University, 744 Motooka, Fukuoka 819-0395, Japan*
[b]*Japan Science and Technology Agency, CREST, 5 Sanbancho, Chiyoda-ku, Tokyo 102-0075, Japan*
nakashima-tcm@mail.cstm.kyushu-u.ac.jp

One of the key issues in the material science of carbon nanotubes (CNTs) is to develop a methodology to solubilize/disperse them in a solvent. In this review articles, we first summarize individual solubilization of single-walled (carbon) nanotubes (SWNTs) in solvents using surfactants and polycylcic aromatic molecules, including porphyrins, deoxyribonucleic acid (DNA), and condensed polymers, as well as their optical properties. We then describe our recent approach toward the applications of novel CNT/polymer functional nanocomposites.

7.1 Introduction

Carbon nanotubes (CNTs) are made of rolled-up graphene sheets with one-dimensional extended *p*-conjugated structures, discovered

---

Luminescence: The Instrumental Key to the Future of Nanotechnology
Edited by Adam M. Gilmore
Copyright © 2014 Pan Stanford Publishing Pte. Ltd.
ISBN 978-981-4241-95-3 (Hardcover), 978-981-4267-72-4 (eBook)
www.panstanford.com

in 1991 by Iijima.[1] They are classified into mainly three types in terms of the number of graphene layers within a CNT, that is, single-walled (carbon) nanotubes (SWNTs), double-walled (carbon) nanotubes (DWNTs), and multiwalled (carbon) nanotubes (MWNTs), which have one, two, and more than three walls, respectively. One of the key issues in the utilization of such a seminal material for basic research, together with the material applications, is to develop a methodology to solubilize/disperse them in solvents (Fig. 7.1)[2–4] since as-synthesized CNTs form tight, bundled structures[5] due to their strong van der Waals interaction.[6] Solubilization/dispersion techniques can be categorized mainly into two methods, namely, chemical and physical modification. Solubilization/dispersion of CNTs based on physical adsorption of dispersant molecules possesses several advantages such as the ease of preparation and maintenance of intrinsic CNT properties, which show a sharp contrast with chemical modification.[7–9]

In this chapter, general strategies for CNT solubilization as well as the applications of solubilized CNTs are described.

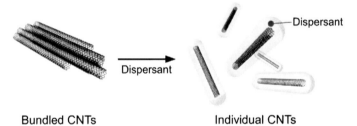

Figure 7.1 Schematic illustration of the solubilization of CNTs through physical adsorption of the dispersant molecules on the surfaces of CNTs.

7.2 Characterization of Dispersion States

The typical procedures for the preparation of an individual solution of CNTs are ultrasonication of the CNTs in a dispersant solution, followed by centrifugation to give gray-transparent supernatant solutions. Individually solubilized/dispersed CNTs are often visualized by an atomic force microscope (AFM)[10,11] and a transmission electron microscope (TEM)[12,13] after casting on substrates. Near-infrared (NIR) absorption and photoluminescence (PL) spectroscopy are strong tools for the direct observation of the dispersion nature of SWNTs associated with the allowed transition (van Hove transition) of SWNTs. Wiseman et al.[14,15] have found that PL in the NIR region can be detected from surfactant-dissolved SWNTs and then have succeeded in the determination of the SWNT chirality indices (n,m) in the solution. Notably only individually dissolved semiconducting SWNTs, and small, bundled SWNTs in some cases[16], exhibit PL because the bundled SWNTs end up with quenching their PL by the metallic SWNTs in the bundle. Thus the PL observation can serve as a good indicator of the individual solubilization of SWNTs in solution as well as in films.[17,18] Recent advances in the NIR detection technique allow us to see PL from a single SWNT swimming in the solution and gel by combining an inversed microscope technique.[19-21] The small-angle neutron scattering (SANS) technique is an even more powerful method to figure out the degree of dispersion states together with the wrapping structures by dispersants.[13,22-27] Even in the absence of PL signals due to the bundling, ultraviolet (UV)-visible (vis) NIR absorption spectroscopy is quite helpful for roughly evaluating the degree of bundling of SWNTs, both in the solution and film states.[28] SWNT bundling renders the red-shifted and broadened features of the absorption spectra compared to that of the isolated SWNTs.[29,30] In some cases, the electron conductivity measurement allows brief estimation of dispersion of the SWNTs in polymer films by evaluating the concentration of the SWNTs at the electron percolation threshold.[31,32] A lower threshold concentration is a consequence of greater dispersion of the SWNTs in the matrices.

7.3 Solubilization by Dispersal Agents

7.3.1 Surfactants

The most convenient and frequently used dispersant for CNTs in aqueous media is a surfactant such as sodium dodecyl sulfate (SDS)[33-35], sodium dodecylbenzene sulfonate (SDBS)[12,36-39], cethyltrimethylammonium bromide (CTAB)[12,40], Brij[12,41], Tween[12,41], and Triton X (Fig. 7.2).[12,36,41,42] An early attempt in preparing CNT dispersions using a surfactant was explored by Bandow et al. for the purification of SWNTs from carbon soot material.[43] The suggested mechanism of the individual dispersion is the encapsulation of

SWNTs in the hydrophobic interiors of the micelles, which results in the formation of a stable dispersion.[29] Among the conventional surfactants, SDBS is one of the most efficient SWNT solubilizer, that is, it has been reported that even in the concentration of 20 mg/mL of SWNTs in an SDBS micelle, no aggregation of the SWNTs occurs for more than three months.[36]

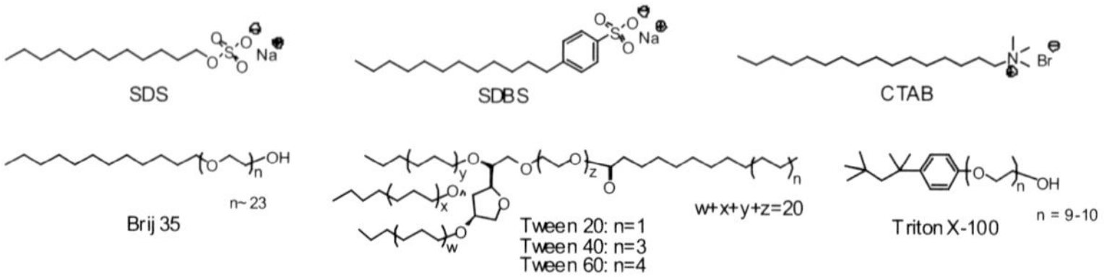

Figure 7.2 Chemical structures of surfactants for CNTs solubilization.

Biological surfactants such as bile salts also act as SWNT solubilizers in water (Fig. 7.3).[41,44,45] We reported individual solubilization of SWNTs using 10 different anionic-, zwitter ionic-, and nonionic steroid biosurfactants and three different sugar biosurfactants. Aqueous micelles of anionic cholate analogs, including sodium cholate (SC), sodium deoxycholate (SDC), sodium taurocholate (STC), sodium taurodeoxycholate (STDC), and sodium glycocholate (SGC), as well as *N,N*-bis(3-D-gluconamidopropyl)cholamide (BIGCHAP) and *N,N*-bis(3-D-gluconamidopropyl)deoxycholamide (deoxy-BIGCHAP), exhibited a high ability to individually dissolve SWNTs. Aqueous micelles of nonionic biosurfactants, including sucrose monocholate (SMC), *n*-octyl-*b*-D-glucoside (OG), *n*-decyl-*b*-D-maltoside (DM), and *n*-decanoyl-N-methylglucamide (MEGA-10), dissolve SWNTs, while the solubilization ability is weaker than that of the anionic cholate analogs, while the solubilization abilities of the zwitter ionic micelles of 3-[(3-cholamidopropyl)dimethylammonio]propanesulfonic acid (CHAPS) and 3-[(3-cholamidopropyl)dimethylammonio]-2-hydroxypropanesulfonic acid (CHAPSO) were very low and almost none for OG. It is evident that the chemical structures, especially, the substituent groups, of the surfactants play an important role in the solubilization of SWNTs.

The contour plots for SWNTs in 1 wt% SC showed the existence of SWNTs with chirality indices of (7,6), (8,4), (8,6), (9,4), and (9,5), whose behavior is virtually identical to that of the SDS-dissolved SWNTs[46] [47] (Fig. 7.4A, left). STC and SGC (both 1 wt%) dissolve

SWNTs with the chirality indices of (7,5), (7,6), (8,4), (8,6), and (9,4), and (7,5), (7,6), (8,4), and (9,4), respectively (Fig. 7.4B,C, left). The micelle of BIGCHAP produced a different pattern, that is, in this solution, the SWNTs with (8,6) and (9,5) are rich (Fig. 7.4D, left). Because of the sugar side chain, the micellar structures of BIGCHAP are different from those of SC, STC, and SGC. Despite the broadness in the NIR absorption spectra in their 1 mM solutions, the NR PL spectra could be detected, indicating that individually dissolved SWNTs exist in these solutions (Fig. 7.4, right columns). The contour plots for SWNTs in 1 wt% and 1 mM solutions of SC were almost identical (Fig. 7.4A, right). On the contrary, the contour plots between micellar (1 wt%) and nonmicellar (1 mM) solutions of STC, SGC, and BIGCHAP are somewhat different (Fig. 7.4B–D). The difference would be due to the difference of solubilization mechanisms. The contour plots of SWNTs in 1 mM SGC are complex, that is, the following three patterns for 1 mM SGC are detected: the (7,5), (7,6), and (8,4)-rich pattern (Fig. 7.4C, right); the (7,6), (8,4), (8,6), (9,4), and (9,5)-rich pattern; and the (6,5) and (8,3)-rich pattern (data not shown). The solubility of SGC in water is low, and this might be related to the solubilization behaviors of the SWNTs.

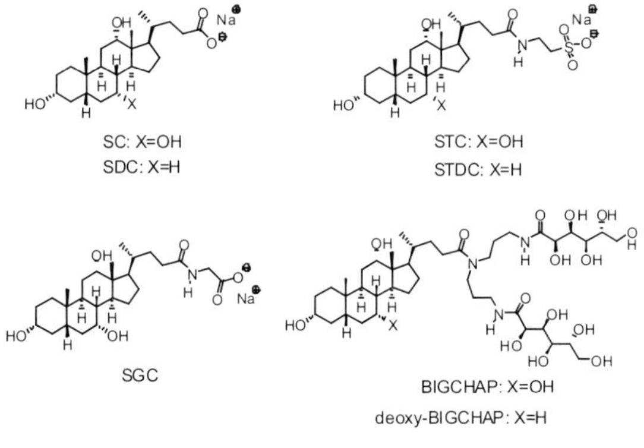

Figure 7.3 Biological surfactants for CNT solubilization.

The mechanism for the solubilization of SWNTs using biosurfactants at concentrations above the critical micelle concentration (cmc) would be similar to that of the so-called "micellar solubilization" using surfactants such as SDS.[29] The solubilization is due to physisorption of the biosurfactants onto the surfaces of the SWNTs since the surfactants have no chemical reactive groups

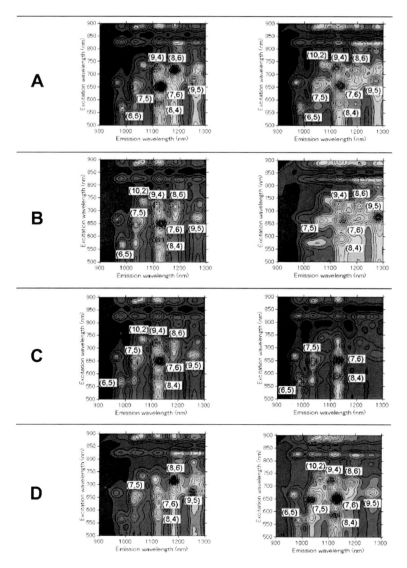

Figure 7.4 Contour plots of PL spectra as a function of excitation wavelength and resultant emission. (A): 1 wt% SC (left) and 1 mM SC (right); (B) 1 wt% STC (left) and 1 mM STC (right); (C) 1 wt% SGC (left) and 1 mM SGC (right); and (D) 1 wt% BIGCHAP (left) and 1 mM BIGCHAP (right).

like SDS to react chemically with the SWNTs. Dialysis experiment for an SWNT aqueous SC solution (1 wt%) using membrane tubing revealed that SC molecules gradually leak from the inside of the tubing to the outer water phase; as a result a precipitate of the SWNTs is produced. Even after two-day-dialysis, about 15% of the SWNTs remained in the tubing as a transparent solution/dispersion. After the dialysis, the NIR absorption spectrum changes to somewhat broad, indicating that the SWNTs form a bundled structure in water after removing the free SC molecules, suggesting the adsorption of SC molecules onto the surfaces of the SWNTs. A possible model for the adsorption of a steroid-surfactant is presented in Fig. 7.5. The steroid biosurfactants have a large, rigid, and planar hydrophobic moiety of a steroid nucleus with two or three hydroxyl groups. This unique structure would be important for the adsorption onto the surfaces of the SWNTs that leads to the preparation of transparent dispersions of the SWNTs in water not containing free steroid molecules.

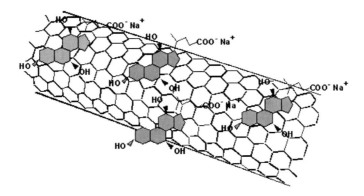

Figure 7.5 A model for the adsorption of a steroid-surfactant onto the surfaces of SWNTs.

7.3.2 Polycyclic Aromatic Compounds

The surfaces of CNTs can be readily functionalized through p–p interactions with compounds having p-electron-rich structures due to the highly delocalized p-electrons of CNTs. The p–p interaction between polycyclic aromatic compounds and CNT sidewalls has been discussed based on both theoretical[48] and experimental[49] approaches. We reported that a pyrene-based ammonium salt (compound **1** in Table 7.1) is able to solubilize SWNTs[50] and fullerene-filled CNTs (so-called peapods)[51] in water. The pyrene-carrying compound acts as an efficient dispersant compared to naphthyl- and phenyl-based ammonium salts.[52] This is due to the strong binding affinity between the pyrene group and the CNT sidewalls. Now pyrene derivatives have been widely recognized as excellent solubilizers for CNTs, as summarized in Table 7.2.[50-58] By taking advantage of the efficient adsorbing capability on CNT surfaces, pyrene derivatives have been used as decent interlinkers to anchor functional materials that can communicate with CNTs, as summarized in Table 7.3.[59-78] Pioneering work demonstrating that the pyrene derivative functioned as an interlinker was carried out by Dai et al.[59] They successfully attached a protein on the surfaces of the SWNTs with the aid of a pyrene-carrying succinidyl compound (compound **8** in Table 7.2). Pyrene-ammonium **1** was also used in many research studies for anchoring anionic functional molecules on the surfaces of SWNTs and MWNTs.[64,65,79] Other polycyclic aromatic moieties such as anthracene[80,81], terphenyl[81,82], perylene[83], triphenylene[84], phenanthrene[85], and pentacene[86] also have affinity for the sidewalls of CNTs, and various solubilizers bearing these molecules have been developed.

Table 7.1   Pyrene-based dispersants for CNTs

	Pyrene	Research targets	Ref.
1		Long-lived charge separation between porphyrin and SWNTs	50, 52
		Solubilization of C70@SWNTs into an aqueous system	51
2		ROMP on SWNTs	53
3		Functionalization of MWNTs in supercritical fluids	54

	Pyrene	Research targets	Ref.
4	(pyrene-OH structure)	Functionalization of MWNTs in supercritical fluids	54
5	(pyrene-fullerene structure)	Electrochemical response of a fullerene/SWNT hybrid	55
6	(pyrene-chlorophyll structure)	Attachment of a pyrene-modified chlorophyll derivative	56
7	(pyrene-N structure)	Selective fictionalization of SWNTs	57, 58

Abbreviation: ROMP, ring-opening metathesis polymerization.

We have reported that a green-tea solution also acts as an excellent SWNT dispersant.[87] Our dissolution scenario is that catechin, a polycyclic aromatic compound, mainly contributes to the dispersion because epigallocatechin gallate also disperses the SWNTs in water.

The degree of the interactions between these polycyclic aromatic moieties and the CNT sidewalls has been accessed based on Raman spectroscopic analysis by monitoring the shift of the radial breathing mode (RBM).[86] The high-performance liquid chromatography (HPLC) technique using CNTs as a stationary phase is a powerful tool to rank the affinity of several dispersants on the surfaces of CNTs at one time.[88,89] Chang et al. evaluated the degree of affinity on the surfaces of SWNTs using the HPLC technique and found the polycyclic aromatic system shows better interaction than a monocyclic compound[88] in accordance with several papers.[52,81] The difference in the affinity worked for CNTs is fascinating in the view of chirality recognition and enrichment of specific SWNTs.[90]

Table 7.2 Pyrene-based interlinkers for immobilizing functional molecules on CNT surfaces

	Pyrenes	Research target	Ref.
8		Immobilization of protein onto the SWNT surface	59
9		Au nanoparticle immobilization on MWNTs	60
10		Layer-by-layer assembly of SWNTs with polyanion	61
11		Photo-induced electron transfer from porphyrin to SWNTs	62–65
		Photo-induced electron transfer between CdTe and MWNTs	66
		Photo-induced electron transfer from polythiophene	67
12		Photo-induced electron transfer from porphyrin to SWNTs	68
13		Immobilization of magnetic particle	69
14		Immobilization of metal nanoparticles onto SWNTs	70
15		Enhancing of bioelectrocatalyzed oxidation	71

	Pyrenes	Research target	Ref.
16	pyrene-CH$_2$-NH$_2$	Immobilization of Au nanoparticle onto the MWNT surface	72
17	pyrene-CH$_2$-NH$_3^+$	Layer-by-layer assembly of polyelectrolytes	73
		DNA adsorption and gene transcription	236
18	pyrene-CO-CH$_2$-imidazolium	Layer-by-layer assembly of polyelectrolytes	73
19	pyrene-CH=N-O-(CH$_2$)$_n$-O-NH$_2$	Immobilization of tobacco mosaic virus onto SWNTs	74
20	pyrene-(CH$_2$)$_3$-CO-NH-C$_6$H$_4$-imidazole	Photo-induced electron transfer from naphthalocyanine to SWNTs	75
21	pyrene-(CH$_2$)$_3$-CO-NH-cyclodextrin	Modification of cyclodextrin on the SWNT surface	76
22	pyrene-(CH$_2$)$_4$-O-CO-(CH$_2$)$_4$-S-S	CdSe immobilization	77
23	pyrene-(CH$_2$)$_3$-CO-NH-glycodendrimers	Binding to cell surface	78

7.3.3 Porphyrins

Porphyrin compounds are able to individually solubilize SWNTs.[91] Zinc protoporphyrin IX (compound **24** in Table 7.3) was used, and it was found that a resulting **24**/SWNT solution is stable even after six months. Fluorescence quenching of the porphyrin in the **24**/SWNT evidenced the adsorption of the porphyrin onto the surfaces of the

Table 7.3 Porphyrin-based CNT solubilizers

	Prophyrin	M	R	Topic	Ref.
24		Zn		First porhyrin-based solubilizer	91
25		2H		Solubilization of SWNTs	92
26		FeCl		Electrochemical response	92
27		Co		EPR study	93
28		2H	3,5-di-t-Bu phenyl	Photo-induced charge Injection	110

#	Structure	M	R	Description	Ref
29	porphyrin	2H	phenyl	Porphyrin-driven supramolecular assembly	103
30	porphyrin	Zn	butyl	Direct observation of adsorbed porphyrin	115
31	porphyrin	2H	tolyl	Photo-induced charge injection	110
32	porphyrin	2H, Zn	tolyl	Nonlinear optical properties	118
33	porphyrin (meso-R)	2H, Zn	phenyl	Interaction study	96
34	porphyrin (meso-R)	2H	3,5-di-t-Bu-phenyl	Photo-induced charge injection from excited porphyrin into SWNTs	110
35	porphyrin (meso-R)	4H^{2+}	$-C_6H_4-SO_3^-$	Solubilization of SWNTs in water	98
36	porphyrin (meso-R)	2H, Zn	$-C_6H_4-O(CH_2)_{11}SAc$	Separation of semiconducting SWNTs	97
37	porphyrin (meso-R)	2H, Zn	$-C_6H_4-O(CH_2)_{11}SAc$	J-aggregation on the SWNTs surface	237
38	porphyrin polymer	Zn	3,5-di($OC_{16}H_{33}$)-4-$OC_{16}H_{33}$-phenyl	Solubilization by conjugated porphyrin polymer	100

(Continued)

Table 7.3 (Continued)

	Prophyrin	M	R	Topic	Ref.
39		Zn		Supramolecular solubilization	99
40			R_1 = t-Bu, t-Bu; CN	Solubilization by fully fused porphyrin	104
41			(R): R_1 = CH$_2$Ph, R_2 = H, R3 = NHCO$_2$t-Bu (S): R_1 = H, R_2 = CH$_2$Ph, R3 = NHCO$_2$t-Bu	Separation of optically active SWNTs	119
42		2H	—SO$_3$·Na$^+$	Long-lived charge separation	105

Abbreviation: EPR, electron paramagnetic resonance.

SWNTs. A series of porphyrin derivatives were tested and revealed that a wide range of porphyrin derivatives, including **25** and **26** (Table 7.3), can also act as effective dispersants for SWNTs.[92,93] The finding led theoretical as well as experimental attempts to understand the interaction between porphyrins and CNTs.[94-96] Importantly, not only the *p–p* interaction but also the charge transfer interaction has been pointed out to serve the adsorption of porphyrin derivatives on the surfaces of CNTs.[96] The center metals in porphyrin affect the degree of solubilization of the CNTs.[96,97] Porphyrin derivatives[97-105] as well as their analog molecules such as phthalocyanines[106-108] and sapphyrin[109] have been reported to serve as dispersants for CNTs.

The combination of porphyrins and CNTs has attracted extensive interest due to the unique photophysical[62,110,111], electrochemical[112-114], electronic[102,110,115,116], and optical[117,118] properties of the composites. Extensive efforts have been carried out on the photo-induced electron transfer from porphyrins to CNTs, including not only physically connected but also covalently bonded porphyrin/CNT hybrids.[110] Dye-sensitize organic solar cells have emerged as a potential application for porphyrin/CNT hybrids. On the other hands, similar to other solubilizers, a porphyrin compound was used for a separation medium for CNTs based on molecular recognition.[97] One of the striking results realized in these studies is the separation of optical active SWNTs reported by Osuka et al.[119] They found that optically active porphyrin dimers (compound **41** in Table 7.3) can pick up SWNTs with a right- or left-handed helicity structure from racemic SWNT mixtures, depending on the chirality of compound **41**. A closer look at the system additionally revealed compound **41** also recognized and enriched the specific diameter of the SWNTs.[120]

7.3.4 DNA

The combination of CNTs and DNA (ribonucleic acid [RNA]) is of fundamental and applied interest in many chemical and biochemical areas. We have reported the finding that double-stranded DNA (dsDNA) molecules dissolve SWNTs in aqueous solutions.[121] We also showed that electrochemical deposition of dsDNA/SWNT complexes by poly(ethylenedioxythiophene) (PEDOT) on an indium tin oxide (ITO) electrode is possible.[122] A number of groups have now used this scheme in their studies of various dsDNA and CNTs structures. Barisci et al.[123] fabricated dsDNA/SWNT fibers that were mechanically strong and conductive and exhibited useful capacitance values up to 7.2 F g^{-1}. Iijima et al.[124] showed high-resolution TEM and scanning tunneling microscopy (STM) images of dsDNA/MWNT. Gladchenko et al.[125] characterized fragmented dsDNA-wrapped SWNTs in aqueous solutions. He and Bayachou[126] described the layer-by-layer fabrication and characterization of dsDNA-wrapped SWNTs particles. Coleman et al.[127] reported spontaneous dispersion of SWNTs by dsDNA and suggested the SWNT-induced denaturing of the dsDNA is the key step for the wrapping of SWNT by dsDNA.[128]

At the same time as our report that dsDNA solubilizes SWNTs, Zheng et al.[129,130] showed that single-stranded DNA (ssDNA) solubilizes SWNTs. An interesting finding is the ability to separate metallic and semiconducting SWNTs using size exclusion chromatography.[131] Molecular modeling for the ssDNA/SWNT composites suggests that ssDNA can bind to SWNTs through p-stacking, resulting in the helical wrapping of SWNTs.[129] Zhao and Johnson[132] reported molecular dynamics simulation of DNA adsorption on an SWNT in an aqueous environment. A binding model of (10,0) SWNTs wrapped by a poly(T) was reported, and the binding of ssDNA to SWNTs was probed by flow linear dichroism.[133,134] Detailed optical properties, including NIR absorption and PL and Raman studies for ssDNA/SWNT hybrids, have been characterized.[135,136] A wide range of the studies focusing on the properties of the ssDNA/SWNT hybrid has been reported. Fagan et al.[137] described length-dependent optical effects in SWNTs dissolved in ssDNA, and (30-mer)/SWNT solutions were used to discuss solution redox chemistry of CNTs[138], photo-induced charge transfer on an Ag$^+$/DNA/CNT complex[139], and electrocatalytic oxidation.[140] Phonon-assisted excitation dynamics for (6,5)-enriched ssDNA-wrapped SWNTs has been discussed theoretically.[141,142] The adsorption behavior of ssDNA-wrapped SWNTs on substrates was examined using reflection absorption Fourier transform infrared spectroscopy (FT-IR), X-ray photoelectron spectroscopy (XPS), and Raman spectroscopy.[143] Strano et al.[144] described the optical detection of DNA conformational polymorphism on SWNTs. Li et al.[145] showed that CNTs selectively destabilized duplex and triplex DNA and induced B-A transition in solution. The DNA-immobilized, aligned CNTs have been shown to be significant for sensing complementary DNA and/or target DNA chains of specific sequences with high sensitivity and selectivity.[146] Fantini et al.[147] characterized GT-DNA oligomer-wrapped SWNTs by Raman and optical spectroscopies and revealed different interactions for semiconducting and metallic SWNTs. Chou et al.[148] described in detail the length characterization of ssDNA-wrapped SWNTs using Raman spectroscopy. We have constructed multilayer assemblies with alternating monolayers of poly(G)-wrapped SWNTs and poly(C)-wrapped SWNTs on quartz based on the complementary base pairing between nucleic bases G and C, which would be applicable in wide fields of nanoscience and technology.[149] ssDNA/SWNT solutions can be used as materials for gene and protein delivery and nanotherapy.[150] Douglas et al. reported ssDNA/SWNT-induced alignment of membrane proteins for nuclear magnetic resonance (NMR) structure determination.[151] Deng et al.[152] described that ssDNA-wrapped SWNTs can serve well as rigid templates for the self-assembly of gold nanoparticles. We described the well-resolved structure of dsDNA-wrapped SWNTs and then the modulation of the bandgaps of individually solubilized SWNTs (high-pressure carbon monoxide conversion [HiPCO] and CoMoCAT) in dsDNA aqueous solutions, which is possible by changing the solution pH, where a very small pH change (from pH = 5.8 to pH = 6.4 in NIR absorption and from pH = 6.4 to pH = 7.4 in PL) causes a dramatic spectral change (Fig. 7.6).[153] The results are explained in terms of reversible SWNT oxidation by analysis of the oxide-induced absorption bleaching and luminescence quenching at low pH, and the differences in the observed pH breakpoints for the NIR absorption bleaching and PL quenching are due to the difference in the number of holes generated on the SWNTs.

Thermodynamical stability of the DNA/SWNT hybrids is another feature of importance. By using the gel permeation chromatography (GPC) technique, we proved that the binding of dsDNA and SWNTs is highly stable, namely, the detachment of the dsDNA from the surfaces of the SWNTs is not detectable for at least one month.[154] Thanks to the formation of stable hybrids between DNA and CNTs, a wide range of research studies has been achieved in view of biological applications, such as the conformation transition monitoring of DNA[144], redox sensing of glucose and hydrogen peroxide[155], hybridization detection between ssDNA and its complementary DNA[156], and uptake estimation of DNA/SWNTs into the cell.[157] As the possibility of DNA/CNTs as a gene delivery carrier increases, the strong demand to avoid DNA damage during sonication is raised. A modified DNA-wrapping protocol[158] using surfactant-dissolved SWNTs followed by the exchange of DNA in a dialysis membrane realized the sonication-free process.[125,155] Stable DNA/SWNT dispersions are quite suitable for the chromatography-based separation of SWNTs. Zheng et al.[130,159–161] developed the enrichment of specific chirality indices as well as the removal of free DNA by anion exchange chromatography for ssDNA/SWNTs. Furthermore, stable ssDNA/SWNT dispersions also enable the length sorting and removal of free DNA by size exclusion chromatography.[131,162] Finally, their dedicated studies led to the excellent result of chromatography-based enrichment of single chirality of SWNTs.[159] Especially, length separation of an ssDNA/SWNT composite is expected to provide a significant opportunity for the precise assessment of biological activity of the composites since the size effects are a general factor in nanomaterials for such studies as cell uptake, retention, and distribution.[163,164] Indeed, Becker et al.[165] reported lengths dependent on the uptake of the ssDNA/SWNT hybrid in the cell.

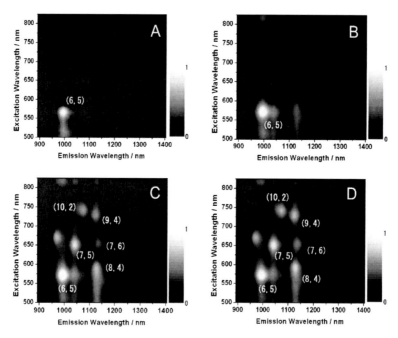

Figure 7.6   PL mapping of the HiPCO-SWNTs dissolved in dsDNA in phosphate buffer at pH = 5.8 (A), pH = 6.4 (B), pH = 7.0 (C), and pH = 8.0 (D).

7.3.5 Condensation Polymers

Many papers describing the nanocomposite formation of CNTs with condensation polymers such as polyesters and polyamides have been published.[166–170] Most of them were prepared by melt mixing, polymer grafting, and in situ polymerization methods by using oxidized CNTs due to the ease of sample preparation. We have reported an extremely efficient individual dissolution of SWNTs by a totally aromatic polyimide (**PI-1** in Fig. 7.7, left).[171] As much as 2.0 mg/mL of the SWNTs is individually dissolved in a 1.0 mg/mL dimethylsulfoxide (DMSO) solution of **PI-1**. The major driving force for the solubilization of SWNTs is attributed to a p–p interaction between the condensed aromatic moieties on the polyimide and the surfaces of SWNTs. Generally speaking, the composite films from the individual dispersion of CNTs would maximize the performance of the materials, such as mechanical properties, with minimum addition of the CNTs. For this reason, precise analysis of the degree of dispersion will become a strong focus of interest also for other polyimide/CNTs[172–182] since polyimides are widely known to possess excellent mechanical strength and heat resistance.[183]

Polybenzimidazole (**PBI** in Fig. 7.7, right) is also recognized as a highly thermal stable polymer and is widely used for firefighters' protective clothing, high-temperature gloves, and astronaut flight suits.[184] Different from typical aromatic polyimides, **PBI** is soluble in common organic solvents such as N,N-dimethylacetamide (DMAC), DMSO, and dimethylformamide (DMF). We have reported that **PBI** acts as a good dispersant for SWNTs due to the p–p interaction between the polymer and SWNT sidewalls. The vis-NIR absorption and PL spectra of a **PBI**/SWNT solution in DMAC clearly show the characteristic absorption peaks and strong PL spots, respectively, derived from the individual SWNTs (Fig. 7.8). Effective dispersion of SWNTs in the **PBI** matrix results in dramatic reinforcement in the composite film. We have found that the addition of very small amounts of SWNTs (0.06 wt%) reinforces the mechanical properties of the original polymer by *ca.* 150% without reducing their thermal stabilities.

Figure 7.7 Chemical structure of **PI-1** and **PBI**.

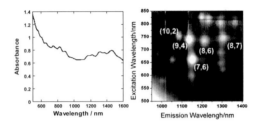

Figure 7.8 Absorption (left) and PL (right) spectra of a **PBI**/SWNT composite solution in DMAC.

7.4 Nanotube/Polymer Composites

7.4.1 Curable Monomers and Nanoimprinting

Heat- and photo-curable resins have been interesting as a promising matrix for CNT composites owing to several advantages:

(i) Most of these monomers are viscose liquids, and principally, there is no need to add any solvents to obtain polymer/CNT composites.
(ii) Mixing with the small monomers is expected to have lower entropic barriers to disperse compared to polymer melt mixing.
(iii) Quick solidification, especially in the photo-curable system, can avoid the reaggregation often occurring during the solvent evaporation process.

Especially, epoxy/CNTs are one of the most extensively researched thermoset composites so far.[185–192] The combination of rheological studies[193] and SANS measurements[25] are strong tools to understand the degree of dispersion in the composite.[193] Quick solidification without the solvent removal process was utilized to keep the CNT alignment formed prior to the polymerization via magnetic- or electronic-induced orientation.[194,195] The combination has been demonstrated to yield good processability.[196] We have reported the mold-assisted photolithography of biacrylate/SWNT composites (**UV-1** in Fig. 7.9) by using a polydimethylsiloxane (PDMS) stamp[197] and clear two-dimensional (2D) patterns with a submicron scale were easily fabricated on a silicon wafer in a few seconds (Fig. 7.10). As expected, the degree of dispersion showed no change upon polymerization, which was proved by the monitoring of vis-NIR absorption spectroscopy. The composite present an extremely low electric percolation threshold (0.05–0.1 wt%) as well as low surface resistance accompanied by a nice dispersion compared to other systems (in the order of 10^{-2} ohms/square), suggesting effective dispersion of SWNTs in the matrix. These patterned polymer/SWNT composites with high conductivity may offer novel potential applications, including an optical waveguide utilizing the nonlinear response of SWNTs[198], a scaffold for cell culture media[199,200], a thin-film transistor composed of an SWNT network in the insulating resin, a separator for fuel cells, and a chemical/biological sensor.[201,202]

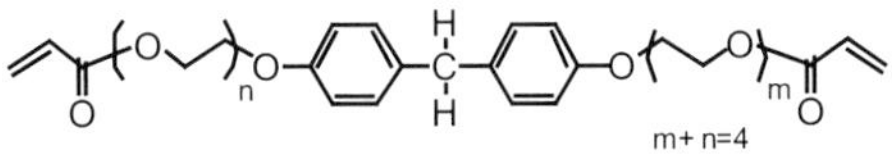

Figure 7.9 Chemical structure of a bisacrylate photocurable monomer (**UV-1**).

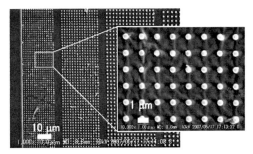

Figure 7.10 SEM images of the nanoimprint patterns prepared from **UV-1**/SWNTs using PDMS stamps. *Abbreviation*: SEM, scanning electron microscopy.

7.4.2 Nanotube/Polymer Gel for NIR-Responsive Materials

CNTs are characterized by their intense absorption in the NIR region, and this absorption gives rise to a potential use for NIR functional materials. Mainly two NIR-responsive materials have been explored. One is the NIR-saturable absorber necessary for solid-state lasers based on the saturable absorption property of CNTs.[203–205] Sakakibara et al. demonstrated that SWNTs composites dispersed in a polyimide matrix are well suited for the reproducible construction of mode-locked fiber lasers and the generation of extremely short pulse durations.[206-208] Homogeneous dispersion of SWNTs in a polyimide matrix serve to minimize the loss of light caused by scattering and to realize such an excellent property. This application is quite unique and gives requisite optical devices such as laser and optical switches for NIR high-speed optical communication systems.

Another unique application of CNTs working in the NIR region is the photon-to-heat convertor utilizing efficient photoabsorption and photothermal conversion of CNTs in the NIR region. Boldor et al. reported that MWNTs showed higher photothermal conversion efficiency than graphite.[209] Among the various light sources, NIR laser light is a fascinating stimulus, especially from the biomedical point of view, because biomedical tissues have only a slight absorption in the NIR region, which enables remote stimulation of the NIR absorbent in the body from the outside. Dai et al. reported an NIR-induced release of ssDNA from an ssDNA/SWNT composite dispersed in an aqueous medium.[150] Photothermal conversion occurred due to the effective nonradiative process of excited SWNTs generating intense heat in a very short period. As a result, the wrapped polymer is dissociated from the composites and the SWNTs start to aggregate through strong van der Waals interactions. They demonstrated that the photothermal conversion of CNTs irradiated by NIR light is effective to kill cancer cells stained with CNTs.[150] Clear unwrapping of the dispersant polymer induced by the NIR photothermal conversion was reported by our group.[210] We described that NIR light irradiation to the SWNTs solubilized with an anthracene-carrying vinyl polymer (Anth-P in Fig. 7.11) caused flocculation of the SWNTs. With increasing irradiation time, black flocculates are generated in the solution (Fig. 7.12), indicating that the photothermal conversion of SWNTs provided intense heat just around them, and as a result, Anth-P was dissociated from the irradiated SWNTs. Furthermore, we have proposed the utilization of photothermal conversion of CNTs to thermoresponsive polymer materials. Poly(N-isopropylacrylamide) (PNIPAM)[211] and its derivatives are well-known thermoresponsive materials, which show a phase transition triggered by external stimuli such as the solvent composition[212], pH[213], ionic strength[213], electric field[214], and light.[215] Upon irradiation with the NIR light centered at 1,064 nm, the PNIPAM/SWNT composite gel (200 mm in diameter) containing the SWNTs in a PNIAPM matrix immediately shrinks to a narrower gel (Fig. 7.13) after 15 seconds. After turning off the irradiation, the shrunken gel gradually swells and becomes around 200 mm in diameter after about 67 seconds. The response time of the volume change is controllable by changing the concentration of the SWNTs as well as the power of the NIR laser light. Amazingly, no notable deterioration of the gel actuation is observed even after the 1,200-cycle operation; namely, the SWNT composite gels are highly durable due to the toughness of the CNTs. In fact, the Raman spectra of the gels before and after the endurance test supports exhibit virtually identical graphite/defect (G/D) ratios, which guarantee that the SWNTs remain structurally intact. Very recently Miyako et al.[216] reported two different kinds of smart polymer gels (agarose and PNIPAM gels) containing SWNTs and single-walled nanohorns (SWNHs) that show marked phase transitions upon NIR irradiation; namely, they found that under NIR laser irradiation (1064 nm), the nanocarbon–agarose gel hybrids exhibit a gel-to-sol transition, whereas control agarose gel (without the nanocarbons) does not show any phase transition. Such NIR actuation of the polymer/CNT composites covers both a soft-gel-type and solid-film materials.[217-223] A wide range of absorption on CNTs provides an opportunity for a "molecular heater" to work at the various wavelengths of the light source.

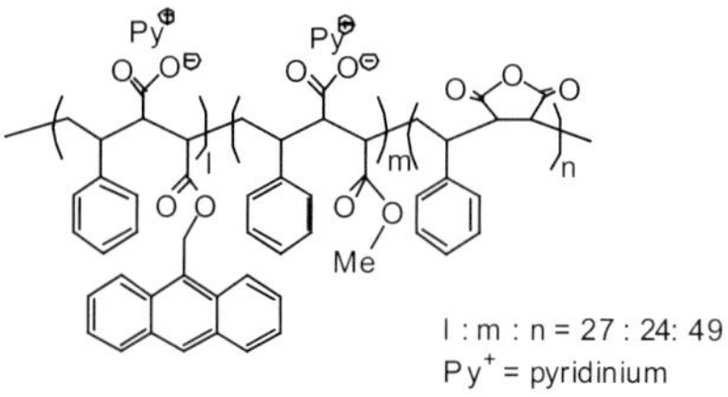

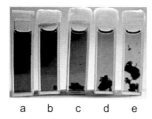

Figure 7.11 Chemical structure of **Anth-P**.

Figure 7.12 Photographs of the **Anth-P**/SWNT solutions in DMF before (a) and after laser (1,064 nm) irradiation for (b) 5 min, (c) 10 min, (d) 30 min, and (e) 60 min.

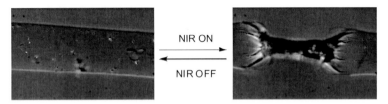

Figure 7.13 PNIPAM/SWNT gel that shows NIR laser-triggered volume phase transition (left) before irradiation and (right) after irradiation.

7.4.3 Conductive Nanotube Honeycomb Film

Honeycomb structures from organic (polymer) and organic/inorganic hybrid materials are of interest due to their unique structures and functions. Since the first report by François et al.[224] that self-organized honeycomb structures are formed from star-shaped polystyrene or poly(styrene)-poly(paraphenylene) block copolymers in carbon disulfide under flowing moist gas, many papers have been published describing the formation of similar honeycomb structures using different kinds of organic (polymer) materials, including symmetric diblock copolymers[225], rod-coil diblock copolymers[226], a coil-like polymer[227], ion-complexed polymers[228], lipid-packaged Pt complexes[229], poly(D,L-lactic-co-glycolic acid)[230], polysulfone[231], amphiphilic poly(*p*-phenylenes)[232], and a poly(ε-caprolactone) amphiphilic copolymer.[233]

We have reported the discovery that the self-assembly of SWNTs with a honeycomb structure is spontaneously formed on glass substrates[234] and transparent plastic films like poly(ethylene terephthalate) (PET), which is a widely used engineering plastic in the industrial field, by a simple solution-casting method using a lipid/SWNT conjugate (complex **1**, Fig. 7.14) as the material, which is an ion complex of shortened SWNTs and tridodecylmethylammonium chloride, a molecular-bilayer-forming amphiphile. Complex **1** is soluble in several organic solvents, including dichloromethane, chloroform, benzene, and toluene. We recently reported the formation of conducting SWNT honeycomb structures on flexible transparent polymer films.[235] As the film, we have chosen PET. This study should be important from the viewpoints of potential applications of conducting SWNTs with honeycomb structures for the fabrication of conducting plastic films with transparent flexible properties. Such films might be useful in many areas of application that require flexible conducting films as materials. The typical SEM images of a honeycomb structure are shown in Fig. 7.15. The sizes of the unit cells are controllable by changing the experimental conditions.

The surface resistivity (R_s) of the cast films of complex **1** with honeycomb structures is insulating (R_s >10^8 ohms/square) due to the coating of the tube surfaces with the ammonium lipid. We developed a method to remove the lipid from the films by employing the "ion-exchange method," as shown by Scheme 7.1. The experimental procedure is very simple: namely, each cast film is immersed overnight in a *p*-toluenesulfonic acid methanol solution and then rinsed with methanol, followed by air-drying. By this procedure, the methylene-stretching vibrations in the FT-IR of the film almost disappear. The Raman spectra of complex **1** before and after the ion exchange are virtually identical. The SWNTs remain intact during all the processes. The SEM images of the cast films after the ion exchange are shown in Fig. 7.16. After the ion exchange, the skeletons with the honeycomb structures become thin due to the removal of the lipid. Higher-magnification SEM measurements show oriented nanotubes along the honeycomb skeletons. After ion exchange, a dramatic change in the R_s values is observed. The R_s values decrease with increasing concentration of complex **1** due to the formation of network structures in larger areas on the films. When the film is prepared from complex **1** = 3.0 mg mL^{-1} in chloroform, the R_s reached a high conducting value, 3.2 × 10^2 ohms/square. A similar behavior is observed when dichloromethane and benzene were used in place of chloroform. Interestingly, after the ion exchange, the R_s values of the films decrease more than 10^4–10^6-fold compared to the original values.

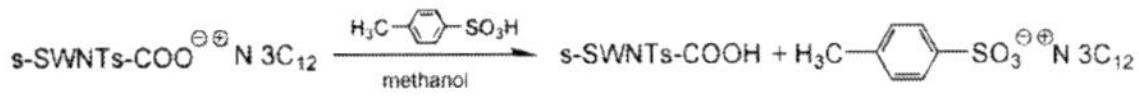

s-SWNTs-COO$^{\ominus}$ $^{\oplus}$N 3C$_{12}$ $\xrightarrow[\text{methanol}]{\text{H}_3\text{C}-\bigcirc-\text{SO}_3\text{H}}$ s-SWNTs-COOH + H$_3$C—⟨ ⟩—SO$_3^{\ominus}$ $^{\oplus}$N 3C$_{12}$

Scheme 7.1

The conductive SWNT honeycomb films on glass substrates and plastic films fabricated by self-organization from nanotube solutions are useful in many areas of nanoscience and technology.

188 | Advances in Dispersal Agents and Methodology for SWNT Analysis

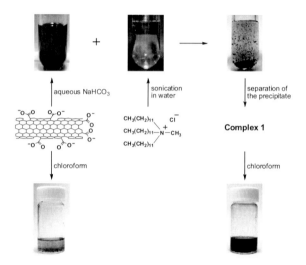

Figure 7.14 Preparation of complex **1**.

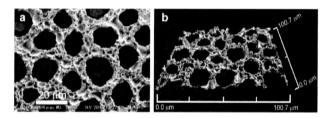

Figure 7.15 SEM images of complex **1** on a glass substrate before the ion exchange.

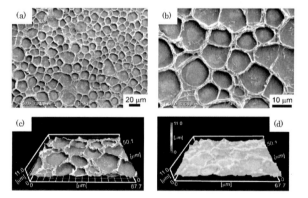

Figure 7.16 SEM images of complex **1** on a glass substrate after the ion exchange.

7.5 Summary

In this chapter, we summarized the recent progress in soluble CNTs based on noncovalent modification and their characterization such as by PL measurement. Individual solubilization of CNTs is necessary for a wide range of science and technology fields because the preparation of individually dissolved SWNTs is the first step toward many practical uses as well as fundamental studies. The individual solubilization based on physical modification maintains CNTs intact and is an attractive route for taking advantage of their intrinsic properties. Tremendous numbers of paper describing the applications of soluble CNTs have been reported, and the some of them are unique for CNT properties. Among a wide range of applications, we here highlighted i) DNA/SWNT hybrids not containing free DNA, ii) NIR-responsive applications based on photothermal conversion properties of CNTs in the NIR range, iii) UV-curable resin/CNT composites having high conductivity and their applications in nanoimprinting, and iv) the formation of conducting CNT-honeycomb films on a transparent plastic film. CNT composites have high potential for applications in the areas of nanomaterials science and technology and bioscience.

References

1. S. Iijima, *Nature*, **354**, 56 (1991).
2. N. Nakashima, *Int. J. Nanosci.*, **4**, 119 (2005).
3. H. Murakami and N. Nakashima, *J. Nanosci. Nanotechnol.*, **6**, 16 (2006).
4. N. Nakashima and T. Fujigaya, *Chem. Lett.*, **36**, 692 (2007).
5. C. A. Dyke and J. M. Tour, *J. Phys. Chem. A,* **108**, 11151 (2004).
6. L. A. Girifalco, M. Hodak, and R. S. Lee, *Phys. Rev. B,* **62**, 13104 (2000).
7. D. Tasis, N. Tagmatarchis, A. Bianco, and M. Prato, *Chem. Rev.,* **106**, 1105 (2006).
8. K. Balasubramanian and M. Burghard, *Small,* **1**, 180 (2005).
9. N. T. V. G. M. P. Dimitrios Tasis, *Chem. Eur. J.,* **9**, 4000 (2003).
10. P. J. Boul, J. Liu, E. T. Mickelson, C. B. Huffman, L. M. Ericson, I. W. Chiang, K. A. Smith, D. T. Colbert, R. H. Hauge, J. L. Margrave, and R. E. Smalley, *Chem. Phys. Lett.,* **310**, 367 (1999).

11. C. A. Furtado, U. J. Kim, H. R. Gutierrez, L. Pan, E. C. Dickey, and P. C. Eklund, *J. Am. Chem. Soc.,* **126**, 6095 (2004).
12. V. C. Moore, M. S. Strano, E. H. Haroz, R. H. Hauge, R. E. Smalley, J. Schmidt, and Y. Talmon, *Nano Lett.,* **3**, 1379 (2003).
13. Y. Dror, W. Pyckhout-Hintzen, and Y. Cohen, *Macromolecules,* **38**, 7828 (2005).
14. S. M. Bachilo, M. S. Strano, C. Kittrell, R. H. Hauge, R. E. Smalley, and R. B. Weisman, *Science,* **298**, 2361 (2002).
15. M. J. O'Connell, S. M. Bachilo, C. B. Huffman, V. C. Moore, M. S. Strano, E. H. Haroz, K. L. Rialon, P. J. Boul, W. H. Noon, C. Kittrell, J. Ma, R. H. Hauge, R. B. Weisman, and R. E. Smalley, *Science,* **297**, 593 (2002).
16. O. N. Torrens, D. E. Milkie, M. Zheng, and J. M. Kikkawa, *Nano Lett.,* **6**, 2864 (2006).
17. Y. Kim, N. Minami, and S. Kazaoui, *Appl. Phys. Lett.,* **86**, 073103/1 (2005).
18. S. Kazaoui, N. Minami, B. Nalini, Y. Kim, and K. Hara, *J. Appl. Phys.,* **98**, 084314/1 (2005).
19. L. Cognet, D. A. Tsyboulski, and R. B. Weisman, *Nano Lett.,* **8**, 749 (2008).
20. D. A. Tsyboulski, J.-D. R. Rocha, S. M. Bachilo, L. Cognet, and R. B. Weisman, *Nano Lett.,* **7**, 3080 (2007).
21. L. Cognet, D. A. Tsyboulski, J.-D. R. Rocha, C. D. Doyle, J. M. Tour, and R. B. Weisman, *Science,* **316**, 1465 (2007).
22. W. Zhou, M. F. Islam, H. Wang, D. L. Ho, A. G. Yodh, K. I. Winey, and J. E. Fischer, *Chem. Phys. Lett.,* **384**, 185 (2004).
23. B. J. Bauer, E. K. Hobbie, and M. L. Becker, *Macromolecules,* **39**, 2637 (2006).
24. H. Wang, W. Zhou, D. L. Ho, K. I. Winey, J. E. Fischer, C. J. Glinka, and E. K. Hobbie, *Nano Lett.,* **4**, 1789 (2004).
25. T. Chatterjee, A. Jackson, and R. Krishnamoorti, *J. Am. Chem. Soc.,* **130**, 6934 (2008).
26. K. Yurekli, C. A. Mitchell, and R. Krishnamoorti, *J. Am. Chem. Soc.,* **126**, 9902 (2004).
27. J. A. Fagan, B. J. Landi, I. Mandelbaum, J. R. Simpson, V. Bajpai, B. J. Bauer, K. Migler, A. R. H. Walker, R. Raffaelle, and E. K. Hobbie, *J. Phys. Chem. B,* **110**, 23801 (2006).
28. Y. Tan and D. E. Resasco, *J. Phys. Chem. B,* **109**, 14454 (2005).

29. M. J. O'Connell, S. M. Bachilo, C. B. Huffman, V. C. Moore, M. S. Strano, E. H. Haroz, K. L. Rialon, P. J. Boul, W. H. Noon, C. Kittrell, J. Ma, R. H. Hauge, R. B. Weisman, and R. E. Smalley, *Science,* **297**, 593 (2002).
30. M. S. Strano, V. C. Moore, M. K. Miller, M. J. Allen, E. H. Haroz, C. Kittrell, R. H. Hauge, and R. E. Smalley, *J. Nanosci. Nanotechnol.,* **3**, 81 (2003).
31. F. Du, C. Guthy, T. Kashiwagi, J. E. Fischer, and K. I. Winey, *J. Polym. Sci., Part B: Polym. Phys.,* **44**, 1513 (2006).
32. F. Du, J. E. Fischer, and K. I. Winey, *Phys. Rev. B: Condens. Matter Mater. Phys.,* **72**, 121404/1 (2005).
33. C. Richard, F. Balavoine, P. Schultz, T. W. Ebbesen, and C. Mioskowski, *Science,* **300**, 775 (2003).
34. G. S. Duesberg, M. Burghard, J. Muster, G. Philipp, and S. Roth, *Chem. Commun.,* 435 (1998).
35. M. Burghard, G. Duesberg, G. Philipp, J. Muster, and S. Roth, *Adv. Mater.,* **10**, 584 (1998).
36. M. F. Islam, E. Rojas, D. M. Bergey, A. T. Johnson, and A. G. Yodh, *Nano Lett.,* **3**, 269 (2003).
37. J. I. Paredes and M. Burghard, *Langmuir,* **20**, 5149 (2004).
38. L. A. Hough, M. F. Islam, B. Hammouda, A. G. Yodh, and P. A. Heiney, *Nano Lett.,* **6**, 313 (2006).
39. M. F. Islam, M. Nobili, F. Ye, T. C. Lubensky, and A. G. Yodh, *Phys. Rev. Lett.,* **95**, 148301/1 (2005).
40. Y. Kim, S. Hong, S. Jung, M. S. Strano, J. Choi, and S. Baik, *J. Phys. Chem. B,* **110**, 1541 (2006).
41. W. Wenseleers, I. I. Vlasov, E. Goovaerts, E. D. Obraztsova, A. S. Lobach, and A. Bouwen, *Adv. Funct. Mater.,* **14**, 1105 (2004).
42. M. O. Lisunova, N. I. Lebovka, O. V. Melezhyk, and Y. P. Boiko, *J. Colloid Interface Sci.,* **299**, 740 (2006).
43. S. Bandow, A. M. Rao, K. A. Williams, A. Thess, R. E. Smalley, and P. C. Eklund, *J. Phys. Chem. B,* **101**, 8839 (1997).
44. A. Ishibashi and N. Nakashima, *Chem. Eur. J.,* **12**, 7595 (2006).
45. A. Ishibashi and N. Nakashima, *Bull. Chem. Soc. Jpn.,* **79**, 357 (2006).
46. S. M. Bachilo, M. S. Strano, C. Kittrell, R. H. Hauge, R. E. Smalley, and R. B. Weisman, *Science,* **298**, 2361 (2002).
47. Y. Miyauchi, S. Chiashi, Y. Murakami, Y. Hayashida, and S. Maruyama, *Chem. Phys. Lett.,* **387**, 198 (2004).

48. K. A. Park, S. M. Lee, S. H. Lee, and Y. H. Lee, *J. Phys. Chem. C,* **111**, 1620 (2007).
49. P. Leyton, J. S. Gomez-Jeria, S. Sanchez-Cortes, C. Domingo, and M. Campos-Vallette, *J. Phys. Chem. B,* **110**, 6470 (2006).
50. N. Nakashima, Y. Tomonari, and H. Murakami, *Chem. Lett.,* **31**, 638 (2002).
51. N. Nakashima, Y. Tanaka, Y. Tomonari, H. Murakami, H. Kataura, T. Sakaue, and K. Yoshikawa, *J. Phys. Chem. B,* **109**, 13076 (2005).
52. Y. Tomonari, H. Murakami, and N. Nakashima, *Chem. Eur. J.,* **12**, 4027 (2006).
53. F. J. Gomez, R. J. Chen, D. Wang, R. M. Waymouth, and H. Dai, *Chem. Commun.,* 190 (2003).
54. L. S. Fifield, L. R. Dalton, R. S. Addleman, R. A. Galhotra, M. H. Engelhard, G. E. Fryxell, and C. L. Aardahl, *J. Phys. Chem. B,* **108**, 8737 (2004).
55. D. M. Guldi, E. Menna, M. Maggini, M. Marcaccio, D. Paolucci, F. Paolucci, S. Campidelli, M. Prato, G. M. A. Rahman, and S. Schergna, *Chem. Eur. J.,* **12**, 3975 (2006).
56. J. S. Kavakka, S. Heikkinen, I. Kilpeläinen, M. Mattila, H. Lipsanen, and J. Helaja, *Chem. Commun.,* 519 (2007).
57. C. Menard-Moyon, N. Izard, E. Doris, and C. Mioskowski, *J. Am. Chem. Soc.,* **128**, 6552 (2006).
58. S. Campidelli, M. Meneghetti, and M. Prato, *Small,* **3**, 1672 (2007).
59. R. J. Chen, Y. Zhang, D. Wang, and H. Dai, *J. Am. Chem. Soc.,* **123**, 3838 (2001).
60. L. Liu, T. Wang, J. Li, Z.-X. Guo, L. Dai, D. Zhang, and D. Zhu, *Chem. Phys. Lett.,* **367**, 747 (2002).
61. A. B. Artyukhin, O. Bakajin, P. Stroeve, and A. Noy, *Langmuir,* **20**, 1442 (2004).
62. V. Sgobba, G. M. A. Rahman, D. M. Guldi, N. Jux, S. Campidelli, and M. Prato, *Adv. Mater.,* **18**, 2264 (2006).
63. C. Ehli, G. M. A. Rahman, N. Jux, D. Balbinot, D. M. Guldi, F. Paolucci, M. Marcaccio, D. Paolucci, M. Melle-Franco, F. Zerbetto, S. Campidelli, and M. Prato, *J. Am. Chem. Soc.,* **128**, 11222 (2006).
64. D. M. Guldi, G. M. A. Rahman, M. Prato, N. Jux, S. Qin, and W. Ford, *Angew. Chem., Int. Ed.,* **44**, 2015 (2005).
65. D. M. Guldi, G. M. A. Rahman, N. Jux, D. Balbinot, N. Tagmatarchis, and M. Prato, *Chem. Commun.,* 2038 (2005).

66. D. M. Guldi, G. M. A. Rahman, V. Sgobba, N. A. Kotov, D. Bonifazi, and M. Prato, *J. Am. Chem. Soc.,* **128**, 2315 (2006).
67. G. M. A. Rahman, D. M. Guldi, R. Cagnoli, A. Mucci, L. Schenetti, L. Vaccari, and M. Prato, *J. Am. Chem. Soc.,* **127**, 10051 (2005).
68. D. M. Guldi, G. M. A. Rahman, N. Jux, D. Balbinot, U. Hartnagel, N. Tagmatarchis, and M. Prato, *J. Am. Chem. Soc.,* **127**, 9830 (2005).
69. V. Georgakilas, V. Tzitzios, D. Gournis, and D. Petridis, *Chem. Mater.,* **17**, 1613 (2005).
70. X. Li, Y. Liu, L. Fu, L. Cao, D. Wei, and Y. Wang, *Adv. Funct. Mater.,* **16**, 2431 (2006).
71. E. Granot, B. Basnar, Z. Cheglakov, E. Katz, and I. Willner, *Electroanalysis,* **18**, 26 (2006).
72. Y.-Y. Ou and M. H. Huang, *J. Phys. Chem. B,* **110**, 2031 (2006).
73. H. Paloniemi, M. Lukkarinen, T. Aeaeritalo, S. Areva, J. Leiro, M. Heinonen, K. Haapakka, and J. Lukkari, *Langmuir,* **22**, 74 (2006).
74. P. G. Holder and M. B. Francis, *Angew. Chem., Int. Ed.,* **46**, 4370 (2007).
75. R. Chitta, A. S. D. Sandanayaka, A. L. Schumacher, L. D'Souza, Y. Araki, O. Ito, and F. D'Souza, *J. Phys. Chem. C,* **111**, 6947 (2007).
76. T. Ogoshi, Y. Takashima, H. Yamaguchi, and A. Harada, *J. Am. Chem. Soc.,* **129**, 4878 (2007).
77. L. Hu, Y.-L. Zhao, K. Ryu, C. Zhou, J. F. Stoddart, and G. Gruner, *Adv. Mater.,* **20**, 939 (2008).
78. P. Wu, X. Chen, N. Hu, U. C. Tam, O. Blixt, A. Zettl, and C. R. Bertozzi, *Angew. Chem., Int. Ed.,* **47**, 5022 (2008).
79. D. M. Guldi, G. M. A. Rahman, N. Jux, N. Tagmatarchis, and M. Prato, *Angew. Chem., Int. Ed.,* **43**, 5526 (2004).
80. J. Zhang, J. K. Lee, Y. Wu, and R. W. Murray, *Nano Lett.,* **3**, 403 (2003).
81. E. Gregan, S. M. Keogh, A. Maguire, T. G. Hedderman, L. O. Neill, G. Chambers, and H. J. Byrne, *Carbon,* **42**, 1031 (2004).
82. T. G. Hedderman, S. M. Keogh, G. Chambers, and H. J. Byrne, *J. Phys. Chem. B,* **108**, 18860 (2004).
83. W. Feng, A. Fujii, M. Ozaki, and K. Yoshino, *Carbon,* **43**, 2501 (2005).
84. T. Yamamoto, Y. Miyauchi, J. Motoyanagi, T. Fukushima, T. Aida, M. Kato, and S. Maruyama, *Jpn. J. Appl. Phys.,* **47**, 2000 (2008).
85. S. Gotovac, Y. Hattori, D. Noguchi, J.-i. Miyamoto, M. Kanamaru, S. Utsumi, H. Kanoh, and K. Kaneko, *J. Phys. Chem. B,* **110**, 16219 (2006).

86. S. Gotovac, H. Honda, Y. Hattori, K. Takahashi, H. Kanoh, and K. Kaneko, *Nano Lett.*, **7**, 583 (2007).
87. G. Nakamura, K. Narimatsu, Y. Niidome, and N. Nakashima, *Chem. Lett.*, **36**, 1140 (2007).
88. Y. X. Chang, L. L. Zhou, G. X. Li, L. Li, and L. M. Yuan, *J. Liq. Chromatogr. Relat. Technol.*, **30**, 2953 (2007).
89. A. Fonverne, F. Ricoul, C. Demesmay, C. Delattre, A. Fournier, J. Dijon, and F. Vinet, *Sensors Actuators, B*, **B129**, 510 (2008).
90. R. Marquis, C. Greco, I. Sadokierska, S. Lebedkin, M. M. Kappes, T. Michel, L. Alvarez, J.-L. Sauvajol, S. H. Meunier, and C. Mioskowski, *Nano Lett.*, **8**, 1830 (2008).
91. H. Murakami, T. Nomura, and N. Nakashima, *Chem. Phys. Lett.*, **378**, 481 (2003).
92. H. Murakami, G. Nakamura, T. Nomura, T. Miyamoto, and N. Nakashima, *J. Porphyrins Phthalocyanines*, **11**, 418 (2007).
93. S. Cambre, W. Wenseleers, J. Culin, S. Van Doorslaer, A. Fonseca, J. B. Nagy, and E. Goovaerts, *ChemPhysChem*, **9**, 1930 (2008).
94. V. A. Basiuk, *J. Comput. Theor. Nanosci.*, **3**, 767 (2006).
95. Y. Yamaguchi, *J. Chem. Phys.*, **120**, 7963 (2004).
96. G. M. A. Rahman, D. M. Guldi, S. Campidelli, and M. Prato, *J. Mater. Chem.*, **16**, 62 (2006).
97. H. Li, B. Zhou, Y. Lin, L. Gu, W. Wang, K. A. S. Fernando, S. Kumar, L. F. Allard, and Y.-P. Sun, *J. Am. Chem. Soc.*, **126**, 1014 (2004).
98. J. Chen and C. P. Collier, *J. Phys. Chem. B*, **109**, 7605 (2005).
99. K. S. Chichak, A. Star, M. V. P. Altoe, and J. F. Stoddart, *Small*, **1**, 452 (2005).
100. F. Cheng and A. Adronov, *Chem. Eur. J.*, **12**, 5053 (2006).
101. A. Satake, Y. Miyajima, and Y. Kobuke, *Chem. Mater.*, **17**, 716 (2005).
102. D. S. Hecht, R. J. A. Ramirez, M. Briman, E. Artukovic, K. S. Chichak, J. F. Stoddart, and G. Gruener, *Nano Lett.*, **6**, 2031 (2006).
103. T. Hasobe, S. Fukuzumi, and P. V. Kamat, *J. Am. Chem. Soc.*, **127**, 11884 (2005).
104. F. Cheng, S. Zhang, A. Adronov, L. Echegoyen, and F. Diederich, *Chem. Eur. J.*, **12**, 6062 (2006).
105. D. M. Guldi, H. Taieb, G. M. A. Rahman, N. Tagmatarchis, and M. Prato, *Adv. Mater.*, **17**, 871 (2005).

106. X. Wang, Y. Liu, W. Qiu, and D. Zhu, *J. Mater. Chem.,* **12**, 1636 (2002).
107. A. Ma, J. Lu, S. Yang, and K. M. Ng, *J. Cluster Sci.,* **17**, 599 (2006).
108. Y. Wang, H.-Z. Chen, H.-Y. Li, and M. Wang, *Mater. Sci. Eng., B,* **B117**, 296 (2005).
109. P. J. Boul, D.-G. Cho, G. M. A. Rahman, M. Marquez, Z. Ou, K. M. Kadish, D. M. Guldi, and J. L. Sessler, *J. Am. Chem. Soc.,* **129**, 5683 (2007).
110. T. Hasobe, S. Fukuzumi, and P. V. Kamat, *J. Phys. Chem. B,* **110**, 25477 (2006).
111. K. Saito, V. Troiani, H. Qiu, N. Solladie, T. Sakata, H. Mori, M. Ohama, and S. Fukuzumi, *J. Phys. Chem. C,* **111**, 1194 (2007).
112. M. Alvaro, P. Atienzar, P. De la Cruz, J. L. Delgado, V. Troiani, H. Garcia, F. Langa, A. Palkar, and L. Echegoyen, *J. Am. Chem. Soc.,* **128**, 6626 (2006).
113. Q. Zhao, Z.-N. Gu, and Q.-K. Zhuang, *Electrochem. Commun.,* **6**, 83 (2004).
114. J. Qu, Y. Shen, X. Qu, and S. Dong, *Electroanalysis,* **16**, 1444 (2004).
115. H. Tanaka, T. Yajima, T. Matsumoto, Y. Otsuka, and T. Ogawa, *Adv. Mater.,* **18**, 1411 (2006).
116. H. Tanaka, T. Yajima, M. Kawao, and T. Ogawa, *J. Nanosci. Nanotechnol.,* **6**, 1644 (2006).
117. Z. Guo, F. Du, D. Ren, Y. Chen, J. Zheng, Z. Liu, and J. Tian, *J. Mater. Chem.,* **16**, 3021 (2006).
118. E. M. N. Mhuircheartaigh, S. Giordani, and W. J. Blau, *J. Phys. Chem. B,* **110**, 23136 (2006).
119. X. Peng, N. Komatsu, S. Bhattacharya, T. Shimawaki, S. Aonuma, T. Kimura, and A. Osuka, *Nature Nanotech.,* **2**, 361 (2007).
120. X. Peng, N. Komatsu, T. Kimura, and A. Osuka, *ACS Nano,* **2**, 2045 (2008).
121. N. Nakashima, S. Okuzono, H. Murakami, T. Nakai, and K. Yoshikawa, *Chem. Lett.,* **32**, 456 (2003).
122. A.-H. Bae, T. Hatano, N. Nakashima, H. Murakami, and S. Shinkai, *Org. Biomol. Chem.,* **2**, 1139 (2004).
123. J. N. Barisci, M. Tahhan, G. G. Wallace, S. Badaire, T. Vaugien, M. Maugey, and P. Poulin, *Adv. Funct. Mater.,* **14**, 133 (2004).
124. M. Iijima, T. Watabe, S. Ishii, A. Koshio, T. Yamaguchi, S. Bandow, S. Iijima, K. Suzuki, and Y. Maruyama, *Chem. Phys. Lett.,* **414**, 520 (2005).

125. G. O. Gladchenko, M. V. Karachevtsev, V. S. Leontiev, V. A. Valeev, A. Y. Glamazda, A. M. Plokhotnichenko, and S. G. Stepanian, *Mol. Phys.*, **104**, 3193 (2006).

126. P. He and M. Bayachou, *Langmuir*, **21**, 6086 (2005).

127. H. Cathcart, S. Quinn, V. Nicolosi, J. M. Kelly, W. J. Blau, and J. N. Coleman, *J. Phys. Chem. C*, **111**, 66 (2007).

128. H. Cathcart, V. Nicolosi, J. M. Hughes, W. J. Blau, J. M. Kelly, S. J. Quinn, and J. N. Coleman, *J. Am. Chem. Soc.*, **130**, 12734 (2008).

129. M. Zheng, A. Jagota, D. Semke Ellen, A. Diner Bruce, S. McLean Robert, R. Lustig Steve, E. Richardson Raymond, and G. Tassi Nancy, *Nat. Mater.*, **2**, 338 (2003).

130. M. Zheng, A. Jagota, M. S. Strano, A. P. Santos, P. Barone, S. G. Chou, B. A. Diner, M. S. Dresselhaus, R. S. McLean, G. B. Onoa, G. G. Samsonidze, E. D. Semke, M. Usrey, and D. J. Walls, *Science*, **302**, 1545 (2003).

131. X. Huang, R. S. McLean, and M. Zheng, *Anal. Chem.*, **77**, 6225 (2005).

132. X. Zhao and J. K. Johnson, *J. Am. Chem. Soc.*, **129**, 10438 (2007).

133. J. Rajendra, M. Baxendale, L. G. D. Rap, and A. Rodger, *J. Am. Chem. Soc.*, **126**, 11182 (2004).

134. J. Rajendra and A. Rodger, *Chem. Eur. J.*, **11**, 4841 (2005).

135. S. G. Chou, H. B. Ribeiro, E. B. Barros, A. P. Santos, D. Nezich, G. G. Samsonidze, C. Fantini, M. A. Pimenta, A. Jorio, F. Plentz Filho, M. S. Dresselhaus, G. Dresselhaus, R. Saito, M. Zheng, G. B. Onoa, E. D. Semke, A. K. Swan, M. S. Uenlue, and B. B. Goldberg, *Chem. Phys. Lett.*, **397**, 296 (2004).

136. H. Kawamoto, T. Uchida, K. Kojima, and M. Tachibana, *J. Appl. Phys.*, **99**, 094309/1 (2006).

137. J. A. Fagan, J. R. Simpson, B. J. Bauer, S. H. Lacerda, M. L. Becker, J. Chun, K. Migler, A. R. H. Walker, and E. K. Hobbie, *J. Am. Chem. Soc.*, **129**, (2007).

138. M. Zheng and B. A. Diner, *J. Am. Chem. Soc.*, **126**, 15490 (2004).

139. M. Zheng and V. V. Rostovtsev, *J. Am. Chem. Soc.*, **128**, 7702 (2006).

140. M. E. Napier, D. O. Hull, and H. H. Thorp, *J. Am. Chem. Soc.*, **127**, 11952 (2005).

141. S. G. Chou, M. F. DeCamp, J. Jiang, G. G. Samsonidze, E. B. Barros, F. Plentz, A. Jorio, M. Zheng, G. B. Onoa, E. D. Semke, A. Tokmakoff, R. Saito, G. Dresselhaus, and M. S. Dresselhaus, *Phys. Rev. B: Condens. Matter Mater. Phys.*, **72**, 195415/1 (2005).

142. S. G. Chou, F. Plentz, J. Jiang, R. Saito, D. Nezich, H. B. Ribeiro, A. Jorio, M. A. Pimenta, G. G. Samsonidze, A. P. Santos, M. Zheng, G. B. Onoa, E. D. Semke, G. Dresselhaus, and M. S. Dresselhaus, *Phys. Rev. Lett.,* **94**, 127402/1 (2005).

143. R. A. Zangmeister, J. E. Maslar, A. Opdahl, and M. J. Tarlov, *Langmuir,* **23**, 6252 (2007).

144. D. A. Heller, E. S. Jeng, T.-K. Yeung, B. M. Martinez, A. E. Moll, J. B. Gastala, and M. S. Strano, *Science,* **311**, 508 (2006).

145. X. Li, Y. Peng, and X. Qu, *Nucleic Acids Res.,* **34**, 3670 (2006).

146. P. He, S. Li, and L. Dai, *Synth. Met.,* **154**, 17 (2005).

147. C. Fantini, A. Jorio, A. P. Santos, V. S. T. Peressinotto, and M. A. Pimenta, *Chem. Phys. Lett.,* **439**, 138 (2007).

148. S. G. Chou, H. Son, J. Kong, A. Jorio, R. Saito, M. Zheng, G. Dresselhaus, and M. S. Dresselhaus, *Appl. Phys. Lett.,* **90**, 131109/1 (2007).

149. A. Ishibashi, Y. Yamaguchi, H. Murakami, and N. Nakashima, *Chem. Phys. Lett.,* **419**, 574 (2006).

150. N. W. S. Kam, M. O'Connell, J. A. Wisdom, and H. Dai, *Proc. Natl. Acad. Sci. U S A,* **102**, 11600 (2005).

151. S. M. Douglas, J. J. Chou, and W. M. Shih, *Proc. Natl. Acad. Sci. U S A,* **104**, 6644 (2007).

152. X. Han, Y. Li, and Z. Deng, *Adv. Mater.,* **19**, 1518 (2007).

153. Y. Noguchi, T. Fujigaya, Y. Niidome, and N. Nakashima, *Chem. Eur. J.,* **14**, 5966 (2008).

154. Y. Noguchi, T. Fujigaya, Y. Niidome, and N. Nakashima, *Chem. Phys. Lett.,* **455**, 249 (2008).

155. Y. Xu, P. E. Pehrsson, L. Chen, R. Zhang, and W. Zhao, *J. Phys. Chem. C,* **111**, 8638 (2007).

156. E. S. Jeng, A. E. Moll, A. C. Roy, J. B. Gastala, and M. S. Strano, *Nano Lett.,* **6**, 371 (2006).

157. N. W. S. Kam, Z. Liu, and H. Dai, *Angew. Chem., Int. Ed.,* **45**, 577 (2006).

158. E. S. Jeng, P. W. Barone, J. D. Nelson, and M. S. Strano, *Small,* **3**, 1602 (2007).

159. M. Zheng and E. D. Semke, *J. Am. Chem. Soc.,* **129**, 6084 (2007).

160. M. S. Strano, M. Zheng, A. Jagota, G. B. Onoa, D. A. Heller, P. W. Barone, and M. L. Usrey, *Nano Lett.,* **4**, 543 (2004).

161. S. R. Lustig, A. Jagota, C. Khripin, and M. Zheng, *J. Phys. Chem. B,* **109**, 2559 (2005).

162. B. J. Bauer, M. L. Becker, V. Bajpai, J. A. Fagan, E. K. Hobbie, K. Migler, C. M. Guttman, and W. R. Blair, *J. Phys. Chem. C,* **111**, 17914 (2007).

163. X. Sun, R. Rossin, J. L. Turner, M. L. Becker, M. J. Joralemon, M. J. Welch, and K. L. Wooley, *Biomacromolecules,* **6**, 2541 (2005).

164. Y. Matsumura and H. Maeda, *Cancer Res.,* **46**, 6387 (1986).

165. M. L. Becker, J. A. Fagan, N. D. Gallant, B. J. Bauer, V. Bajpai, E. K. Hobbie, S. H. Lacerda, K. B. Migler, and J. P. Jakupciak, *Adv. Mater.,* **19**, 939 (2007).

166. Y. Lin, M. J. Meziani, and Y.-P. Sun, *J. Mater. Chem.,* **17**, 1143 (2007).

167. J. N. Coleman, U. Khan, and Y. K. Gun'ko, *Adv. Mater.,* **18**, 689 (2006).

168. J. N. Coleman, U. Khan, W. J. Blau, and Y. K. Gun'ko, *Carbon,* **44**, 1624 (2006).

169. R. Yerushalmi-Rozen, and I. Szleifer, *Soft Matter,* **2**, 24 (2006).

170. M. Moniruzzaman, and K. I. Winey, *Macromolecules,* **39**, 5194 (2006).

171. M. Shigeta, M. Komatsu, and N. Nakashima, *Chem. Phys. Lett.,* **418**, 115 (2006).

172. P. T. Lillehei, C. Park, J. H. Rouse, and E. J. Siochi, *Nano Lett.,* **2**, 827 (2002).

173. C. Park, R. E. Crooks, E. J. Siochi, J. S. Harrison, N. Evans, and E. Kenik, *Nanotechnology,* **14**, L11 (2003).

174. C. Park, Z. Ounaies, K. A. Watson, R. E. Crooks, J. Smith, S. E. Lowther, J. W. Connell, E. J. Siochi, J. S. Harrison, and T. L. St. Clair, *Chem. Phys. Lett.,* **364**, 303 (2002).

175. Z. Ounaies, C. Park, K. E. Wise, E. J. Siochi, and J. S. Harrison, *Compos. Sci. Technol.,* **63**, 1637 (2003).

176. L. Qu, Y. Lin, D. E. Hill, B. Zhou, W. Wang, X. Sun, A. Kitaygorodskiy, M. Suarez, J. W. Connell, L. F. Allard, and Y.-P. Sun, *Macromolecules,* **37**, 6055 (2004).

177. X. Jiang, Y. Bin, and M. Matsuo, *Polymer,* **46**, 7418 (2005).

178. Y. Lin, B. Zhou, R. B. Martin, K. B. Henbest, B. A. Harruff, J. E. Riggs, Z.-X. Guo, L. F. Allard, and Y.-P. Sun, *J. Phys. Chem. B,* **109**, 14779 (2005).

179. C. Park, J. H. Kang, J. S. Harrison, R. C. Costen, and S. E. Lowther, *Adv. Mater.,* **20**, 2074 (2008).

180. S.-M. Yuen, C.-C. M. Ma, and C.-L. Chiang, *Compos. Sci. Technol.,* **68**, 2842 (2008).

181. W.-J. Chou, C.-C. Wang, and C.-Y. Chen, *Compos. Sci. Technol.,* **68**, 2208 (2008).

182. K. J. Sun, R. A. Wincheski, and C. Park, *J. Appl. Phys.,* **103**, 023908/1 (2008).
183. K. L. Mittal, *Polyimides and Other High Temperature Polymers: Synthesis, Characterization and Applications*, Brill Academic, Vol. 2 (2003).
184. T.-S. Chung, *Plastics Eng.*, **41**, 701 (1997).
185. A. S. dos Santos, T. d. O. N. Leite, C. A. Furtado, C. Welter, L. C. Pardini, and G. G. Silva, *J. Appl. Polym. Sci.,* **108**, 979 (2008).
186. L.-Q. Liu and H. D. Wagner, *Composite Interfaces,* **14**, 285 (2007).
187. E. Bekyarova, E. T. Thostenson, A. Yu, H. Kim, J. Gao, J. Tang, H. T. Hahn, T. W. Chou, M. E. Itkis, and R. C. Haddon, *Langmuir,* **23**, 3970 (2007).
188. A. Moisala, Q. Li, I. A. Kinloch, and A. H. Windle, *Compos. Sci. Technol.,* **66**, 1285 (2006).
189. R. Guzman de Villoria, A. Miravete, J. Cuartero, A. Chiminelli, and N. Tolosana, *Composites, Part B: Eng.*, **37B**, 273 (2006).
190. N. Li, Y. Huang, F. Du, X. He, X. Lin, H. Gao, Y. Ma, F. Li, Y. Chen, and P. C. Eklund, *Nano Lett.,* **6**, 1141 (2006).
191. J. Zhu, H. Peng, F. Rodriguez-Macias, J. L. Margrave, V. N. Khabashesku, A. M. Imam, K. Lozano, and E. V. Barrera, *Adv. Funct. Mater.,* **14**, 643 (2004).
192. J. Zhu, J. Kim, H. Peng, J. L. Margrave, V. N. Khabashesku, and E. V. Barrera, *Nano Lett.,* **3**, 1107 (2003).
193. Y. S. Song and J. R. Youn, *Carbon,* **43**, 1378 (2005).
194. V. N. Bliznyuk, S. Singamaneni, R. L. Sanford, D. Chiappetta, B. Crooker, and P. V. Shibaev, *Polymer,* **47**, 3915 (2006).
195. C. Park, J. Wilkinson, S. Banda, Z. Ounaies, K. E. Wise, G. Sauti, P. T. Lillehei, and J. S. Harrison, *J. Polym. Sci., Part B: Polym. Phys.,* **44**, 1751 (2006).
196. J. Sandoval, K. Soto, L. Murr, and R. Wicker, *J. Mater. Sci.,* **42**, 156 (2007).
197. T. Fujigaya, S. Haraguchi, T. Fukumaru, and N. Nakashima, *Adv. Mater.,* **20**, 2151 (2008).
198. T. R. Schibli, K. Minoshima, H. Kataura, E. Itoga, N. Minami, S. Kazaoui, K. Miyashita, M. Tokumoto, and Y. Sakakibara, *Opt. Express,* **13**, 8025 (2005).
199. N. Aoki, A. Yokoyama, Y. Nodasaka, T. Akasaka, M. Uo, Y. Sato, K. Tohji, and F. Watari, *J. Biomed. Nanotechnol.,* **1**, 402 (2005).

200. N. Aoki, A. Yokoyama, Y. Nodasaka, T. Akasaka, M. Uo, Y. Sato, K. Tohji, and F. Watari, *Chem. Lett.*, **35**, 508 (2006).
201. B. L. Allen, P. D. Kichambare, and A. Star, *Adv. Mater.*, **19**, 1439 (2007).
202. K. Balasubramanian and M. Burghard, *Anal. Bioanal. Chem.*, **385**, 452 (2006).
203. Y. C. Chen, N. R. Raravikar, L. S. Schadler, P. M. Ajayan, Y. P. Zhao, T. M. Lu, G. C. Wang, and X. C. Zhang, *Appl. Phys. Lett.*, **81**, 975 (2002).
204. S. Tatsuura, M. Furuki, Y. Sato, I. Iwasa, M. Tian, and H. Mitsu, *Adv. Mater.*, **15**, 534 (2003).
205. Y. Sakakibara, S. Tatsuura, H. Kataura, M. Tokumoto, and Y. Achiba, *Jpn. J. Appl. Phys., Part 2*, **42**, L494 (2003).
206. Y. Sakakibara, A. G. Rozhin, H. Kataura, Y. Achiba, and M. Tokumoto, *Jpn. J. Appl. Phys., Part 1*, **44**, 1621 (2005).
207. T. R. Schibli, K. Minoshima, H. Kataura, E. Itoga, N. Minami, S. Kazaoui, K. Miyashita, M. Tokumoto, and Y. Sakakibara, *Opt. Express*, **13**, 8025 (2005).
208. N. Nishizawa, Y. Seno, K. Sumimura, Y. Sakakibara, E. Itoga, H. Kataura, and K. Itoh, *Opt. Express*, **16**, 9429 (2008).
209. D. Boldor, N. M. Gerbo, W. T. Monroe, J. H. Palmer, Z. Li, and A. S. Biris, *Chem. Mater.*, **20**, 4011 (2008).
210. K. Narimatsu, Y. Niidome, and N. Nakashima, *Chem. Phys. Lett.*, **429**, 488 (2006).
211. S. Hirotsu, Y. Hirokawa, and T. Tanaka, *J. Chem. Phys.*, **87**, 1392 (1987).
212. S. Katayama, Y. Hirokawa, and T. Tanaka, *Macromolecules*, **17**, 2641 (1984).
213. T. Tanaka, D. Fillmore, S.-T. Sun, I. Nishio, G. Swislow, and A. Shah, *Phys. Rev. Lett.*, **45**, 1636 (1980).
214. T. Tanaka, I. Nishio, S. T. Sun, and S. U.-Nishio, *Science*, **218**, 467 (1982).
215. A. Suzuki and T. Tanaka, *Nature*, **346**, 345 (1990).
216. E. Miyako, H. Nagata, K. Hirano, and T. Hirotsu, *Small*, **4**, 1711 (2008).
217. S. V. Ahir and E. M. Terentjev, *Nat. Mater.*, **4**, 491 (2005).
218. S. V. Ahir and E. M. Terentjev, *Phys. Rev. Lett.*, **96**, 133902/1 (2006).
219. S. V. Ahir, A. M. Squires, A. R. Tajbakhsh, and E. M. Terentjev, *Phys. Rev. B: Condens. Matter Mater. Phys.*, **73**, 085420/1 (2006).
220. S. Lu and B. Panchapakesan, *Nanotechnology*, **16**, 2548 (2005).
221. S. Lu and B. Panchapakesan, *Nanotechnology*, **18**, 305502 (2007).

222. L. Yang, K. S. A. Li, and S. G. J. Chen, *Adv. Mater.,* **20**, 2271 (2008).
223. H. Koerner, G. Price, N. A. Pearce, M. Alexander, and R. A. Vaia, *Nat Mater,* **3**, 115 (2004).
224. G. Widawski, M. Rawiso, and B. Francois, *Nature,* **369**, 387 (1994).
225. Z. Li, W. Zhao, Y. Liu, M. H. Rafailovich, J. Sokolov, K. Khougaz, A. Eisenberg, R. B. Lennox, and G. Krausch, *J. Am. Chem. Soc.,* **118**, 10892 (1996).
226. S. A. Jenekhe and X. L. Chen, *Science,* **283**, 372 (1999).
227. M. Srinivasarao, D. Collings, A. Philips, and S. Patel, *Science,* **292**, 79 (2001).
228. H. Yabu, M. Tanaka, K. Ijiro, and M. Shimomura, *Langmuir,* **19**, 6297 (2003).
229. C.-S. Lee and N. Kimizuka, *Proc. Natl. Acad. Sci. U S A,* **99**, 4922 (2002).
230. X. Zhao, Q. Cai, G. Shi, Y. Shi, and G. Chen, *J. Appl. Polym. Sci.,* **90**, 1846 (2003).
231. Y. Xu, B. Zhu, and Y. Xu, *Polymer,* **46**, 713 (2005).
232. M. H. Nurmawati, R. Renu, P. K. Ajikumar, S. Sindhu, F. C. Cheong, C. H. Sow, and S. Valiyaveettil, *Adv. Funct. Mater.,* **16**, 2340 (2006).
233. K. Arai, M. Tanaka, S. Yamamoto, and M. Shimomura, *Colloids Surf., A,* **313–314**, 530 (2008).
234. H. Takamori, T. Fujigaya, Y. Yamaguchi, and N. Nakashima, *Adv. Mater.,* **19**, 2535 (2007).
235. N. Wakamatsu, H. Takamori, T. Fujigaya, and N. Nakashima, *Adv. Funct. Mater.,* **19**, 311 (2009).

Chapter 8

Time Domain Luminescence Instrumentation

Graham Hungerford, Kulwinder Sagoo, and David McLoskey
HORIBA Jobin Yvon IBH Ltd Skypark 5, 45 Finnieston Street, Glasgow G3 8JU, Scotland, UK
graham.hungerford@horiba.com

The main principles of contemporary time domain measurements using time-correlated single-photon counting have not changed since the early 1960s. However, the advances in instrumentation and measurement techniques have now allowed for a great improvement in time resolution and measurement of time. From what were once measurements involving substantial equipment performed in specialized research laboratories, time domain fluorescence techniques are becoming commonplace tools to study a variety of areas, ranging from biology to materials science. Advances in software have also allowed accessibility and ease of use of the technique. Time-resolved fluorescence has the advantage over steady-state measurements that the luminescence lifetime is independent of the sample concentration. Using single-photon counting (only available in the time domain) it is even possible to detect single-

---

Luminescence: The Instrumental Key to the Future of Nanotechnology
Edited by Adam M. Gilmore
Copyright © 2014 Pan Stanford Publishing Pte. Ltd.
ISBN 978-981-4241-95-3 (Hardcover), 978-981-4267-72-4 (eBook)
www.panstanford.com

molecule fluorescence. These factors make fluorescence well suited to the study of nanomaterials. Although a major effort has been to improve time resolution, it should not be forgotten that there is still an important area of investigation using "longer" timescales (μs and ms), for example, using lanthanides, with potential applications such as biological labels, nonlinear optical materials, and optical devices.

The instrumentation has evolved over the years from using "simple" spark sources to femtosecond laser systems, producing an improvement in time resolution of several orders of magnitude. The advent of semiconductor light-emitting diodes and laser diodes, with wavelengths toward the ultraviolet region of the spectrum, means that relatively inexpensive and compact instruments are available to study diverse areas (e.g., protein fluorescence, semiconductor properties) at picosecond time resolution. Furthermore, application of time-tag techniques means that dynamic events can be followed.

8.1 Introduction

The main principles of contemporary time domain measurements making use of time-correlated single-photon counting (TCSPC) have not significantly changed since the first use of a time-to-amplitude converter in the early 1960s by Bollinger and Thomas.[1] Their system was employed to measure the scintillation response of crystals and glasses upon excitation with different forms of radiation. This approach was consolidated during the following decade and is now more usually associated with the measurement of fluorescence decay parameters after optical excitation. The principles of TCSPC have remained largely unaltered, but major advances have been made in improved instrumentation. Several comprehensive texts provide details concerning the main concepts behind this technique[2-4], so only a brief overview will be given here. The major reason for the popularity of TCSPC is the great sensitivity that it affords depending on the detection of the arrival times of individual photons after an optical excitation pulse. At low signal levels the probability of detecting more than one photon in the detection window is small, and upon repeated excitation-collection cycles a histogram can be accumulated, which is indicative of the decay process. This is shown schematically in Fig. 8.1.

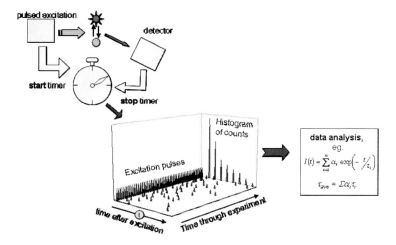

Figure 8.1 Schematic representation of TCSPC.

As with all techniques TCSPC has its advantages and drawbacks. On the plus side it offers the high sensitivity of single-photon counting, which is only possible in the time domain. The histograms obtained give an immediate, qualitative indication of the lifetime and can show the presence of more than one decay pathway. However, it suffers from an inherent disadvantage—to be certain of acquiring histograms representative of the decay process there should not be more than one photon detected per excitation-collection cycle. This restriction mainly relates to the instrumental dead time and is to avoid the possibility of a photon arriving while the equipment is busy and not able to time it. In practice this has meant limiting the rate of detection to less than 2% of the excitation frequency—a value that means distortion linked to undetected fluorescence photons will be less than other experimental errors.[5] Historically with the use of low-repetition-rate excitation sources this has implied that to obtain statistically meaningful numbers of fluorescence photons, long run times were necessary. This factor has favored the use other fluorescence lifetime measurement techniques, such as those in the frequency domain[6], when short acquisition times were necessary. Laterally, advances in excitation source technology, especially the advent of diode lasers, have improved matters significantly. Much faster acquisition rates are now possible, allowing measurement times of only a few milliseconds. Although the basis of the technique has not noticeably changed since the early 1960s, successive improvements in instrumentation have combined to produce an extremely powerful and popular method for the determination of excited-state molecular kinetics.

8.2 Overview

The main components of a TCSPC system are shown schematically in Fig. 8.2. From the user's point of view, the important aspects are probably the resolution of the equipment plus the time to make a measurement. Obviously the capability to measure a large range of decay times is also advantageous. In this section we will briefly consider the history of state of the art and future development of the parts of the instrumentation that have the biggest influence on performance.

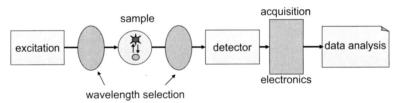

Figure 8.2 Basic components of a TCSPC measurement system.

In general the instrumentation can be divided into:

(1) light sources;
(2) detectors;
(3) electronics; and
(4) optical components.

It is the first three that we concern ourselves with in this section as their development, especially of 1 and 2, has proved to be the major factor in the instrument performance. Ideally for TCSPC the excitation source should have the form of a δ function.[2,3] However, experimentally the measured response of a time domain lifetime system depends on the timing jitter of the component parts, as indicated by Eq. 8.1.[3]

$$\Delta t_m \sim \left[\Delta t_{exc}^2 + \Delta t_{det}^2 + \sum_i \Delta_i^2 \right]^{1/2} \qquad (8.1)$$

where Δt_m is the measured full width at half maximum (FWHM), Δt_{exc} is the FWHM of the excitation, Δt_{det} is that associated with the detector, and Δt_i is related to the response of the ith component in the system. The excitation source and detector with their associated electronics are by far the greatest influences on the measured response, and Δt_m can be approximated as just being dependent on these two components. In TCSPC an assumption is made of instantaneous (delta function) excitation, which because of the time response of the electrical components is clearly not the case. This leads to the need to measure the instrumental response (normally made with a scattering solution), as well as the decay and the use of reconvolution analysis[7] to recover the decay parameters. Generally speaking the shortest lifetime that can be determined with this approach is approximately a tenth of the instrumental FWHM.

8.3 Light Sources

The capabilities of light sources have improved dramatically since the first application of the technique. The initial three choices of excitation sources, flashlamps[8-10], mode-locked laser[11], and synchrotron radiation[12], have been augmented and superseded as (mainly semiconductor) technology has advanced. Today the favored excitation sources for TCSPC include femtosecond lasers[13] (such as the Coherent Chameleon and the Lambda Physik Mai Tai), semiconductor laser diodes, and light-emitting diodes (LEDs)[14-18] and more recently supercontinuum lasers[19-21], such as those from Fianium. Here we will summarize capabilities of these sources. The role of the flashlamp is now limited and has largely been replaced as the entry-level TCSPC source by LEDs, which have recently been developed in the ultraviolet (UV) region of the spectrum. However, because of the large wavelength range possible with a flashlamp and the number still in operation it will be discussed later.

8.3.1 Flashlamps

Although now largely superseded by semiconductor sources, flashlamps are still in use, and their advantage has been their wavelength range. This type of source first appeared in the 1960s, and their usage in TCSPC became widespread in the 1970s. They remained the primary excitation source for TCSPC into the 1980s. An example of a successful commercial version is the coaxial flashlamp (shown in Fig. 8.3).[10] The principle of operation involves the application of a high voltage to produce a spark in a gas-filled enclosure. The coaxial transmission line geometry was designed to minimize electrical reflections and reduce radio frequency interference resulting from the electrical discharge. The voltage is gated, although less stable, free-running designs were also produced, to give the required pulse repetition rate. This is typically several tens of kilohertz, which is nowadays considered a limitation as it can lead to extended data collection times. The wavelength range is dependent on the filler gas, which also can influence the FWHM of the optical pulse. The narrowest pulse widths are associated with the use of hydrogen (typically producing a system FWHM of 0.6 to 1.5 ns), and this has a useful continuous wavelength range until ~500 nm, with the advantage that there is little wavelength dependency on the temporal response. It is also possible to employ deuterium, which increases the output intensity but also equally augments the FWHM. More intense excitation can be obtained using nitrogen as a filler gas, but this is at the expense of a doubling of the FWHM (compared with using hydrogen) and is more or less restricted to the spectral line emissions at 316 nm, 337 nm, and 358 nm.

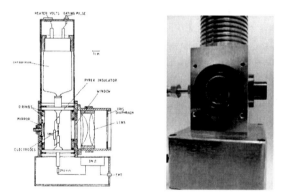

Figure 8.3 Coaxial flashlamp schematic and close-up of the gas enclosure.

The wavelength range has been demonstrated to be usable up to ~800 nm by the use of argon as the filler gas. However, this produces a wide temporal response but can be narrowed to produce an FWHM

< 1 ns by combining argon with hydrogen in a 1:4 mixture.[22]

Although the large tuning range, from UV to near infrared (NIR), can be advantageous, there are several drawbacks that have led to reduced usage of this type of excitation source. These include the requirement of routine cleaning and refilling, combined with the (now) relatively low repetition rate.

8.3.2 Dye Laser Systems

The limitations of the flashlamp in terms of repetition rate and pulse width and the associated restriction of the shortest lifetime that could be determined (see Eq. 8.1) meant that there was a need for "faster" sources. Starting in the 1980s the usage of cavity-dumped dye lasers (see the schematic in Fig. 8.4) increased. They became the system of choice for the measurement of shorter fluorescence lifetimes, as they could provide pulse widths of several picoseconds at repetition rates of many hundreds of kilohertz or even megahertz. This led to an obvious increase in time resolution accompanied by a decrease in data collection time.

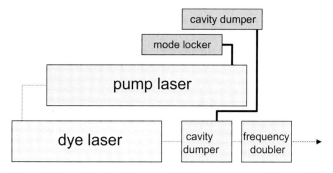

Figure 8.4 Schematic of a picosecond dye laser system (adapted from Ref. 4).

The main types of pump laser used have been mode-locked argon ion lasers, making use of the emission line at 514 nm, or neodymium yttrium aluminum garnet (YAG) lasers, with their output frequency doubled to give 532 nm or even tripled to give 355 nm. This output is then fed to the dye laser, whose tunable output will depend on the dye employed and the option of frequency doubling. The advantages of these systems over the flashlamp are those of shorter pulse width (typically ~5 ps), higher repetition rates (up to many MHz), and high pulse intensities (mW to W, depending on wavelength) over a wavelength range of ~300 nm to 1 μm. However, these are relatively large and complex systems requiring routine maintenance. In the past, this fact coupled with their expense, limited their widespread usage in comparison to flashlamp-based fluorometers.

8.3.3 LEDs and Laser Diodes

Recent advances in semiconductor technology have enabled this class of excitation source to become the cost-effective choice for routine time-resolved measurements. Although lacking the tuning range of the flashlamp, the combination of the different wavelength–emitting sources covers an extensive wavelength range, from the UV[16–18] (from 250 nm at the time of writing) up to the NIR (1,310 nm). This is exemplified in Fig. 8.5.

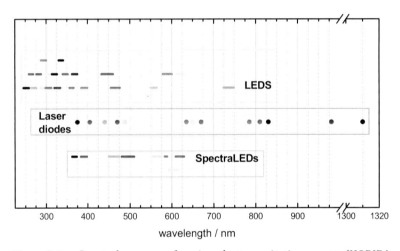

Figure 8.5 Spectral coverage of semiconductor excitation sources (HORIBA Scientific). The line length signifies the typical spectral FWHM of the source.

In addition to the good wavelength coverage, the repetition rate of these sources can be easily varied and range from one shot to several tens of MHz and the LEDs exhibit pulse widths from *ca.* 500 ps (for the UV variants) to ~1.5 ns (visible/NIR). Laser diodes typically have pulse widths < 200 ps, and recently an instrumental response of 32 ps was obtained using a DeltaDiode 405 nm laser diode. This means that the optical pulse was less than 30 ps. The performance of these sources is more than sufficient for many applications. A comparison of a BODIPY[23,24] in ethanol demonstrates the difference between using an LED and a laser diode source (in this case an N-495 and N-485L from HORIBA Scientific) is shown in Fig. 8.6. In both measurements the same lifetime, within error, was obtained.

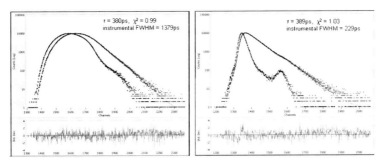

Figure 8.6 Fluorescence decay and instrumental response, along with fitted function and weighted residuals, for a BODIPY in ethanol. These decays were measured using an LED (left) and a laser diode (right) for excitation and monitored with a TBX picosecond detection module.

Making use of the rough guide that the minimum lifetime that can be determined is ~1/10 that of the instrumental FWHM, it is clear that both of these excitation sources are capable of measuring the BODIPY lifetime. However, employing a laser diode should allow lifetimes about six times shorter to be determined. However, other factors, such as the detector and timing electronics, require consideration.

Another factor favoring the use of semiconductor-based excitation sources is the repetition rate. In a recent development, DeltaDiode laser diodes have the capability to run up to 100 MHz, which coupled with low-dead-time timing electronics (such as the FluoroHub-Bi) enable one of the historic disadvantages of TCSPC to be tackled. This relates to the fact that traditionally the data collection rate has been limited to 2% of the excitation rate to avoid data distortion, with the implication that most of the fluorescence photons are wasted. The obvious ramification of an increase in the excitation rate is a decrease in the data collection time—an important factor if these components are to be employed in microscope systems to perform fluorescence lifetime imaging (FLIM). It has previously been reported that only ~185 photons are required to fit a monoexponential decay.[25] It has been empirically shown that the combination of a DeltaDiode at a repetition rate of 100 MHz, coupled to a FluoroHub-Bi (dead time ≈ 10 ns) and a TBX detection module, enable the fast acquisition of decay data[26,27], illustrated in Fig. 8.7. This figure shows the effect of different data collection rates (up to 8% stop rate). The effect of a pileup was negligible, allowing efficient photon collection, and enabled the collection time for 200 counts (sufficient for determining a monoexponential decay) to be obtained. An example measurement for a BODIPY derivative with this number of counts, collected in 60 μs, is also shown.

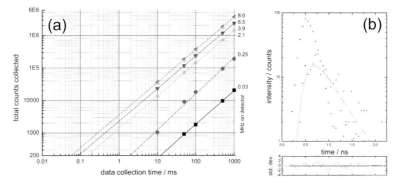

Figure 8.7 (a) Influence of the data collection time on the number of counts in the measured decay. (b) Decay data for a BODIPY derivative measured with 100 MHz excitation frequency with a 7.8 MHz collection rate, measured in 60 μs. The recovered lifetime (336 ps) was in agreement with that expected.

The fact that the LED and laser diode heads can be mechanically interchangeable (so-called plug-and-play) along with their cost, plus the fact that they rarely require maintenance, has made them ideal choices for routine fluorescence lifetime determination. This is enhanced by recent developments increasing their repetition rate coupled with low-dead-time electronics, allowing faster data acquisition rates.

8.3.4 Femtosecond Lasers

In terms of temporal resolution titanium:sapphire (Ti:sapphire) lasers, such as those from Coherent and Lambda Physik, constitute the present state of the art. Their usage in TCSPC measurements really goes back to the last decade of the previous century. In addition to being capable of producing pulses in the order of 100 fs for TCSPC, their high photon flux allows for two and three photon excitation measurements to be performed.[28] However, the high repetition rate (~80 MHz) of the Ti:sapphire laser is problematic in fluorescence lifetime determination for all but the shortest of decays, as it can lead to the re-excitation of the sample before the fluorescence has fully decayed. At a rate of 80 MHz there will be an excitation pulse every 12.5 ns, and usually a period between pulses is chosen to be at least 10 times that of the fluorescence lifetime. Hence for most measurements the repetition rate needs to be reduced. This cannot be accomplished using the Ti:sapphire laser itself, which runs at a fixed frequency and requires the use of a pulse picker (or cavity dumper working in pulse-picking mode) to discard a selected fraction of the output pulses (see the schematic in Fig. 8.8). Typically a frequency around 4 MHz is employed to excite the fluorescence.

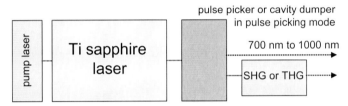

Figure 8.8 Schematic of a Ti:sapphire laser system.

The spectral output of these lasers can be tuned and, combined with the second or third harmonic generation, covers a large spectral range, although it should be noted that this range is not continuous. The downside to this type of excitation sources is that they are relatively expensive, are bulky (e.g., space required for power supply, pump laser, cooling, beam diagnostics), and can be quite labor intensive to use, although newer models are simpler for users to operate.

8.3.5 Supercontinuum Lasers

Supercontinuum or "white light" laser sources have only recently started to be applied to fluorescence lifetime determination in the time domain. The potential of this type of laser was demonstrated in last the last 20 years[19], and their high repetition rate (~MHz) and short pulse duration (fs to ps) appear well suited for TCSPC. These lasers take coherent laser light from a short pulsed laser and route it through a nonlinear device such as a photonic crystal fiber (see Fig. 8.9 for a schematic), which allows strong nonlinear interactions to produce a broad spectral output in the range 400 nm to 2,100 nm.

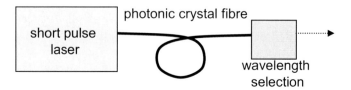

Figure 8.9   Scheme of a supercontinuum laser.

One of the areas where this type of excitation source has found application is in microscopy[29,30], where using an acousto-optic tunable filter (AOTF) allows flexibility for excitation.[31] It is in microscopy and imaging where at present supercontinuum lasers have been employed to measure fluorescence lifetimes[32], but this type of laser has potential to be one of the first choice sources for TCSPC in the visible to NIR range.

8.3.6   Sources for Longer-Lived Decays

When considering the properties for an excitation source to use for measurement of luminescence lifetimes in the micro- to millisecond time range and longer, factors such as narrow temporal pulse width and high repetition rate are not so significant. A more important factor is that of large pulse energy. This is because of the fact that the lifetimes will be considerably longer than the pulse width, which negates the need for reconvolution analysis. It should be kept in mind that the time range should be about 10 times the lifetime to be measured and hence the repetition rate should be kept to ~1/time range to avoid sample re-excitation before it has fully decayed. As the efficiency of TCSPC measurements depends on the repetition rate, longer timescale measurements tend to use multichannel scalers. Their multistop capability means that the 2% stop limit no longer applies, although there are other constraints.

The excitation sources are similar to those used for the shorter timescale measurements. Typically xenon-filled flashlamps provide a broadly tunable emission from *ca.* 200 nm to the visible spectral region. These sources have the advantage of having a high energy per pulse; however, their low repetition rate can be problematic, and their long temporal tail can cause difficulties in interpreting decays with lifetimes less than 1 ms. Lasers, for example, N_2 and YAG, with their discrete wavelength output have overcome some of the drawbacks of flashlamp sources. As with fluorescence measurements, the introduction of sources based on LEDs has proved beneficial. Figure 8.5 shows some of the wavelengths for which they are available. Figure 8.10 shows an example of the time-resolved luminescence decay, made with a SpectraLED source, of a lanthanide-containing polyoxometalate in the presence of increasing quantities of protein.[33]

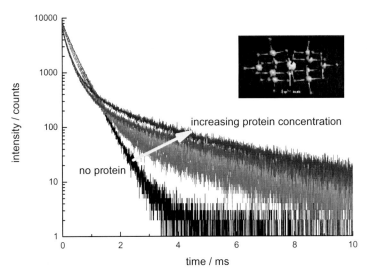

Figure 8.10 Time-resolved decay for a europium-containing poloxometalate in the presence of bovine serum albumin excited with a SpectraLED emitting at 395 nm.

These SpectraLED sources have variable pulse widths from 100 ns to dc, with near instantaneous on-off transitions. Computer control of their repetition rate and duration is advantageous, as it can be tuned to match the lifetime and hence optimize the data collection time. Increasing the pulse duration allows the intensity to be controlled. As the lifetime is many times longer than the instrumental response, fitting of the data is possible without reconvolution.

8.4 Detectors

The ideal choice of detector for TCSPC should exhibit most of the following features[34]:

- Large spectral coverage
- Fast time response (typically <1.5 ns rise time)
- Low transit time jitter across the device's detecting area (ideally <20 ps mm^{-1})
- High gain (typically in excess of five million)
- Negligible wavelength dependency of the temporal response (ideally <0.5 ps nm^{-1})
- Low noise (dominated by dark count and ideally <100 cps)
- Few detector generated artefacts, for example, secondary pulses
- Sensitive area compatible with other optical components

However, some of these ideals are conflicting, such as decrease in noise performance with increasing gain.

A major factor when choosing a detector is its wavelength range. The lower limit is, in large, determined by any window material. The composition of the photocathode material responsible for converting the incoming light into an electrical signal determines the longer wavelength response. As longer wavelengths correspond to lower energies the photocathode is required to have a lower work function (i.e., less energy required to liberate a photoelectron). This is problematic as ambient thermal energy can be sufficient to cause the liberation of these electrons and hence the subsequent secondary electrons causing an output signal. As this does not correspond to an incident light pulse, it is commonly referred to as noise. Thus, detectors that are more sensitive to the red end of the spectrum tend to have higher noise counts. This can, in part, be counteracted by cooling the detector. However, even with cooling, photomultiplier tubes with sensitivity over 1,000 nm may exhibit a large number (several tens of thousands) of noise counts. Another important factor to take into account is by how much the incoming signal is spread during its passage through the detector. This is called the transit time spread (TTS) and is a good measure of the detector's temporal resolution. The following table (Table 8.1) gives an indication of the typical TTS of four types of representative detectors.

Table 8.1 Comparison of the typical TTS for four TCSPC detectors

Detector	R928P[a]	TBX	SPAD	MCP (R3809U)
TTS (ps)	1,200	≤200	<50	≤25

[a]R928P—side on PMT; TBX—picosecond detection module; SPAD—avalanche photodiode; MCP—microchannel plate PMT.

Abbreviations: SPAD, single-photon avalanche diode; MCP, microchannel plate; PMT, photomultiplier tube.

Traditionally PMTs[35] have been a popular choice of detector for TCSPC, and significant improvements in their performance have been made in recent years. The detector of choice for measurement of short fluorescence lifetimes remains the MCP[36] photomultiplier, although it is limited in its count rate (recommended <20,000 cps for normal operation[37]) before damage can occur. Semiconductor devices, namely, SPADs[38,39], are also becoming more popular, although they are limited by having a small active area, which requires the fluorescence emission to be well focused. They exhibit a wavelength-dependent response, which can cause complications when using reconvolution analysis.

8.4.1 Photomultiplier Tubes

The main purpose of a photomultiplier is to convert incoming light into an amplified electrical signal. This is done via the photoelectric effect, with light interacting with the photocathode and liberating an electron. This electron is accelerated toward a dynode structure, where secondary electrons are generated and current (electron) amplification occurs. The amplification is typically in excess of $\sim 10^6$ at the end of this process. The trend over the last 20 years has been to move to smaller detectors. Initially larger linear-focused PMTs (such as the Philips XP2020 series) were the mainstay of fluorescence lifetime measurement systems; these gave way to the usage of a cheaper side on PMTs (e.g., Hamamatsu R2949 and R928P). Relatively recently, very compact metal package PMTs have shown promise as fast detectors for use with full photocathode illumination. Packages also exist combining PMTs with associated electronics to give a fast response and "black box" operation for single-photon counting applications, for example, the TBX picosecond detection module from HORIBA Scientific. This combines amplifier, discriminator, and high voltage supply on one board, eliminating an analog connection from the detector to the timing electronics.

It should be mentioned here that for a large number of fluorescence studies the detectors used are sensitive from the UV to the visible spectral region. However, some studies require detection into the NIR, and detectors are available for use up until 1,700 nm, for example, the Hamamatsu H10330 and R5509 series. As previously mentioned PMTs capable of usage in the NIR are prone to high noise counts and require cooling. In combination with TCSPC this may actually confer a limitation in the lifetimes that can be measured.

To obtain distortion-free data the start-to-stop ratio should ideally be kept below 2%. Some NIR tubes exhibit noise levels approaching 200,000 cps, so on certain time ranges noise alone may account for this. This means that care has to be exercised in the determination of fluorescence lifetimes in this spectral region. The source repetition rate, measurement time range (relating to sample lifetime), and the selection of the detector need to be accounted for to ensure that a distortion-free measurement is possible.

Figure 8.11 shows the possible contribution of correlated noise to the start-to-stop ratio for TCSPC over different time ranges for different detector noise values. Combining the data given in this figure along with the fact that usually a time range approximately 10 times that of the lifetime should be employed for a measurement shows that using TCSPC with very noisy detectors only allows shorter lifetimes to be determined free of distortion.

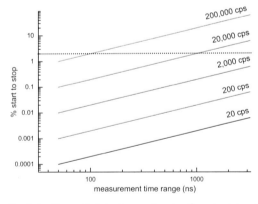

Figure 8.11 Contribution of detector noise to the start-to-stop ratio for different measurement time ranges. The dotted line shows the 2% value.

8.4.2 Microchannel Plate Detectors

This form of photomultiplier, rather than using a dynode structure for signal amplification, uses continuous semiconductor-coated glass multiplier tubes (*ca*. 10 μm in diameter). The multiplied electrons are collected at the anode in the same way as a conventional PMT. This provides a faster time response and lower TTS because of the compactness. Most often an external amplifier is required to produce an increase in the gain prior to the timing device. The photocathode compositions are similar to those used in conventional PMTs, so equivalent wavelength responses are obtained.

Although the favored device for the determination of short lifetimes, the sensitivity of MCPs limits the amount of light that they can be exposed to. Therefore care has to be taken to avoid exposure to high count rates (for normal operation <20,000 cps[37]) and to ambient light. MCPs are not recommended for phosphorescence or steady-state measurements, which means that they are commonly used in conjunction with another general-purpose TCSPC detector.

8.4.3 Avalanche Photodiodes

These semiconductor devices are based on a reverse-biased *p-n* junction. Incident photons create electron-hole pairs, which are multiplied in the depletion layer to produce gain with a small TTS (<50 ps). A major restriction of the use of this type of detector is their small active area, which requires good focusing of the fluorescence emission and limits their usage with monochromators. This restriction can be circumvented for applications in microscope- and laser-based fluorescence systems. These detectors also exhibit some wavelength dependency in the temporal response, caused by longer-wavelength light penetrating further into the device. This can be problematic using reconvolution analysis when there is a large wavelength separation between excitation and emission.

The associated electronics depend on the specific application of these devices, and photon-counting modules are available and typically are sensitive in the wavelength range of 400 nm to 1,000 nm.

8.5 Data Acquisition Electronics

8.5.1 TCSPC Electronics

Largely because of historical reasons, TCSPC electronics have been based on nuclear instrumentation modules (NIMs), developed for the nuclear physics area. These modules have been employed with little customization to fluorescence lifetime determination. The start and stop timing pulses (see Figs. 8.1 and 8.12) are usually "cleaned up" using a discriminator, commonly a constant fraction discriminator (CFD) as it counteracts fluctuating pulse heights in the signals by producing a common timing point with low timing jitter. The "start" and "stop" signals are fed to a timing device, typically a TAC, shown in Fig. 8.12, which when triggered ramps up the voltage on a capacitor. This process is demonstrated in Fig. 8.13, and the arrival of a "stop" pulse then halts this increase, and the voltage output is fed to an ADC, which outputs the time to a histogram bin.

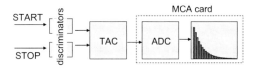

Figure 8.12 Scheme of TCSPC timing electronics. *Abbreviations*: TAC, time-to-amplitude converter; ADC, analog-to-digital converter; MCA, multichannel analyzer.

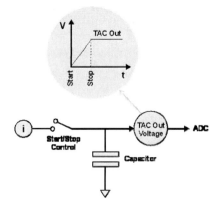

Figure 8.13 Schematic representation of the voltage buildup in a TAC used for the timing of photon events.

Histogram sizes can vary depending on application. Typically they can be up to 8,192 channels for high-resolution measurements or only 256 channels if data is to be acquired quickly at the expense of resolution, for example, in fluorescence lifetime microscopy. The width of the bins can be as little as 700 fs. The time width of the histogram bins is dependent on the time range selected and the number of points in the histogram. Typically the time range should be at least 10 times that of the lifetime to be measured. The number of histogram points is normally a trade-off between resolution and the data collection time, as doubling the number points, for example, will double the time to measure the data. Typically at least 1,024 channels in the histogram are advisable, although this is highly application dependent.

Because of the nature of TCSPC the timing electronics may be triggered many times by a "start" pulse without any data output expected. When using high-repetition-rate excitation sources this means that the equipment dead time becomes significant. One method used to overcome this is to run the equipment in "reverse mode," with the TAC triggered using the signal from the detector and the voltage ramp halted by the correlated signal from the excitation source. As the TAC is started fewer times a reduction in equipment dead time is achieved.

Further efficiencies in TAC usage and hence in the TCSPC technique can be made by the use of several detectors or a multicathode device coupled via a multiplexing unit working in conjunction with the TAC. The reports on the theory[40] and first reported usage[41] have shown the ability to record simultaneous fluorescence decays. This has been applied to the study of self-absorbing samples in solution[42] and a sol-gel medium.[43]

Timing devices for fluorescence lifetime determination historically started as stand-alone devices, with the histogram data finally being read to a computer for treatment and analysis of the data. With advances in computing technology there was a move to integrate the collection hardware (MCA) within the computer along with a move away from the NIM format. However, computers are "noisy" environments in which to process the small signals originating from the detectors and advances in computer technology have made some older MCA cards obsolete, as they will not physically couple within newer computers. These two factors have led to the return of more stand-alone hardware with control and connection achieved via a universal serial bus (USB) connection to a computer.

Combining the functionality of TACs and ADCs in newer devices called time-to-digital converters (TDCs)[44] has also occurred, and a newer representation of the electronics is given in Fig. 8.14. A practical example is the HORIBA Scientific FluoroHub-Bi, where all electronics are in a stand-alone module in communication via a USB link to a computer running control and analysis software.

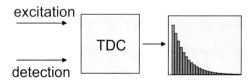

Figure 8.14 Scheme of TCSPC timing electronics incorporating a TDC.

The traditional histograms just contain information concerning when a detection event happens in relation to the excitation pulse ("microtime" events). The histograms give no information, however, as to when during the experiment the detection event occurred. Nevertheless, there is importance in knowing this "macrotime" information in the study of dynamic processes, such as diffusion and fluorescence correlation spectroscopy.[45] Again advances in computer technology have allowed this to be realized, and it is possible to label or tag each detection event in terms of its macrotime (time during the experiment) and microtime (time after excitation pulse)—a process referred to as time tagging (see Fig. 8.15). In one respect this may seem a step backward in a technological sense. Online "histogramming" was necessary when disk sizes were small and computer interfaces slow. This is no longer the case, so data can simply be streamed to disk and "histogrammed" in the software as part of postprocessing and analysis. This means that data of excitation-collection cycles can be continuously streamed, reconstructed, and visualized in terms of the macrotime. There is one major drawback, however: this is that large amounts of data are generated. Typically this can be in the region of hundreds of megabits compared with a few kilobits for the usual measurements.

Thus, the methodologies in the treatment of the data have changed from the "reductionist" summing of detection events in stand-alone NIM modules via computer integration on to integrated modules linked to computers by USB connections, where the complete dynamical information of the fluorescence process is preserved.

However, because of the large data files generated, at present, this should only be employed when necessary.

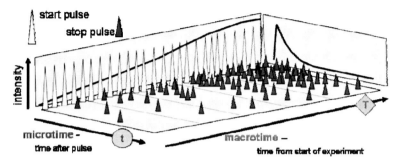

Figure 8.15 Schematic view of a time-tagged measurement.

8.5.2 Longer Timescale Measurements

Up until this point we have been mainly concerned with the fluorescence timescale and the use of the TCSPC technique. Much of the instrumentation can also be applied to the determination of lifetimes on the phosphorescence timescale. More traditionally, rather than using TAC-based systems, multichannel scaling (MCS)[46] is employed.

MCS works by counting all events occurring within a specified time interval t to $t + \Delta t$ (the dwell time or bin width). The time t is divided into i histogram points (number of bins), and t is quantized by the relation $t = i\Delta t$. When the measurement starts it inputs counting events into the first channel; at the end of the dwell time, selected counting is commenced in the next channel, and so on (see the schematic in Fig. 8.16). This continues until all (i) channels have been utilized (called a sweep). This process can then be repeated for a prerequisite number of sweeps, specified time, or peak number of counts.

This type of operation means that the device is acting in a multiple stop mode as it collects all the counts that are occurring within a specified time interval. Thus, the luminescence collection rate, in relation to that of the excitation, can be higher than that of a TCSPC-based system (as these only allow for one stop). However, this enhanced collection rate is not true for all time ranges, as the electronic dead time effects dominate at shorter timescales. This form of operation is also advantageous when it comes to treatment of the detector noise, allowing access to different time ranges for the previously mentioned NIR tubes.

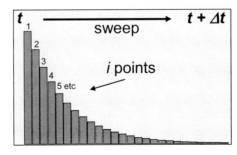

Figure 8.16 Representation of an MCS measurement.

8.6 Time-Resolved Measurement System Considerations

There are many aspects that require consideration when choosing a system to measure time-resolved luminescence decays. The major factor is the sample that is to be measured, and concerns its spectral properties and luminescence lifetime. This influences the source and detector options, as well as the electronics required for data acquisition. For example, the use of a narrow temporal pulse width excitation source operating at a high repetition rate is erroneous for the measurement of microsecond luminescence lifetimes. Also the combination of a femtosecond laser system with a side on a photomultiplier would not produce optimum performance in terms of time resolution as this would be detector limited. Here, it is worth referring back to Eq. 8.1. Thus, it is important to match the performance of the component parts of the time-resolved measurement system, as the overall performance is limited by the worse-performing part.

Improved and higher performance comes at a cost. A system containing a flashlamp will be limited in terms of time resolution by the source, just as a side on a PMT could be a limitation on the detector side. The highest time resolution systems, although probably not the fastest in terms of data acquisition, would probably consist of a femtosecond laser system with MCP detection. However, factors like the electronics, wavelength range, and ease of use should not be overlooked; the faster system is not always the best for the researchers' need. For a majority of applications a solid-state laser diode or LED source and picosecond detection module are more than sufficient and easier to use and maintain.

So far the above discussion of instrumentation has concentrated on systems for "cuvette" measurements, or rather where the position of the emitting molecule within the sample has not been the major consideration. Obviously if position is of importance then a microscope-based system is required. Most of the factors encountered so far are also applicable to microscope-based systems. It is not the purpose of this chapter to go into details concerning microscopy, but typically a microscope can be used to take point measurements, analogous to the time-resolved measurements taken with a "cuvette"-based instrument or to perform FLIM. Several point measurements can be combined to form a fluorescence map of a designated area. Imaging can produce a "picture" containing both spatial and temporal information.

Several companies produce instrumentation covering a range of fluorescence techniques and measurement methods. The following picture (Fig. 8.17) shows examples of the time domain systems available from HORIBA Scientific; these range from the compact filter base TemPro Fluorescence Lifetime System, to the MCP-based FluoroCube Ultrafast 01, to the DynaMyc fluorescence microscope with confocal and fluorescence lifetime-mapping capabilities.

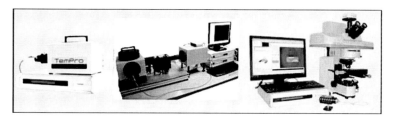

Figure 8.17 Examples of time-resolved fluorescence lifetime measurement systems.

8.7 Summary

Although the advancement of technology has allowed great improvements in time resolution, measurement times, and the power of data acquisition, the fundamentals of the TCSPC technique have remained unaltered since its conception in the latter part of the last century. However, this progress has also helped reduce the costs of these pieces of apparatus, which in turn has made the technique more accessible, and its application has gone from restricted to a few specialized instrumentation centers to individual laboratories as the potential of fluorescence lifetime determination is realized.

References

1. L. M. Bollinger, G. E. Thomas, *Rev. Sci. Instrum.*, **32**, 1044 (1961).
2. D. V. O'Connor, D. Phillips, *Time-Correlated Single Photon Counting*, London: Academic Press, 1984.
3. D. J. S. Birch, R. E. Imhof, in J. R. Lakowicz, ed., *Topics in Fluorescence Spectroscopy*, Vol. 1, New York: Plenum Press, 1.
4. J. R. Lakowicz, *Principles of Fluorescence Spectroscopy*, 2nd ed., New York: Kulwer Academic/Plenum Publishers, 1999.
5. C. C. Davis, T. A. King, *J. Phys. A: Gen. Phys.*, **3**, 101 (1970).
6. J. R. Lakowicz, I. Gryczynski, in J. R. Lakowicz, ed., *Topics in Fluorescence Spectroscopy*, Vol .1, New York: Plenum Press, 293.
7. D. J. S. Birch, R. E. Imhof, *Anal. Instrum.*, **14**, 293 (1985).
8. J. H. Malmberg, *Rev. Sci. Instrum.*, **28**, 1027 (1957).
9. S. S. Brody, *Rev. Sci. Instrum.*, **28**, 1021 (1957).
10. D. J. S. Birch, R. E. Imhof, *Rev. Sci. Instrum.*, **52**, 1206 (1981).
11. C. K. Chan, D. Phillips, *Spectra-Physics Laser Technical Bulletin No 8*, (1978).
12. L. Linqvist, R. Lopez-Delgado, M. M. Martin, A. Treamer, *Opt. Commun.*, **10**, 283 (1974).
13. A. Volkmer, D. A. Hatrick, D. J. S. Birch, *Meas.Sci. Technol.*, **8**, 1319 (1997).
14. T. Imasaka, A. Yoshitake, K. Hirata, Y. Kawabata, N. Ishibashi, *Anal. Chem.*, **57**, 947 (1985).
15. W. J. O'Hagan, M. McKenna, D. C. Sherrington, O. J. Rolinski, D. J. S. Birch, *Meas. Sci. Technol.*, **13**, 84 (2002).
16. C. D. McGuinness, A. Macmillan, K. Sagoo, D. McLoskey, D. J. S. Birch, *Appl. Phys. Lett.*, **89**, 063901 (2006).
17. C. D. McGuinness, K. Sagoo, D. McLoskey, D. J. S. Birch, *Meas. Sci. Technol.*, **15**, L19 (2004).

18. C. D. McGuinness, K. Sagoo, D. McLoskey, D. J. S. Birch, *Appl. Phys. Lett.*, **86**, 261911 (2005).
19. G. McConnell, *Opt. Express.*, **12**, 2844 (2004).
20. C. F. Kaminski, R. S. Watt, A. D. Elder, J. H. Frank, J. Hult, *Appl. Phys. B*, **92**, 367 (2008).
21. J. Y. Ye, C. J. Divin, J. R. Baker, T. B. Norris, *Opt. Express*, **15**, 10439 (2007).
22. D. J. S. Birch, G. Hungerford, R. E. Imhof, *Rev. Sci. Instrum.*, **62**, 2405 (1991).
23. M. K. Kuimova, G. Yahioglu, J. A. Levitt, K. Suhling, *J. Am. Chem. Soc.*, **130**, 6672 (2008).
24. G. Hungerford, A. Allison, D. McLoskey, M. K. Kuimova, G. Yahioglu, K. Suhling, *J. Phys. Chem. B*, **113**, 12067 (2009).
25. M. Köllner, J. Wolfrum, *Chem. Phys. Lett.*, **200**, 199 (1992).
26. M. Toury, L. Chandler, A. Allison, D. Campbell, D. McLoskey, A. S. Holmes-Smith, G. Hungerford, *Proc. SPIE*, **7903**, 79031W (2011).
27. D. McLoskey, D. Campbell, A. Allison, G. Hungerford, *Meas. Sci. Technol.*, **22**, 067001 (2011).
28. S. H. Lin, Y. Fujimura, H. J. Neusser, E. W. Schlag, *Multiphoton Spectroscopy of Molecules*, Orlando: Academic Press, 1984, 7.
29. D. Wildanger, E. Rittweger, L. Kastrup, S. Hell, *Opt. Express.*, **16**, 9614 (2008).
30. G. McConnell, J. M. Girkin, S. M. Ameer-Beg, P. R. Barber, B. Vojnovic, T. Ng, A. Banerjee, T. F. Watson, R. J. Cook, *J. Microsc.*, **225**, 126 (2007).
31. J. H. Frank, A. D. Elder, J. Swartling, A. R. Venkitaraman, A. D. Jeyasekharan, C. F. Kaminski, *J. Microsc.*, **227**, 203 (2007).
32. D. M. Owen, E. Auksorius, H. B. Manning, C. B. Talbot, P. A. A. De Beule, C. Dunsby, M. A. A. Neil, P. M. W. French, *Opt. Lett.*, **32**, 3408 (2007).
33. G. Hungerford, K. Suhling, M. Green, *Photochem. Photobiol. Sci.*, **7**, 734 (2008).
34. G. Hungerford, D. J. S. Birch, *Meas. Sci. Technol.*, **7**, 121 (1996).
35. B. H. Candy, *Rev. Sci. Instrum.*, **56**, 183 (1985).
36. S. Dhawan, *IEEE Trans. Nucl. Sci.*, **28**, 672 (1981).
37. Hamamatsu Photonics K. K., *MCP-PMTs R3809U-50 Series. Note TPMH1067E08*, 2007.
38. S. Cova, A. Ghioni, G. Ripamonti, T. Louis, *Rev. Sci. Instrum.*, **60**, 1104 (1989).

39. H. Dautet, P. Deschamps, A. D. MacGregor, D. MacSween, R. J. McIntyre, C. Trottier, P. P. Webb, *Appl. Opt.*, **32**, 3894 (1993).
40. K. Suhling, D. McLoskey, D. J. S. Birch, *Rev. Sci. Instrum.*, **67**, 2238 (1996).
41. D. J. S. Birch, A. S. Holmes, J. R. Gilcrest, R. E. Imhof, S. M. Al Alawi, B. Nadolski, *J. Phys. E: Sci. Instrum.*, **20**, 471 (1987).
42. D. J. S. Birch, D. McLoskey, A. Sanderson, K. Suhling, A. S. Holmes, *J. Fluoresc.*, **4**, 91 (1994).
43. G. Hungerford, M. R. Pereira, J. A. Ferreira, T. M. R. Viseu, A. F. Coelho, M. I. C. Ferreira, K. Suhling, *J. Fluoresc.*, **12**, 397 (2002).
44. A. Kirichenko, S. Sarwana, D. Gupta, I. Rochwarger, O. Mukhanov, *IEEE Trans. Appl. Superconduct.*, **13**, 454 (2003).
45. S. Felekyan, R. Kühnemuth, V. Kudryavtsev, C. Sandhagen, W. Becker, C. A. M. Seidel, *Rev. Sci. Instrum.*, **76**, 083104 (2005).
46. M. Zieliński, K. Karasek, R. S. Dygdala, *Rev. Sci. Instrum.*, **67**, 3325 (1996).

Chapter 9

Key Approaches to Linking Nanoparticle Metrology and Photoluminescence

Yu Chen, Jan Karolin, and David J. S. Birch
Centre for Molecular Nanometrology, University of Strathclyde, Department of Physics, SUPA, John Anderson Building, 107 Rottenrow, Glasgow G4 0NG, Scotland, UK
djs.birch@strath.ac.uk

We illustrate some of the links between nanoparticle dimensions and luminescence. Properties, application, and techniques for studying luminescence in two distinct types of nanoparticles are described. Those in silica colloids require extrinsic labeling, while gold nanoparticles possess intrinsic luminescence. Firstly, extrinsically doped luminescence is typified in terms of its use to determine nanoparticle size by means of the rotational correlation time, which describes the fluorescence anisotropy decay due to Brownian rotation of a fluorescent dye attached to amorphous silica nanoparticles. Measurement optimization is described with respect to both dye and particle and robustly illustrated using LUDOX* colloids despite the complex fluorescence anisotropy decay observed. Good agreement of 4.0 ± 0.4 nm, 6.4 ± 0.5 nm, and 11.0

---

*LUDOX is a registered trade mark of DuPont Corporation.

Luminescence: The Instrumental Key to the Future of Nanotechnology
Edited by Adam M. Gilmore
Copyright © 2014 Pan Stanford Publishing Pte. Ltd.
ISBN 978-981-4241-95-3 (Hardcover), 978-981-4267-72-4 (eBook)
www.panstanford.com

± 1.6 nm with expected values was found for LUDOX SM30, AM30, and AS40, respectively. This approach has topical application in addressing the need for inexpensive and in situ techniques for nanoparticle metrology. Secondly, gold nanoparticles have unique optical properties arising from surface plasmon resonance. Strong fluorescence via two-photon excitation has been observed from gold nanorods resulting from the enhanced local field. The gold nanoclusters less than 2 nm have unique optical, electronic, and chemical properties significantly different from larger nanoparticles. In this nanoscopic length scale, further confinement of the electrons results in discrete energy states with strong intrinsic fluorescence. These molecule-like properties of highly polarizable and emissive noble metal clusters open new opportunities for biological labels, energy-transfer pairs, and light-emitting sources on the nanoscale. Moreover, the properties of these nanoscopic systems can be tuned as their atomic arrangement and electronic structure changes when adding or removing atoms. Here we summarize how the luminescent properties of gold nanoclusters are size dependent.

9.1 Introduction

Nanoparticles present new challenges and opportunities for quantum understanding, spectroscopic measurement, and application. While there is much in common between traditional nanoparticles (e.g., silica and carbon) and the new varieties (e.g., noble atomic clusters and semiconductor quantum dots), there are also important differences, which we will try to illustrate in this chapter.

Requirements common to all types of nanoparticles include the urgent need for reliable and practical methods of measuring nanoparticle size and the establishment of appropriate internationally recognized standards with which to assess and compare data. Techniques for monitoring structures in situ and fabrication during bottom-up synthesis, and comparisons with simple standards, would be particularly helpful when designing and optimizing nanotechnologies. Moreover, recent reports[1-3] have stressed the need for identifying, characterizing, and managing the potential risks to human health and the environment posed by nanotechnologies. Within the ever-increasing range of nanotechnologies, of particular note is the present lack of clarity concerning the potential toxicological effects of nanoparticles.[3]

Methods used for nanoparticle metrology include small-angle X-ray scattering (SAXS)[4,5], small-angle neutron scattering (SANS)[4], transmission electron microscopy (TEM)[5], light scattering[6], spectroturbidimetric methods[7], and electrospray.[8]

However, all of these have drawbacks, including SAXS and SANS being expensive and complex, TEM not being in situ, and light scattering being increasingly difficult below a 10 nm radius, which is a size that includes many of the new nanoparticles, such as quantum dots, fullerenes, and carbon nanotubes. Recently a ten-laboratory international round-robin produced a preliminary comparison of nanoparticle size using conventional measurement methods for radii ≈ 10 nm and 50 nm. For the polystyrene nanoparticles studied the results were found to be in general consistence with the manufacturer's values, but for the silver nanoparticles studied discrepancies were observed.[9] In this chapter we start to address the gap in nanoparticle standards in the important 1 nm to 10 nm range, illustrated by means of both traditional (silica) and new (gold) types of nanoparticles.

Fluorescence and luminescence techniques in general have much to offer in nanoparticle characterization and application. For example, previous work in our laboratory demonstrated a new model for interpreting fluorescence anisotropy decay in sol-gels in terms of dye molecules binding to nanoparticles, and this led directly to an inexpensive and yet high-resolution method of measuring silica nanoparticle size during the sol-gel transition in silica hydrogels and alcogels.[10–13]

In the case of hydrogels, primary particles with a mean radius r of ~1.5 nm were found to be initially formed before subsequently aggregating to give secondary particles with $r \approx 4.5$ nm. This growth was resolved with close to angstrom resolution, and the technique we have introduced directly addresses the need for low-cost and in situ nanoparticle size measurement below the limit where light scattering can be reliably performed in what might be termed the "ultranano" region, that is, $r < 5$ nm. The fluorescence anisotropy decay technique is generic in the context of nanoparticles, and in principle there is no reason why it cannot be adapted to measure the size of other nanoparticles such as semiconductor quantum dots, metal colloids, titanium dioxide, and carbon black. In the first demonstration, a near-infrared dye, JA120[10,11], with a fluorescence decay time $\tau_f \approx 2.0$ ns, was attached to the nanoparticle and thereby the Brownian rotation observed in terms of the rotational correlation time τ_r (equivalent to the fluorescence anisotropy decay time for an isotropic rotor), which then leads to the determination of r using the Stokes–Einstein equation. However, many silica colloids (e.g., the DuPont LUDOX range of stable silica colloids) have r much greater than 5 nm, and although in this case the measurable lower particle size limit is determined by the size of the dye, the upper limit is determined by the ratio of τ_r to τ_f, the extreme case being when the particle rotational time is substantially longer than the fluorescence decay time such that an immeasurably small change in depolarization of the fluorescence occurs during the fluorescence lifetime. The ideal situation for unbound dye molecules of $\tau_r \approx \tau_f$ was identified over half a century ago by Weber[14], and Fig. 9.1A illustrates the problem. Hence a question arises regarding the magnitude of the fluorescence decay time f required for measuring r > 5 nm and indeed what nanoparticle standard might be used for calibration purposes. Recently we have shown how LUDOX silica colloids provide a useful metrology standard[15] and that sufficient measurement statistical precision can be achieved for $\tau_f \approx 25$ ns to enable the measurement of a mean radius $r \approx 11$ nm, corresponding to $\tau_r \approx 1.4$ s, that is, an τ_r/τ_f ratio of ~50. The fluorophore used was based on the 6-MQ core, as shown in Fig. 9.1B.

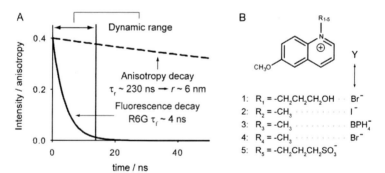

Figure 9.1 (A) Illustration of the dynamic range over which the decay of fluorescence anisotropy can be observed. (B) Structure of various 6-MQ derivatives. *Abbreviation:* 6-MQ, 6-methoxy quinolinium.

The photoluminescence from bulk gold was first reported by Mooradian[16] in 1969. It was found that the emission peak was centered near the interband absorption edge of the metal and therefore was attributed to direct radiative recombination of the excited electrons in conduction band states below the Fermi level with holes in the d bands. However, luminescence from noble metals has a very low efficiency because metals do not have bandgaps and the nonradiative decay can proceed all the way back down to the ground state. The quantum yield for photoluminescence from bulk gold and copper films is on the order of 10^{-10}.

Boyd et al. compared the luminescence from smooth and rough surfaces and found a rise in the intensity of multiphoton-excited luminescence from a rough gold surface.[17] The multiphoton luminescence is believed to be emitted predominantly from the surface atoms of protrusion on the rough surface with localized plasmon resonance. The influence of surface roughness on the luminescence is largely attributable to local-field enhancement in the rough surface protrusions. Roughing metal surfaces can be thought of as creating a collection of randomly oriented hemispheroids of nanometer dimensions on the smooth surfaces. These hemispheroids show a surface plasmon resonance similar to the gold nanorods, and therefore the incoming and outgoing electric fields are amplified by the local field created around the hemispheroids by the plasmon resonances. This kind of enhancement has been proposed for the observed luminescence of gold nanorods. The luminescence efficiency is found to increase by six orders of magnitude because of this lightning rod effect. In the last decade, gold nanorods have attracted intensive interest because of their shape-dependent optoelectronic properties, such as strong surface-enhanced Raman scattering[18-20], fluorescence[21,22], and anisotropic chemical reactivity[23], and have become promising candidates for biological sensing or imaging.[24-27] Several review articles provide thorough surveys on the latest development in this area[28-30], which interested readers can refer to.

By reducing the size and dimensionality of a material, electronic properties change drastically as the density of states and the spatial length scale of the electronic motion are reduced with decreasing size. A transition from the bulk band structure to individual localized energy levels occurs in nanoclusters of subnanometer to nanometer size. Closely related to size-induced changes in the electronic structure are the optical properties of nanoparticles.

Large Au nanoparticles have optical properties dominated by a surface plasmon when their conduction electrons are confined to dimensions comparable to the electron mean free path length (ca. 20 nm). However, Au nanoparticles smaller than 2 nm no longer possess plasmon resonance and Mie's theory no longer can be applied.[31] In fact small metal nanoclusters have sizes comparable to the Fermi wavelength of electrons (ca. 0.7 nm), which results in molecule-like properties, including size-dependent fluorescence[32–34] and discrete size-dependent electronic states.[35–37]

The fluorescent properties of Au nanoclusters make them potential labels for biological sensing and imaging. Compared to commonly used semiconductor quantum dots, such as CdSe/ZnS, which possess size-dependent fluorescence in cases where the particle size is smaller than the exciton Bohr radius (about 4–5 nm for CdSe)[38], Au nanoclusters do not contain toxic heavy metals and have smaller sizes.[39] Thus, they have potential for biological labeling and light-emitting sources in a new generation of optical devices.

9.2 Fluorescence Anisotropy Theory

The theory and application of time-resolved fluorescence anisotropy decay of fluorophores have been widely reported[40–42], and so here we offer only a brief summary in the context of nanometrology.

By recording vertically and horizontally polarized fluorescence decay curves, $I_V(t)$ and $I_H(t)$, orthogonal to vertically polarized excitation, an anisotropy function $R(t)$ can be generated, that is,

$$R(t) = \frac{I_V(t) - I_H(t)}{I_V(t) + 2I_H(t)} \tag{9.1}$$

For a spherical rigid rotor in an isotropic medium the decay of $R(t)$, which describes depolarization of the fluorescence due to Brownian rotation, can be expressed as:

$$R(t) = R_0 \exp\left(-\frac{t}{\tau_r}\right) \tag{9.2}$$

where R_0 is the initial anisotropy with a maximum value of 0.4 for one-photon excitation and, in the simplest case, τ_r is described by the Stokes–Einstein equation:

$$\tau_r = \frac{\eta V}{kT} \tag{9.3}$$

where η is the microviscosity, V the hydrodynamic volume = $4\pi r^3/3$ prescribed by the rotor, T temperature, and k Boltzmann constant. In the case where the probe is in two environments

$$R(t) = (1-f)R_0 \exp\left(-\frac{t}{\tau_{r1}}\right) + fR_0 \exp\left(-\frac{t}{\tau_{r2}}\right) \quad (9.4)$$

where the fraction of total fluorescence f and $1-f$ represent the probe in each environment. In silica sol-gels f can, by careful choice of probe, be interpreted as being due to probe molecules bound to silica particles rotating with a correlation time τ_{r2} and $1-f$ to probe molecules solvated and unbound in the sol rotating faster with a correlation time τ_{r1}.[10-13]

If in Eq. 9.4 we expand $\exp(-t/\tau_{r2})$ and put $\tau_{r1} \ll \tau_{r2}$ to reflect free probe molecules rotating much faster than those that are bound to nanoparticles, then in the case where the fluorescence decay time $\tau_f \ll \tau_{r2}$, a simplified expression to Eq. 9.4 is obtained, namely:

$$R(t) = (1-f)R_0 \exp\left(-\frac{t}{\tau_{r1}}\right) + fR_0 \quad (9.5)$$

It should be noted that only if the fluorescence quantum yield of the probe is the same when unbound and bound to nanoparticles does the fraction f exactly describe the probe partition ratio between the nanoparticle and fluid. Also, τ_f for an unbound and a bound probe also needs to be similar for Eqs. 9.4 and 9.5 to be valid. Moreover a single rotational correlation time description is only strictly valid for a spherical rotor of a fixed radius of rotation. Hence in the case of a nonspherical or a distribution of nanoparticle sizes the analysis becomes more complicated and the above treatment becomes more approximate. Nevertheless, notwithstanding these constraints, the interpretation of fluorescence anisotropy we have proposed for probes bound to nanoparticles has been validated for the initial stages of silica gelation. In this case it has been shown to be consistent with what is known about the growth of silica nanoparticles from a variety of other techniques[4-7,43,44] and indeed is already finding broader application such as in studying the surface modification of silica.[45,46]

In the case of a small probe totally bound to a larger nanoparticle with none free, but wobbling within a cone angle with a rotational time τ_{r1} as the particle rotates with a rotational correlation time τ_{r2} such that $\tau_{r1} \ll \tau_{r2}$, the anisotropy decay can be described to a good approximation by:[40]

$$R(t) = b_1 \exp\left(-\frac{t}{\tau_{r1}}\right) + b_2 \exp\left(-\frac{t}{\tau_{r2}}\right) \qquad (9.6)$$

where $b_1 + b_2 = R_0$. If in addition there are probe molecules bound to aggregates of the "fundamental nanoparticles" and of similar composition, but a much larger size then we can similarly approximate this by expressions such as

$$R(t) = (1-f-g)R_0 \exp\left(-\frac{t}{\tau_{r1}}\right) + fR_0 \exp\left(-\frac{t}{\tau_{r2}}\right) + gR_0 \qquad (9.7)$$

where $gR_0 = R_\infty$ can also effectively describe a very slow rotation (i.e., an exponential of long time constant, c.f. Eq. 9.5), g is the fraction of fluorescence derived from the aggregate and f describes free dye or bound dye wobbling on a nanoparticle or a compromise between the two.

9.3 Experimental

9.3.1 Instrumentation

Time-resolved fluorescence and anisotropy decay time measurements were performed using the time-correlated single-photon counting (TCSPC)[47,48] technique on a FluoroCube (Horiba Jobin Yvon IBH Ltd, Glasgow). The compact instrument configuration occupies less than 1 m², and a schematic specified for the present application is shown in Fig. 9.2. The excitation source was a NanoLED source of appropriate wavelength (e.g., 370 nm for the methoxy quinolinium dyes) with a 1 MHz repetition rate and an instrumental full width half maximum of 370 ps. The fluorescence rate was kept to below 1% of the source repetition rate to eliminate data pileup.[47,48] In recent years we have formed a favorable assessment of light-emitting diodes (LEDs) as a lower-cost, simpler, and more appropriate alternative to mainframe lasers for the vast majority of fluorescence decay measurements. Flashlamps had hitherto dominated the fluorescence decay time field for over 20 years[49], and although LEDs have a comparable peak (~1 mW) and average power (~1 W) they offer higher stability, reliability, and ease of use. We have successfully demonstrated a wide range of LEDs in fluorescence decay studies from the visible[50] to the near ultraviolet down to 265 nm[51-53], thus making the whole spectrum of aromatic dyes accessible as fluorescence decay time probes. LEDs emit light over a broader spectrum in comparison to laser diodes and can sometimes generate weaker emissions at longer wavelengths[52] than their fundamental wavelength band. For these reasons the excitation as well as the fluorescence wavelength (typically 430–440 nm for the methoxy quinolinium dyes) was selected using a monochromator (Seya–Namioka geometry). Both thin-film and prism optical polarizers were used successfully on different occasions to select vertically polarized excitation light and an identical rotating polarizer on the emission was toggled between vertical and horizontal orientations to obtain the anisotropy decay (Eq. 9.1). Motor-driven lenses enable the excitation and emission focusing to be automatically computer optimized to maximize the data collection rates for samples of differing optical density. Fluorescence photons were detected using a TBX detection system incorporating a compact photomultiplier in a T0-8 package (Hamamatsu Model R7400) and fluorescence decay data accumulated using the FluoroHub electronics incorporating a time-to-amplitude converter (operated in reverse mode to minimize dead time) and multichannel analyzer with the operation of all functions fully controlled under Windows software.

Data analysis was performed using nonlinear least squares with the DAS 6 (Horiba Jobin Yvon IBH Ltd, Glasgow) iterative reconvolution library. The anisotropy data was analyzed using impulse reconvolution by first fitting to the fluorescence decay given by the denominator in Eq. 9.1 ($I_V(t) + 2I_H(t)$). The difference curve $I_V(t) - I_H(t)$ (from Eq. 9.1) was then fitted using different models for the anisotropy decay $R(t)$ in order to obtain the best-fitting rotational parameters, and selection of the most appropriate model was aided using the χ^2 goodness of fit criterion.[47] Errors shown are 1 std. dev. unless stated otherwise.

The steady-state absorption measurements were performed on a Perkin–Elmer Lambda 2 spectrometer. Spectral measurements of intrinsic fluorescence were performed on a Perkin–Elmer LS50B luminescence spectrometer, and extrinsic probe fluorescence was studied on a Jobin Yvon SPEX Fluoromax.

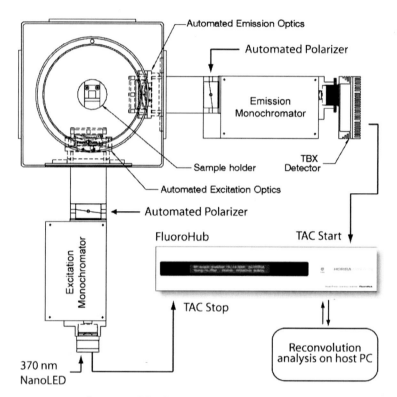

Figure 9.2 Schematic of the fluorometer used to perform the fluorescence and anisotropy decay time measurements.

9.3.2 Choice of Dyes and Nanoparticles and Sample Preparation

The three colloidal silicas selected for investigation from a wider range available from DuPont were chosen to reflect a range of sizes and surface chemistry, namely, LUDOX SM30, AM30, and AS40 of radius 3.5 nm, 6 nm, and 11 nm, respectively. SM30 has a negative particle charge at pH 10.2, contains 30% SiO_2 and a 0.58% Na_2O stabilizer, and at 25°C the macroviscosity is 5 cP; AM30 has a negative particle charge at pH 8.9, contains 30% SiO_2, trivalent aluminum ions substituted for some of the tetravalent silicon ions on the surface, and a 0.24% Na_2O stabilizer, and at 25°C the macroviscosity is 11 cP; AS40 has a negative particle charge at pH 9.2 and contains 40% SiO_2, and at 25°C the macroviscosity is 9 cP.[54]

All the LUDOX samples studied have free hydroxyl groups tethered on the outside of the nanoparticle and capable of binding to water, potentially increasing the measured hydrodynamic radius. LUDOX is quite stable, but its repeated exposure to atmospheric CO_2 has the effect of lowering the pH, eventually resulting in gelation. A field emission scanning electron microscope (Sirion 200 supplied by FEI) image of dried LUDOX AS40 shown in Fig. 9.3 verifies the spherical size to be of a ~20 nm diameter and usefully provides a cross-check of the in situ fluorescence anisotropy measurements. The size distribution is small, with a standard deviation of 2.5 nm in the diameters measured for 100 spheres, and unlikely to be evident in the anisotropy analysis other than contributing to the final uncertainty. Although at first sight a nanoparticle standard based on fluorescence anisotropy with LUDOX might seem trivial by virtue of LUDOX's expected simple spherical rigid rotation to give a single rotational correlation time (Eq. 9.2), the results in the next section show there to be other factors that need to be taken into account before the fluorescence anisotropy decay is meaningful when used for nanometrology.

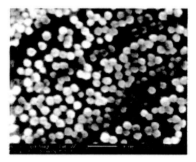

Figure 9.3   Secondary electron image of LUDOX AS40 diluted to 2% w/w and dried at 40°C.

6-MQ dyes were chosen as fluorescent probes because of their cationic properties for ease of binding to the negatively charged nanoparticles without covalent attachment and, in comparison to many other dyes, a relatively long fluorescent lifetime, stability at pH 10, small size, and compatibility with low-cost LEDs. The materials used to synthesize the fluorescent probes were purchased from Aldrich Chemical Co. Ltd. The detailed synthesis of the fluorescent probes has been reported elsewhere.[55,56]

The fluorescence probes were added directly to the LUDOX without dilution unless otherwise indicated. Typical probe concentrations used were 10 M, and all the samples were studied in cuvettes at 90° between the excitation and emission axes. Despite fears of the effect of strain-induced birefringence on polarization disposable plastic cuvettes (4 cm³ from Hughes & Hughes Ltd) were found to give comparable results to fused silica cuvettes. All cuvettes were sealed with parafilm to prevent autogelation. It is informative to consider how the choice of dye came about with respect to alternatives. Figure 9.4 illustrates the measurement principle deployed where free and bound dyes coexist for one type of LUDOX AM30. Figure 9.5 shows how the nonbinding of the anionic fluorescein to the anionic nanoparticle is reflected in little change in the anisotropy decay when the nanoparticles are added to the dye solution. Figure 9.6 shows how the cationic rhodamine 6G binds to the nanoparticle but has too fast a fluorescence decay time (~4 ns) to track the depolarization as the nanoparticle rotates. Figure 9.7 (~25 ns) shows that the binding and longer fluorescence decay time of the 6-MQ dye are just right for revealing the particle rotation and size.

Figure 9.4 Illustration of dye free in solution and bound to a silica nanoparticle.

Gold clusters can be prepared through the wet-chemical process or physical gas-phase deposition. Synthesis in organic solvents includes direct synthesis of small nanoparticles and formation of clusters through etching the surface of a larger preformed particle. The two-phase reduction method for the preparation of alkyl-thiol monolayer-protected gold particles was first introduced by Schiffrin et al.[57] This process has been extensively used to synthesize small clusters.[58] In a typical reaction, monolayer-protected gold clusters are synthesized by two-phase (e.g., water–methanol) reduction of $AuCl_4^-$ by sodium borohydride in the presence of surfactant molecules.

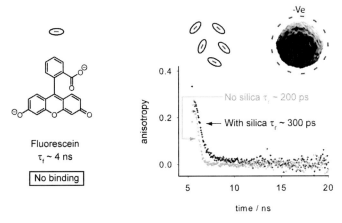

Figure 9.5 Comparison of fluorescence anisotropy decays recorded on fluorescein in a neat solution and in a silica colloid (LUDOX).

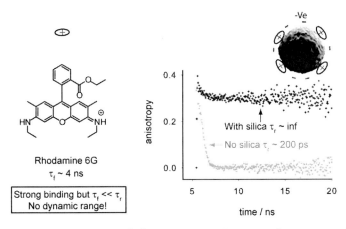

Figure 9.6 Comparison of fluorescence anisotropy decays recorded on rhodamine 6G in a neat solution and in a silica colloid (LUDOX).

Lin et al.[59] synthesized Au nanoclusters based on precursor-induced Au nanoparticle etching in organic phase and ligand exchange with reduced lipoic acid (dihydrolipoic acid [DHLA]) to transfer the particles into aqueous solution. DDAB-stabilized gold nanoparticles are etched by the addition of Au precursors (HAuCl$_4$ or AuCl$_3$) to smaller nanoclusters. The hydrophobic AuNC@DDAB become water soluble upon ligand exchange with dihydrolipoic acid (AuNC@DHLA). TEM images reveal articles size change from 5.55 ± 0.68 nm to 3.17 ± 0.38 nm after the etching process. The ligand exchange results in further reduction in the particle size down to 1.56 ± 0.3 nm. It is found that only after phase transfer to aqueous solution do nanoclusters exhibit a pronounced red photoluminescence.

Whereas chemical synthesis offers a versatile, cost-effective approach and has been used broadly to prepare gold nanoparticles with various passivating ligands, gas-phase growth provides a beneficial alternative to create nanoparticle arrays.[60] In particular, size-selected nanoclusters with narrow size distribution (5%) can be produced and deposited onto a variety of supports[61–64], creating the potential for the investigation of single-nanoparticle fluorescence in dilute arrays.

Compared to colloidal nanoparticles synthesized through chemical processes, the nanoclusters thus formed have well-controlled and tunable sizes ranging from subnanometer to a few nanometers. This size region crosses the gap between gold nanoparticles with surface plasmon–dominated properties and small ones with molecular fluorescent behavior. It provides the possibility to investigate size-dependent fluorescence behavior and the evolution of surface plasmons as the atom number increases.

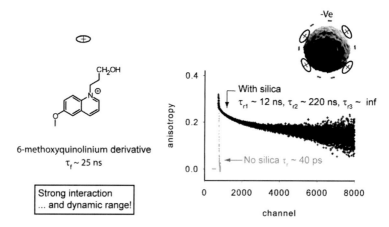

Figure 9.7 Comparison of fluorescence anisotropy decays recorded on a 6-MQ derivative in a neat solution and in a silica colloid (LUDOX).

A state-of-the-art magnetron sputtering gas aggregation cluster source, which incorporated a novel time-of-flight mass selector, has been used to produce Au nanoclusters with a size distribution of 5%. The construction and principle of this cluster source has been described previously.[62] The TEM image taken from Au nanoclusters containing 8860 ± 443 atoms deposited on an amorphous carbon film at a kinetic energy of 1.5 KeV, displayed in Fig. 9.8a, reveals monodispersed nanoclusters scattered on supports. Figure 9.8b shows the distribution of the aspect ratio. The majority of nanoclusters have a nonspherical shape with aspect ratios peaked at 1.4. Elongated nanoclusters with an aspect ratio larger than 2 have also been observed. Figure 9.8c shows a nanocluster with a width of 3.8 nm and an aspect ratio of about 2. Nanoclusters containing 15000 ± 750 gold atoms were also created and deposited on supports. Elongated particles of an aspect ratio of over 3 have been observed. Figure 9.8d displays a gold nanorod with an aspect ratio of 3.3. Gold nanoclusters were deposited on glass substrates for a fluorescence microscopy study.

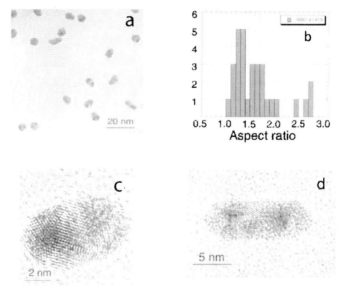

Figure 9.8 (a) TEM image taken from Au_{8600} nanoclusters deposited on an amorphous carbon film at 1.5 KeV, (b) distribution of aspect ratios, (c) an elongated nanocluster with an aspect ratio of 2, and (d) TEM image of a $Au1_{5000}$ nanorod with an aspect ratio of 3.3.

9.4 Results and Discussions

9.4.1 Ludox Labeled with Extrinsic Probes

In the types of LUDOX studied the emission spectra of the 6-MQ dyes show a hypsochromic shift of ~6–9 nm and the absorption shows a bathochromic shift of ~2–4 nm compared to the dyes in water. Such typical spectra are shown in Fig. 9.9 for dye 1 in LUDOX AS40 and water. These shifts are indicative of the dye attaching to the silica nanoparticle and the associated interaction as no change in the emission maxima is observed when the dyes are in sodium hydroxide solution at the relevant pH.

The fluorescence decay times for the 6-MQ dyes have already been investigated in a range of solvents and found to consistently display biexponential components.[55] In the present study time-resolved fluorescence decays were collected from solutions of the dyes in water and LUDOX. In both environments the fluorescence decay could also be adequately described by a biexponential decay function giving typically decay components of ~18 and 27 ns in aqueous solution and ~10 and 25 ns when in LUDOX. For both environments, the long lifetime component was the main contributor to the fluorescence decay.

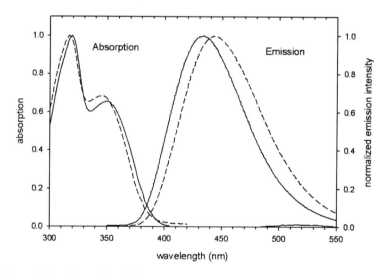

Figure 9.9 Normalized absorption and emission spectra of dye 1 in H_2O (- - - - -) and in LUDOX SM30 (———) at 2 % (w/w) SiO_2.

The rotational times determined from the anisotropy decays were used, in conjunction with Eq. 9.3, to determine the silica particle sizes. For these calculations the fluid viscosity, η, was taken as 1 cP (0.001 Pa· S), the viscosity of water at 20°C, rather than the bulk viscosity of the LUDOX investigated. An alternative approach would be to measure the rotation of the fraction of dye unbound to particles and hence make a correlation with the microviscosity using Eq. 9.3. The microviscosity, that is, the viscosity surrounding dye-silica particles, rather than the bulk viscosity, has previously been used to calculate silica particle sizes;[10-12] however, in the present study free dye rotation could not be unambiguously identified. Although we do not know exactly the microviscosity pertaining to the present measurements, because $r \propto \eta^{-1/3}$, a 30% deviation in viscosity estimation would be required for a 10% error in the calculated radius to be incurred. Similar scaling applies to $\tau_r \sim r^3$. In the particle size determination no allowance was made for any additional size of the dye molecule ($r \approx < 0.5$ nm). Although the orientation of the dye on the nanoparticle is not known the r^3 dependence of τ_r should in any case ensure the dye has negligible effect on the measured rotational correlation time even when bound to the smallest nanoparticle of ~3.5 nm radius.

Table 9.1 6-MQ derivative attached to different LUDOX silicas. Fluorescence anisotropy data indicates three components

LUDOX	r (SEM*)	$R_0 \neq 0.4$ Free dye? R_0	Wobbling? Surface diffusion? $\tau_{r1}/$ ns	Particle rotational diffusion? $\tau_{r2}/$ns	r_{calc}	Particle aggregation R_∞
SM30	3.5 nm	0.24	24	65 ± 15	4.0 ± 0.4	0.12
AM30	6 nm	0.28	17	273 ± 60	6.4 ± 0.5	0.06
AS30	11 nm	0.28	10	1400 ± 590	11.0 ± 1.6	0.07

*Data from scanning electron microscopy (SEM).

Considering first the results for LUDOX AM30 and AS40, anisotropy data for dyes 1, 2, and 4 with ~10,000 counts in the channel of the numerator of Eq. 9.1 containing the maximum number of counts was typically acquired in one hour. For both colloids two rotational times (i.e., Eq. 9.4 or 9.6, rather than Eq. 9.5 with one rotational time plus a residual anisotropy), with the faster component typically at the 10% level, were found to offer an acceptable description of the data in terms of the χ^2 goodness-of-fit criterion and a reasonable agreement with DuPont's data, finding $r \approx 7.7 \pm 0.1$ nm for AM30 (DuPont's $r \approx 6$ nm) and $r \approx 11.8 \pm 0.4$ for AS40 (DuPont's $r \approx 11$ nm radius). (Note that since the DuPont data is based on the dried colloid and our data on the hydrodynamic radius for the wet colloid in situ, they might not necessarily be identical). Table 9.1 shows the analysis outcomes for a typical data set that was part of a much wider investigation using different 6-MQ dyes and LUDOX colloids.[15] Generally in photophysics a low R_0 value (≤ 0.3) below the expected 0.4[40–42] reflects a rapid depolarization such as that due to intramolecular realignment between the absorption and emission dipoles. The long rotational time, τ_{r2}, was used to calculate the particle radii of the colloidal silicas using Eq. 9.3. However the faster component ($\tau_{r1} > 1$ ns) is at least an order of magnitude too large to represent free dye rotation.[10,11] The absence of a significant fraction of free dye is not too surprising, given we have an anionic colloid and a cationic dye; however, it is interesting, and possibly revealing, to discuss the possible origins of τ_{r1}. For example, even though the dye may be electrostatically attached to the nanoparticle it may still be mobile and undergo surface contour diffusion and thereby depolarize the fluorescence by this mechanism, as previously discussed for micelles and membranes.[65–67] Assuming the emission dipole is normal to the surface of the nanoparticle the corresponding correlation times ($\sim r^2/6D$)[67] associated with surface diffusion on AS40, AM30, and SM30 nanoparticles for a surface diffusion constant[68] $D \approx 10^{-10}$ m²s⁻¹ are ~200 ns, 60 ns, and 20 ns, respectively. Although these values are all significantly less than their respective τ_{r2} values they could nevertheless have an influence on the accuracy and interpretation of both τ_{r1} and τ_{r2}. Another possibility is that τ_{r1} reflects the wobbling of the bound dye tethered on the particle surface, thus sensing vicinal rather than the bulk microviscosity. In our view it is most likely that τ_{r1} and indeed τ_{r2} represent weighted averages of several depolarization mechanisms, for example, aggregated nanoparticles, surface contour diffusion, and free and tethered dyes. The two-rotational time model is in itself open to different interpretations, depending on the circumstances. Indeed the interpretation we offer contrasts with our previous work for the xanthene dye JA120 in silica hydrogels[10,11] (where the dominance of free and bound dyes was clearly established) and the work in which similar techniques were used to examine the adsorption of rhodamine B on a different LUDOX colloid HS30.[69] In the latter case Smith et al. were able to extract a fast component of ~300 ps as being consistent with a fluorophore free in solution, while fixing constant a slow component of ~685 ns corresponding to the probe bound to the 18 nm diameter particles in LUDOX HS30. The anisotropy function showed a decay and then a rise consistent with the greater-than-factor-of-two difference in the fluorescence lifetime for rhodamine B in the free and bound states. In the studies reported here the anisotropy function showed no evidence of a rise-time.

Repeat measurements[15] suggested the mean radius values determined for both LUDOX AM30 and AS40 were slightly larger than the values quoted by DuPont, which reports mean particle diameters of 6 nm and 11 nm, respectively.[54] Although the colloids are stable and marketed by DuPont as a single size, which is consistent with our SEM results showing a narrow distribution, some aggregation seems likely to occur with time (albeit to a lesser extent for fresh materials) since the colloids, as already mentioned, do eventually gel when exposed to air. A small degree of aggregation might indeed be reflected in the slightly higher radii than DuPont specifies that we find for AM30 and AS40. The evidence for nanoparticle aggregation is stronger when we consider the results for the particular batch of SM30 we studied in detail.

Given the complexity of the anisotropy decay and to better distinguish between the alternative models we found it necessary to measure all three colloids to a higher statistical precision (~10^5 counts in the peak of the difference curve for a typical measurement time of ~10 hr) than is common practice. Figure 9.10a shows the difference curve, the weighted residuals, the autocorrelation function and the anisotropy decay at this higher level of precision for AS40, which, at 11 nm, is the largest radius studied. It is notable that the agreement between DuPont's dried colloid data and our wet colloid data is within the error of measurement. For SM30, AM30, and AS40 we measure radii of 4.0 ± 0.4 nm, 6.4 ± 0.5 nm, and 11.0 ± 1.6 nm corresponding to DuPont's values of 3.5 nm, 6 nm, and 11 nm, respectively. The level of agreement suggests the assumptions made in the analysis are reasonable. However, to obtain good fits to the data we have had to use a residual anisotropy term gR_0, in addition to two exponentials (Eq. 9.7), and this is to take account of what we suggest is a relatively small fraction of larger aggregates of the fundamental nanoparticle in LUDOX. (A third rotational time of a very long time constant has the same effect in this respect as adding a residual anisotropy). To investigate the presence of nanoparticle aggregates we analyzed data for a dilute colloid when fresh and then aged for 500 hours. If we fix one rotational time equivalent to the DuPont value for the radius we would expect the χ^2 value for a narrow nanoparticle size distribution to be minimized at that radius. Figure 9.10b for AS40 shows that in fact this is only the case for the fresh sample suggesting that larger particles are formed even after only 500 hours.

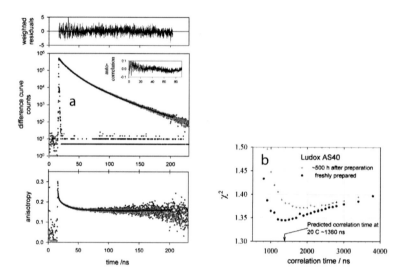

Figure 9.10 (a) Fluorescence anisotropy data recorded on a LUDOX AS40 diluted to 2% w/w SiO_2. The number of counts in the peak of the difference curve is 5×10^5. The measurement time was 10 hours. The anisotropy data was fitted to two exponential decay functions and a limiting anisotropy value, a total of six fitting parameters, including a time-shift parameter. The quality of the fit was judged by inspecting the weighted residuals, upper graph, the autocorrelation function, inserted graph, and reduced $\chi 2$. (b) χ^2 dependence of the longest correlation time.

9.4.2 Fluorescence from Au Nanoparticles

Fluorescence lifetime imaging microscope (FLIM) has been employed to study the fluorescence behavior of size-selected Au_{8600}

nanoclusters deposited on a glass slide. Figure 9.11 shows a typical FLIM image obtained with emission at a 535–590 nm range excited with a 750 nm pulsed laser from a Chameleon Ti:Sapphire laser source (Coherent). The scan area is 210 m × 210 m. The fluorescence lifetime is found in the time scale to be about 80 ps, in line with that measured from gold nanorods synthesized via the seeded growth process under the same excitation. It implies surface plasmon–dominated optical properties and a "lightning" effect originating from an enhanced local field at asymmetric nanoparticles.

Figure 9.11 Fluorescence lifetime image taken from Au_{8600} nanoclusters deposited on a glass slide.

Wilcoxon et al. have investigated the fluorescence characteristic of gold colloids and observed an emission at 440 nm from very small spherical gold clusters (<5 nm) excited at 230 nm.[70] The emission was attributed to sp to d interband transitions with an enhanced quantum yield of ~10^{-4}–10^{-5} in comparison with the 10^{-10} value of bulk gold. They also found that larger 15 nm clusters were not photoluminescent.[70] Recently, water-soluble gold nanoparticles were synthesized through the reduction of a gold salt in the presence of a designed polymer ligand (pentaery-thritol tetrakis (3-mercaptopropionate)-terminated polymethacrylic acid, PTMP-pMAA). The size of nanoparticles can be tuned by varying systematically the polymer-to-gold ratio. A transition from nonfluorescent to fluorescent nanoparticles is observed for core diameters between 1.7 nm and 1.1 nm. The most fluorescent nanomaterials have a 3% quantum yield, a 1.1 nm gold core, and a 6.9 nm hydrodynamic radius.[34]

Previous work on Au nanoclusters of 1.7 nm reported near-infrared photoluminescence when excited at 1064 nm. The quantum yield is about 4.4 10^{-5} at room temperature. This photoluminescence is attributed to sp to sp-like transitions, analogous to intraband transitions in bulk gold. However, the exact mechanism is unknown.[71] Later, Murray reported a visible fluorescence from water-soluble monolayer-protected gold clusters.[58] The quantum yield of the luminescence of these clusters with 1.8 nm diameter cores is estimated to be 0.003. It was found that the efficiency and wavelength of the luminescence varied with the ligands. The mechanism for the luminescence is hypothesized to be associated with interband transitions between the filled $5d^{10}$ band and $6(sp)^1$ conduction band. The visible photoluminescence was also found from small gold clusters protected by glutathione monolayers. The optical absorption spectra are highly structured with clear absorption onsets, which shift toward higher energies with reduction of the core size.[72]

Au38(PhC2S)24 was synthesized and transferred into other Au38 nanoparticles by exchanging the monolayer ligands with different thiolate ligands.[73] Voltammetry of the Au38 nanoparticles in CH_2Cl_2 reveals a highest occupied molecular orbital (HOMO)–lowest unoccupied molecular orbital (LUMO) gap energy *ca.* 1.33 eV, in line with an optical absorbance edge. The ligand-exchanged nanoparticle Au38(PEG135S)13(PhC2S)11, where PEG135S is $-SCH_2CH_2OCH_2CH_2OCH_3$, exhibits a broad (1.77–0.89 eV) near-infrared photoluminescence band resolvable into maxima at 902 nm (1.38 eV) and 1,025 nm (1.2 eV).

Chemically prepared gold clusters composed of 28-atom cores show a distinct absorption onset at 1.3 eV corresponding to the opening of an electronic gap within the conduction band (HOMO-LUMO gap). A broad luminescence extending over visible to infrared spectral range was observed.[74] Two bands of the luminescence with maxima around 1.5 eV and 1.15 eV, as shown in Fig. 9.12, indicate a radiative recombination between ground state and two distinctively different excited states. The total quantum yield of the luminescence was measured as about 3.5×10^{-3} with a lifetime in the microsecond range. This unusually high luminescence quantum yield suggested that the two luminescence bands could be assigned to the fluorescence and phosphorescence from excited singlet and triplet states, respectively, in analogy to the photophysical properties of a molecule.[74]

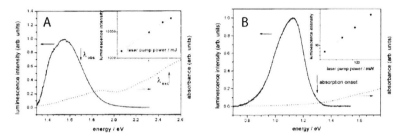

Figure 9.12 Luminescence spectrum (solid line, left axis) of Au$_{28}$(SG)$_{16}$ clusters in D$_2$O (a) excited at 500 nm and recorded on a gated CCD camera; (b) excited at 510 nm and recorded on a camera with a gated CCD. (Reprinted with permission from Ref. 74. Copyright 2002 American Chemical Society.) *Abbreviation*: CCD, charge-coupled device.

Monolayer-protected gold clusters containing 11 atoms were found to exhibit semiconductor electronic characteristics with a bandgap of about 1.8 eV. Photoluminescence in the visible range was observed from the peak position at 840 nm, where surface chemistry plays a vital role in determining the electronic structure.[75] Fluorescence has also been observed from Au nanoclusters created through physical approaches. It was found that Au dimers and trimers embedded in a gas matrix emit light.[76] Fluorescence has also been observed from other noble metal nanostructures fabricated by various methods. Electroluminescence was observed from spatially resolvable gold nanoclusters, formed between electrodes through a controlled electron migration process, in response to pulsed external electric fields from these spatially resolvable individual nanoclusters.[77]

Metal networks have also been created through thermal simultaneous decomposition of gold mercaptides with fluorescence spectra measured.[78]

9.4.3 Size-Dependent Fluorescence

On the nanoscopic length scale, further confinement of the electrons results in discrete energy states with strong intrinsic fluorescence.

These molecule-like properties of these nanoscopic systems can be tuned as their atomic arrangement and electronic structure changes when adding or removing atoms.

Theoretical simulation by density-functional theory found that the energetic and electronic properties of the small gold clusters are strongly dependent on sizes and structures.[79] The size dependence also appears in the optical absorption spectra for Au clusters calculated from first principles, showing transitions with the largest oscillator strength at 3.9, 3.22, and 2.71 eV for Au5, Au8, and Au13 respectively.[80] The theoretical studies suggest that Au nanoclusters in this size region are expected to have size-dependent fluorescence with wavelength covering visible and infrared regions.

Experimental studies have observed size-dependent blue emission and strong fluorescence from Au$_8$ nanoclusters prepared within biocompatible poly(amidoamine) dendrimer hosts. The quantum yield was measured to be 41% in aqueous solution and 52% in methanol, which are quite comparable to aromatics dyes such as those we have used in the silica work. The time dependence of the emission shows that there are two lifetime components. The short lifetime component is 7.5 ns, which is dominant in the emission and likely arises from singlet transitions between low-lying d bands and excited sp bands of gold nanoclusters. The long lifetime component (2.8 ms) may be due to a triplet-singlet intraband transition.[81] Further work from the same group found that these water-soluble, highly fluorescent gold nanoclusters from PAMAM-encapsulated Au behave as multielectron artificial atoms with size-tunable, discrete electronic transitions. The PAMAM-encapsulated gold clusters containing 5, 8, 13, and 23 atoms present discrete absorption and fluorescence that are size dependent from the ultraviolet to the near-infrared region. The well-defined, tunable, discrete excitation and emission suggest that these gold nanoclusters may find utility as energy transfer pairs, a task not so suitable for semiconductor quantum dots because of their extremely broad excitation spectra and larger dimensions.[33] The size-dependent fluorescence properties relate fluorescence to the number of atoms in the clusters, suggesting the possibility of developing fluorescence-based size measurement on an ultrananoscopic length scale. Future research is expected to explore further the relationship between the electronic structure and the size of molecular metal nanoclusters, as well as the size-dependent fluorescence behavior.

9.5 Conclusions

We have described how silica nanoparticles can be sized using extrinsic doping with aromatic fluorophores and how gold nanoparticles possess intrinsic luminescence, thus both offering the potential for sensing and biological imaging. Moreover the size-dependent luminescent properties of both types of nanoparticles offer opportunities for their use as standards.

Acknowledgments

We would like to thank the Engineering and Physical Sciences Research Council, the Scottish Funding Council, and PQ Corporation for financial support.

References

1. Royal Commission on Environmental Pollution, *27th Report Novel Materials in the Environment: The Case of Nanotechnology (TSO)*, 2008.
2. Royal Academy of Engineering & Royal Society, *Nanoscience and Nanotechnologies: Opportunities and Uncertainties*, 2004.
3. Department of Environment, Food and Rural Affairs, *Characterising the Potential Risk Posed by Engineered Nanoparticles—A 2nd Government Research Report (DEFRA)*, 2007.
4. R. Winter, D. W. Hua, P. Thiyagarajan, J. Jonas, *J. Non-Cryst. Solids*, **108**(2), 137–142 (1989).
5. A. van Blaaderen, A. P. M. Kentgens, *J. Non-Cryst. Solids*, **149**, 161–178 (1992).
6. H. Gratz, A. Penzkofer, P. Weidner, *J. Non-Cryst. Solids*, **189**(1–2), 50–54 (1995).
7. J. Antalik, M. Liska, J. Vavra, *Ceram.-Silik.*, **38**(1), 31–34 (1994).
8. I. W. Lenggoro, B. Xia, K. Okuyama, J. F. de la Mora, *Langmuir*, **18**(12), 4584–4591 (2002).
9. C. Y. Wang, W. E. Fu, H. L. Lin, G. S. Peng, "Preliminary study on nanoparticle sizes under the APEC technology cooperative framework," in *Preliminary Study on Nanoparticle Sizes under the APEC Technology Cooperative Framework*, series *Preliminary Study on Nanoparticle Sizes*

under the *APEC Technology Cooperative Framework*, IOP, 2006, 487–495.

10. D. J. S. Birch, C. D. Geddes, *Phys. Rev. E*, **62**(2), 2977–2980 (2000).
11. C. D. Geddes, D. J. S. Birch, *J. Non-Cryst. Solids*, **270**(1–3), 191–204 (2000).
12. C. D. Geddes, J. Karolin, D. J. S. Birch, *J. Phys. Chem. B*, **106**(15), 3835–3841 (2002).
13. J. Karolin, C. D. Geddes, K. Wynne, D. J. S. Birch, *Meas. Sci. Technol.*, **13**(1), 21–27 (2002).
14. G. Weber, *Biochem. J.*, **52**(2), 145–155 (1952).
15. K. Apperson, J. Karolin, R. W. Martin, D. J. S. Birch, *Meas. Sci. Technol.*, **20**(2), 11 (2009).
16. A. Mooradian, *Phys. Rev. Lett.*, **22**(5), 185 (1969).
17. G. T. Boyd, Z. H. Yu, Y. R. Shen, *Phys. Rev. B*, **33**(12), 7923–7936 (1986).
18. X. M. Qian, X. H. Peng, D. O. Ansari, Q. Yin-Goen, G. Z. Chen, D. M. Shin, L. Yang, A. N. Young, M. D. Wang, S. M. Nie, *Nat. Biotechnol.*, **26**(1), 83–90 (2008).
19. C. J. Orendorff, L. Gearheart, N. R. Jana, C. J. Murphy, *Phys. Chem. Chem. Phys.*, **8**(1), 165–170 (2006).
20. X. G. Hu, W. L. Cheng, T. Wang, Y. L. Wang, E. K. Wang, S. J. Dong, *J. Phys. Chem. B*, **109**(41), 19385–19389 (2005).
21. M. B. Mohamed, V. Volkov, S. Link, M. A. El-Sayed, *Chem. Phys. Lett.*, **317**(6), 517–523 (2000).
22. K. Imura, T. Nagahara, H. Okamoto, *J. Phys. Chem. B*, **109**(27), 13214–13220 (2005).
23. N. R. Jana, L. Gearheart, S. O. Obare, C. J. Murphy, *Langmuir*, **18**(3), 922–927 (2002).
24. N. J. Durr, T. Larson, D. K. Smith, B. A. Korgel, K. Sokolov, A. Ben-Yakar, *Nano Lett.*, **7**(4), 941–945 (2007).
25. A. Brioude, X. C. Jiang, M. P. Pileni, *J. Phys. Chem. B*, **109**(27), 13138–13142 (2005).
26. C. J. Murphy, T. K. San, A. M. Gole, C. J. Orendorff, J. X. Gao, L. Gou, S. E. Hunyadi, T. Li, *J. Phys. Chem. B*, **109**(29), 13857–13870 (2005).
27. T. B. Huff, L. Tong, Y. Zhao, M. N. Hansen, J. X. Cheng, A. Wei, "Hyperthermic effects of gold nanorods on tumor cells," in *Hyperthermic Effects of Gold Nanorods on Tumor Cells*, series *Hyperthermic Effects of Gold Nanorods on Tumor Cells*, Future Medicine, 2006, 125–132.

28. C. Burda, X. B. Chen, R. Narayanan, M. A. El-Sayed, *Chem. Rev.,* **105**(4), 1025–1102 (2005).
29. J. Perez-Juste, I. Pastoriza-Santos, L. M. Liz-Marzan, P. Mulvaney, "Gold nanorods: synthesis, characterization and applications," in *Gold Nanorods: Synthesis, Characterization and Applications*, series *Gold Nanorods: Synthesis, Characterization and Applications*, Elsevier Science Sa, 2004, 1870–1901.
30. M. Grzelczak, J. Perez-Juste, P. Mulvaney, L. M. Liz-Marzan, *Chem. Soc. Rev.,* **37**(9), 1783–1791 (2008).
31. S. Link, M. A. Ei-Sayed, *Annu. Rev. Phys. Chem.,* **54**, 331–366 (2003).
32. C. C. Huang, Z. Yang, K. H. Lee, H. T. Chang, *Angew. Chem., Int. Ed.,* **46**(36), 6824–6828 (2007).
33. J. Zheng, C. W. Zhang, R. M. Dickson, *Phys. Rev. Lett.,* **93**(7), 4 (2004).
34. N. Schaeffer, B. Tan, C. Dickinson, M. J. Rosseinsky, A. Laromaine, D. W. McComb, M. M. Stevens, Y. Q. Wang, L. Petit, C. Barentin, D. G. Spiller, A. I. Cooper, R. Levy, *Chem. Commun.,* (34), 3986–3988 (2008).
35. S. W. Chen, R. S. Ingram, M. J. Hostetler, J. J. Pietron, R. W. Murray, T. G. Schaaff, J. T. Khoury, M. M. Alvarez, R. L. Whetten, *Science,* **280**(5372), 2098–2101 (1998).
36. P. Apell, R. Monreal, S. Lundqvist, *Phys. Scr.,* **38**(2), 174–179 (1988).
37. Z. Y. Lin, R. P. F. Kanters, D. M. P. Mingos, *Inorg. Chem.,* **30**(1), 91–95 (1991).
38. A. P. Alivisatos, *Science,* **271**(5251), 933–937 (1996).
39. C. Kirchner, T. Liedl, S. Kudera, T. Pellegrino, A. M. Javier, H. E. Gaub, S. Stolzle, N. Fertig, W. J. Parak, *Nano Lett.,* **5**(2), 331–338 (2005).
40. J. R. Lakowicz, *Principles of Fluorescence Spectroscopy*, New York: Springer-Verlag, 2006.
41. R. F. Steiner, "Fluorescence anisotropy: theory and applications," in J. R. Lakowicz, ed., *Topics in Fluorescence Spectroscopy: Principles*, Vol. 2, New York: Plenum Press, 1991, 1–51.
42. M. Eftink, *Fluorescence Techniques for Studying Protein Structure*, New York: John Wiley & Sons, 1991.
43. H. H. Paradies, *Colloids Surf., A,* **74**(1), 57–69 (1993).
44. Y. N. Xu, P. L. Hiew, M. A. Klippenstein, Y. Koga, *Clays Clay Miner.,* **44**(2), 197–213 (1996).
45. D. Tleugabulova, A. M. Duft, M. A. Brook, J. D. Brennan, *Langmuir,* **20**(1), 101–108 (2004).

46. D. Tleugabulova, A. M. Duft, Z. Zhang, Y. Chen, M. A. Brook, J. D. Brennan, *Langmuir*, **20**(14), 5924–5932 (2004).
47. D. J. S. Birch, R. E. Imhof, "Time-domain fluorescence spectroscopy using time-correlated single-photon counting," in J. R. Lakowicz, ed., *Topics in Fluorescence Spectroscopy: Techniques*, Vol. 1, New York: Plenum Press, 1991, 1–95.
48. D. V. O'Connor, D. Phillips, *Time-Correlated Single Photon Counting*, London, Orlando: Academic Press, 1984.
49. D. J. S. Birch, R. E. Imhof, *Rev. Sci. Instrum.*, **52**(8), 1206–1212 (1981).
50. W. J. O'Hagan, M. McKenna, D. C. Sherrington, O. J. Rolinski, D. J. S. Birch, *Meas. Sci. Technol.*, **13**(1), 84–91 (2002).
51. C. D. McGuinness, K. Sagoo, D. McLoskey, D. J. S. Birch, *Appl. Phys. Lett.*, **86**(26), 3 (2005).
52. C. D. McGuinness, K. Sagoo, D. McLoskey, D. J. S. Birch, *Meas. Sci. Technol.*, **15**(11), L19–L22 (2004).
53. C. D. McGuinness, A. M. Macmillan, K. Sagoo, D. McLoskey, D. J. S. Birch, *Appl. Phys. Lett.*, **89**(6), 3 (2006).
54. DuPont, *Ludox Colloidal Silica: Properties, Uses, Storage and Handling: Data Sheets*, 1987.
55. C. D. Geddes, K. Apperson, D. J. S. Birch, *Dyes Pigm.*, **44**(2), 69–74 (2000).
56. C. D. Geddes, K. Apperson, J. Karolin, D. J. S. Birch, *Anal. Biochem.*, **293**(1), 60–66 (2001).
57. M. Brust, M. Walker, D. Bethell, D. J. Schiffrin, R. Whyman, *J. Chem. Soc., Chem. Commun.*, (7), 801–802 (1994).
58. T. Huang, R. W. Murray, *J. Phys. Chem. B*, **105**(50), 12498–12502 (2001).
59. C. A. J. Lin, T. Y. Yang, C. H. Lee, S. H. Huang, R. A. Sperling, M. Zanella, J. K. Li, J. L. Shen, H. H. Wang, H. I. Yeh, W. J. Parak, W. H. Chang, *ACS Nano*, **3**(2), 395–401 (2009).
60. Y. Chen, J. A. Preece, R. E. Palmer, "Processing and characterization of gold nanoparticles for use in plasmon probe spectroscopy and microscopy of biosystems," in *Processing and Characterization of Gold Nanoparticles for Use in Plasmon Probe Spectroscopy and Microscopy of Biosystems*, series *Processing and Characterization of Gold Nanoparticles for Use in Plasmon Probe Spectroscopy and Microscopy of Biosystems*, Blackwell, 2007, 201–206.
61. C. Leung, C. Xirouchaki, N. Berovic, R. E. Palmer, *Adv. Mater.*, **16**(3), 223–+ (2004).

62. R. E. Palmer, S. Pratontep, H. G. Boyen, *Nat. Mater.*, **2**(7), 443–448 (2003).
63. S. Pratontep, P. Preece, C. Xirouchaki, R. E. Palmer, C. F. Sanz-Navarro, S. D. Kenny, R. Smith, *Phys. Rev. Lett.*, **90**(5), 4 (2003).
64. M. Di Vece, N. P. Young, Z. Y. Li, Y. Chen, R. E. Palmer, *Small*, **2**(11), 1270–1272 (2006).
65. B. W. Vandermeer, K. H. Cheng, S. Y. Chen, *Biophys. J.*, **58**(6), 1517–1526 (1990).
66. S. Y. Chen, K. H. Cheng, B. W. Vandermeer, J. M. Beechem, *Biophys. J.*, **58**(6), 1527–1537 (1990).
67. M. M. G. Krishna, R. Das, N. Periasamy, R. Nityananda, *J. Chem. Phys.*, **112**(19), 8502–8514 (2000).
68. C. Hellriegel, J. Kirstein, C. Brauchle, V. Latour, T. Pigot, R. Olivier, S. Lacombe, R. Brown, V. Guieu, C. Payrastre, A. Izquierdo, P. Mocho, *J. Phys. Chem. B*, **108**(38), 14699–14709 (2004).
69. T. A. Smith, M. Irwanto, D. J. Haines, K. P. Ghiggino, D. P. Millar, *Colloid Polym. Sci.*, **276**(11), 1032–1037 (1998).
70. J. P. Wilcoxon, J. E. Martin, F. Parsapour, B. Wiedenman, D. F. Kelley, *J. Chem. Phys.*, **108**(21), 9137–9143 (1998).
71. T. P. Bigioni, R. L. Whetten, O. Dag, *J. Phys. Chem. B*, **104**(30), 6983–6986 (2000).
72. Y. Negishi, Y. Takasugi, S. Sato, H. Yao, K. Kimura, T. Tsukuda, *J. Am. Chem. Soc.*, **126**(21), 6518–6519 (2004).
73. D. Lee, R. L. Donkers, G. L. Wang, A. S. Harper, R. W. Murray, *J. Am. Chem. Soc.*, **126**(19), 6193–6199 (2004).
74. S. Link, A. Beeby, S. FitzGerald, M. A. El-Sayed, T. G. Schaaff, R. L. Whetten, *J. Phys. Chem. B*, **106**(13), 3410–3415 (2002).
75. Y. Y. Yang, S. W. Chen, *Nano Lett.*, **3**(1), 75–79 (2003).
76. S. Fedrigo, W. Harbich, J. Buttet, *J. Chem. Phys.*, **99**(8), 5712–5717 (1993).
77. J. I. Gonzalez, T. Vosch, R. M. Dickson, *Phys. Rev. B*, **74**(23), 8 (2006).
78. A. Longo, G. P. Pepe, G. Carotenuto, A. Ruotolo, S. De Nicola, V. I. Belotelov, A. K. Zvezdin, *Nanotechnology*, **18**(36), 5 (2007).
79. X. B. Li, H. Y. Wang, X. D. Yang, Z. H. Zhu, Y. J. Tang, *J. Chem. Phys.*, **126**(8), 8 (2007).
80. J. C. Idrobo, W. Walkosz, S. F. Yip, S. Ogut, J. Wang, J. Jellinek, *Phys. Rev. B*, **76**(20), 12 (2007).

81. J. Zheng, J. T. Petty, R. M. Dickson, *J. Am. Chem. Soc.,* **125**(26), 7780–7781 (2003).

Chapter 10

Nanometer-Scale Measurements Using FRET and FLIM Microscopy

Margarida Barroso,[a] **Yuansheng Sun,**[b,c] **Horst Wallrabe,**[c] **and Ammasi Periasamy**[b,c]

[a]*Albany Medical College, Center for Cardiovascular Sciences, Albany, NY 12208, USA*
[b]*W. M. Keck Center for Cellular Imaging, University of Virginia, Charlottesville, VA 22904, USA*
[c]*Department of Biology, University of Virginia, Charlottesville, VA 22904, USA*
barrosm@mail.amc.edu, ap3t@virginia.edu

Förster resonance energy transfer (FRET) microscopy is a valuable methodology to infer distances of 1–10 nm in living specimens, greatly surpassing the resolution of light at ~200 nm and doing so with virtually any light microscopy system. The availability of different types of fluorophores and imaging methodologies allows for a tailored approach to specific research goals. This chapter describes the use of the three main fluorophore categories: organic dyes, conjugated to a protein of interest; GFP-type fusion proteins expressed in living specimens; and semiconductor nanocrystals, widely known as quantum dots (QDs), conjugated to target proteins. Each category presents its own challenges to the researcher, and we

---

Luminescence: The Instrumental Key to the Future of Nanotechnology
Edited by Adam M. Gilmore
Copyright © 2014 Pan Stanford Publishing Pte. Ltd.
ISBN 978-981-4241-95-3 (Hardcover), 978-981-4267-72-4 (eBook)
www.panstanford.com

stress therefore the importance of the judicious choice of a FRET pair by considering its spectral and other properties. FRET microscopy methodologies cover intensity-based emission, filter-based or spectral-based confocal imaging, and fluorescence lifetime imaging (FLIM-FRET)-based approaches. We present actual research data with extensive background information for each of the fluorophore choices and imaging methodologies, with emphasis on QDs.

10.1 Introduction

Förster resonance energy transfer (FRET) is a valuable methodology to conduct measurements at the nanometer scale in the life sciences. FRET imaging assays the distance between fluorophore-labeled molecules within 1–10 nm. Live and fixed cells—grown in suspension or attached to culture plates—microorganisms, and tissue specimens, as well as purified molecules and other types of biological samples have been used in FRET imaging. A wide variety of FRET-based biological experiments have been developed to determine proximities within macromolecular assemblies, such as signaling complexes and receptor-ligand clusters in cellular membrane compartments.[1-5] Depending on the research objectives, a wide range of different microscopy systems can be used for FRET: widefield, confocal, spectral, fluorescence lifetime imaging (FLIM), and anisotropy. The increasing availability of pH- and photo-stable organic dyes with high quantum yields, the ever-growing family of green fluorescent protein (GFP)-type fluorescent fusion proteins, and latterly the development of semiconductor nanocrystals, widely known as quantum dots (QDs), have accelerated the use and widened the utility of FRET microscopy. The recent introduction of spectral imaging to measure FRET events has increased even more the choice of fluorophores and FRET pairs, opening the door to the development of multiplex FRET imaging. Nevertheless, the selection of a fluorophore FRET pair remains a careful consideration that will be influenced by the research goals and availability of suitable instruments; specimen and imaging parameter optimization steps are also important to get the full benefit of FRET microscopy. QDs in particular are still at an early development stage for live-specimen microscopy, and their use should be considered in light of both their advantages and pitfalls, as described in this chapter. Therefore, FRET microscopy is a particularly useful technique to take advantage of the ever-increasing opportunities to conduct nanometer-scale measurements in live biological specimens. Since there are many excellent publications that cover FRET microscopy in detail[2,6–9], we will focus here on the use of different types of fluorophore FRET pairs, such as, organic dyes, GFP-like fluorescent proteins (FPs), and QDs in biological applications.

10.2 FRET Microscopy

In the late 1940s, Förster proposed the theory of FRET, which described how energy could be nonradiatively transferred directly from a fluorophore in the excited state (the donor, D) to a nonidentical neighbor fluorophore (the acceptor, A) (10–12). The energy from a donor molecule can be transferred directly to an acceptor molecule under the following four conditions: (i) when the emission spectrum of the donor overlaps significantly with the absorption spectrum of the acceptor, (ii) when the two fluorophores are within ~1 to ~10 nm of each other, (iii) when the donor emission and acceptor absorption dipole moments are in favorable mutual orientations, and (iv) when the donor has a reasonably high quantum yield. For a FRET system composed of a single donor and a single acceptor, the efficiency of the energy transfer from the donor to the acceptor (E) is described as the ratio of energy transfer rate (k_T) to the sum of all deactivation rates of the excited state of the donor ($k_T + k_D$), where k_D is the sum of its deactivation rates other than FRET (Eq. 10.1):

$$E = k_T/(k_T + k_D) \tag{10.1}$$

E varies inversely with the sixth power of the distance separating the donor and acceptor molecules (r), as described by Eq. 10.2 (12;13):

$$E = R_0^6/(R_0^6 + r^6) \leftrightarrow r = R_0[(1/E) - 1]^{1/6} \tag{10.2}$$

R_0 is the Förster distance for the FRET pair, where E is equal to 50%, and it is described by Eq. 10.3:

$$R_0 = 0.211 \cdot \{k^2 \cdot n^{-4} \cdot QY_D \cdot J\}^{1/6}, J = \varepsilon_A \frac{\int_0^\infty f_D(\lambda)f_A(\lambda)\lambda^4 d\lambda}{\int_0^\infty f_D(\lambda)d\lambda} \tag{10.3}$$

where κ^2 ranging from 0 to 4 is the dipole orientation factor and usually assumed to be 2/3 (the average value integrated over all possible angels), n is the refractive index of the medium, QY_D is the donor quantum yield, and J expresses the degree of spectral overlap between the donor emission and the acceptor absorption (14). Moreover, ε_A is the extinction coefficient of the acceptor at its peak absorption wavelength, λ is the wavelength, $f_D(\lambda)$ is the donor emission spectrum, and $f_A(\lambda)$ is the normalized acceptor absorption spectrum (peak absorbance = 1).

Although the distance over which FRET can occur is limited to the 1–10 nm range, in practice robust energy transfer can only be detected when the distance between donor- and acceptor-labeled molecules lies within the 2–8 nm range, depending on the microscopy system's capabilities and on the donor–acceptor R_0 values. This FRET-based inferred proximity between two labeled cellular components considerably surpasses the resolution of general light microscopy, which at best can resolve distances of ~200 nm. Since in the majority of biological applications, donor and acceptor molecules are not separated by a fixed distance at a 1:1 ratio, nanoscale FRET measurements can only provide relative distances between donor- and acceptor-labeled molecules generated from averages of spatial distribution maps. Nevertheless, relative distances are extremely valuable information when following proximity changes within molecular assemblies, since the distance between interacting proteins falls within the 2–8 nm FRET distance range.

Given a suitable donor–acceptor fluorophore pair, FRET can be executed on different microscopy systems using distinct methodological approaches with the goal of determining the relative measurement of nanometer-scale distances between donor- and acceptor-labeled molecules. Many FRET microscopy techniques have been developed and are generally categorized into intensity-based and lifetime-based methods.

Steady-state intensity-based measurements performed under identical imaging conditions are typically used to measure the "apparent" energy transfer efficiency ($E\%$) as shown in Eq. 10.4—called "apparent," because most of intensity-based FRET microscopy techniques cannot differentiate between the donor molecules that participate in FRET events (FRET donors) and those that do not (non-FRET donors). In Eq. 10.4, qD represents the average intensities of the quenched donor in the presence of the acceptor (i.e., samples containing donor- and acceptor-labeled molecules) and D is the average intensities of the unquenched donor in the absence of the acceptor (i.e., samples containing only donor-labeled molecules).

$$E\% = 100 \times (1 - qD/D) \quad (10.4)$$

Two main intensity-based approaches are routinely applied in FRET studies using most of the standard microscope systems available, such as wide-field, confocal, spectral, or total internal reflection fluorescence (TIRF). One is acceptor photobleaching (AP) or donor de-quenching FRET microscopy, where the $E\%$ (Eq. 10.4) is directly estimated by measuring the donor intensities before (quenched, qD) and after (unquenched, D) photobleaching the acceptor in samples containing donor- and acceptor-labeled molecules (15;16). This method is straightforward and can be used to collect measurements of FRET events in live cells under certain conditions, including reduced diffusion of the donor and acceptor molecules, quick and efficient bleaching of the acceptor molecules, and reduced acceptor–donor photoconversion, which has been shown to occur even in the well-known CFP-EYFP FRET pair (17). In the second intensity-based FRET methodology, including filter-based and spectral FRET, the unquenched donor signal (D) cannot be directly measured from samples containing donor- and acceptor-labeled molecules; instead it is usually determined by adding the energy transfer levels estimated from the acceptor intensities upon donor excitation (acceptor sensitized emission) to the measured quenched donor signal (qD) to calculate $E\%$ (Eq. 10.4). However, accurate quantification of the energy transfer levels requires the removal of the spectral bleedthrough (SBT) contaminations, which are a direct result of the spectral overlap required for FRET events. Algorithms have been developed for various microscopy techniques to identify and remove the SBT contaminations, allowing the accurate measurement of the $E\%$. In filter-based FRET microscopy, the donor, acceptor, and FRET signals are measured in channels separated using band-pass filters (18–25). In contrast, spectral FRET microscopy uses a spectral detector to measure signals acquired over a continuous emission spectrum, followed by the use of spectral linear unmixing to separate the signals emitted from the donor and acceptor (26–29). Both filter-based and spectral FRET microscopy can be used for time-lapse live-cell FRET imaging involving relatively complex postcollection image processing. The spectral FRET approach has the advantage of allowing more flexibility in the choice of fluorophores, such as QDs (30). Moreover, spectral imaging can also be used to develop multiplexed FRET approaches to follow multiple signaling events in live cells by using more than one donor–acceptor pair (31).

In lifetime-based FRET microscopy, FRET events can be identified by measuring the reduction in the donor lifetime that results from quenching, and the energy transfer efficiency (E) is estimated from the donor lifetimes determined in the absence (τ_D – unquenched lifetime) and the presence (τ_{DA} – quenched lifetime) of the acceptor based on Eq. 10.5 (14):

$$E = 1 - (\tau_{DA}/\tau_D) \tag{10.5}$$

The fluorescence lifetime refers to the average time a molecule stays in its excited state before emitting a photon and is an intrinsic property of a fluorophore; it can be measured in different ways, and the experimental techniques are generally divided into time and frequency domain methods (32). In the time domain, a fluorophore is usually excited with a pulsed light source and its decay profile is directly measured at different time windows using high-speed detectors plus fast synchronization electronics; the fluorescence lifetime of the fluorophore is estimated from analyzing the recorded decay profile. In the frequency domain, the excitation light is modulated at different frequencies, which are chosen upon the lifetime scale of a fluorophore to be measured (megahertz for nanosecond decays). At each frequency, the phase shift and amplitude attenuation (modulation) of the fluorescence emission relative to the phase and amplitude of the excitation light are measured. Analyzing the recorded phase shifts or amplitude attenuations or both can estimate the fluorescence lifetime of the fluorophore. In FLIM-FRET microscopy, SBT is not an issue since only the donor signals are monitored, and the fluorescence lifetime is not influenced by probe concentration, excitation light intensity, or light scattering; these facts make FLIM-FRET microscopy an accurate method to measure nanoscale distance distributions. However, fluorescent lifetimes are sensitive to environmental conditions. Thus, in live cells, for example, fluorophores may be exposed to different pH values and therefore may show multiple lifetime values, increasing the complexity of FRET measurements using FLIM. Changes in the fluorescence lifetime of fluorescence proteins have also been detected (33) and should be considered when using FLIM to image FRET between fluorescence proteins in live cells.

10.3 Choosing FRET Pairs

Considering only the fluorescence properties to choose fluorophores for FRET is not a trivial decision. Quantum yields, extinction coefficients, degree of donor–acceptor spectral overlap, and SBT levels, all play a role to be considered and to be optimized. Nevertheless, the final selection of the right donor–acceptor pair should also include the actual biological question to be addressed, the type of biological specimen to be imaged, and the instrument available to measure FRET levels. After selection, optimization steps should be performed to attain sufficient intensity/lifetime levels of donor and acceptor molecules that should result in E levels above background/negative controls. This would entail experimenting with different cell lines, fixed or live specimens, fluorescence intensity levels of donor- and acceptor-labeled proteins due to different expression or labeling levels, different imaging conditions such as laser power, gain, and image collection speed, and other variables. Here we will review FRET methodological approaches from the point of view of different donor–acceptor pairs instead of the more standard approach that considers different microscopy-based approaches.

Fluorophores broadly fall into three categories: organic dyes, FPs, and inorganic semiconductor nanocrystal QDs. Organic dyes are small and well established, and the more recent Alexa Fluor (AF) and Cy families of fluorophores are quite photostable at a wide range of pH and temperature. FPs are normally overexpressed by the cells, and new, brighter varieties with higher quantum yields are added constantly. The newer inorganic semiconductor nanocrystal QDs have been used mostly for in vitro FRET biological assays. QDs are still in the early application phases of biomedical FRET imaging but hold considerable promise and are described here in greater detail. We will focus on in vivo FRET applications using organic dye pairs (e.g., AF488–AF555), new FPs variants (e.g., mTFP–KO2), and QD–organic dye pairs (e.g., QD566–AF568 and QD580–AF594).

An important FRET property that should considered before the donor–acceptor pair selection is their Förster distances (R_0), which can be estimated from the photophysical properties of the donor and acceptor dyes based on Eq. 10.3. Different FRET pairs have different R_0 with the majority being ~5–6 nm (30;34;35). However, it is difficult to obtain the accurate R_0 value in some cases. For example, different values of quantum yield of the particular dye–protein conjugate may alter the R_0 value (36), which is normally estimated based on the quantum yield of the dye itself. Moreover, the dipole orientation factor assumed to be 2/3 may not reflect the truth (14), although it has been recently demonstrated that the dipole–dipole interactions mechanism used in R_0 calculations accurately describes the energy transfer process in QD–dye FRET pairs and that the point dipole approximation is correct for QD donors in FRET reactions (35). As mentioned above, the practical experimental range of FRET measurements is ~2–8 nm, depending on the microscopy system's capabilities and on the R_0 values of the selected FRET pairs. Therefore, the choice of a FRET pair and its R_0 can play a role in extending the distance range that will be assayed in a particular FRET-based assay. When choosing a FRET pair, one needs to keep in mind that a larger R_0 will increase the likelihood of a FRET event. Increasing R_0 may be achieved by using a donor with a higher quantum yield, an acceptor with a larger extinction coefficient, and a pair with a larger spectral overlap. Here, we have used three different types of donor–acceptor pairs with increasing R_0 values: mTFP–mKO2 with R_0 = 5.29 nm (37); QD566–AF568 and QD580–AF594 with R_0 = 6.0 nm (30); and AF488–AF555 with R_0 = 7.0 nm (http://www.invitrogen.com/site/us/en/home/References/Molecular-Probes-The-Handbook/tables/R0-values-for-some-Alexa-Fluor-dyes.html) (Fig. 10.1). These donor–acceptor FRET pairs can assay different ranges of donor–acceptor distances.

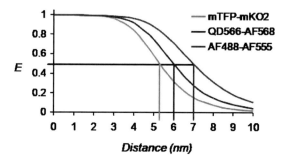

Figure 10.1 Relationship between FRET efficiency and the distance separating the donor and the acceptor.

Therefore, the careful selection of a donor–acceptor pair should match its R_0 value to the nanoscale distances occurring in the FRET-based assay. Importantly, changes in the donor–acceptor pair may lead to an increased range of distances measured in the FRET-based assay.

10.4 Organic Dye Donor–Acceptor FRET Pair: AF488–AF555

Organic dyes, such as fluorescein isothiocyanate (FITC), Rhodamine, and Alexa- and Cy-dye families, may be introduced into biological specimens via their conjugation with target molecules that can cross the plasma membrane or that can enter cells either through fluid-phase or receptor-mediated endocytosis (5;25). Alternatively antibodies labeled with organic fluorophores can be used in FRET reactions in fixed cells and tissues using immunofluorescence-based approaches (38;39). Here we will focus on in vivo FRET applications using the organic dye pair AF488–AF555, one of the most well-known organic dye FRET pairs (5;25;40).

10.4.1 Filter-Based FRET Microscopy

The most commonly used FRET methodology to assay organic dye donor–acceptor pairs, such as AF488–AF555, is the filter-based FRET method since it can be applied in most optical microscopy imaging systems and allows the measurement of discrete effects of pharmacological or other interventions during the course of experimentation in two or three dimensions. However, the high SBT levels between the FRET donor and acceptor channels, which are a direct result of the spectral overlap required for FRET events (Fig. 10.2i), is an obstacle to this method, needing to be overcome by appropriate corrections. These SBT contaminations of the FRET signal result from the donor emission that is detected in the FRET channel, known as the donor SBT (DSBT; Fig. 10.2c), and the acceptor SBT (ASBT; Fig. 10.2f), which is caused by the direct excitation of the acceptor at the donor excitation wavelength. Figure 10.2 shows an illustration of both SBTs using the donor–acceptor AF488–AF555 pair as an example. Both SBTs must be removed for accurate FRET measurements. Commonly, three types of specimens are used: cells containing AF488 alone (Fig. 10.2a–c) or AF555 alone (Fig. 10.2d,e) as a reference specimen that will be used to remove SBT from cells containing both AF488 and AF555 (Fig. 10.2g–i). Images are collected from these types of cells using donor and acceptor excitation wavelengths and specific donor and acceptor emission band-pass filters. Three types of imaging conditions (donor, acceptor, and FRET channel settings) are used. The donor channel (Fig. 10.2a,d,g) includes the use of a donor excitation wavelength (Dex; 488 nm argon laser) and a collection of donor emission spectra (Dem). The acceptor channel (Fig. 10.2b,e,h) includes the use of an acceptor excitation wavelength (Aex; 543 Green HeNe laser) and a collection of acceptor emission spectra (Aem). The FRET channel (Fig. 10.2c,f,i) includes the use of a donor excitation wavelength (Dex) and a collection of acceptor emission spectra (Aem). In Fig. 10.2, the green-shaded rectangle represents the donor emission filter (BP500–530) and the red-shaded rectangle the acceptor emission filter (590LP).

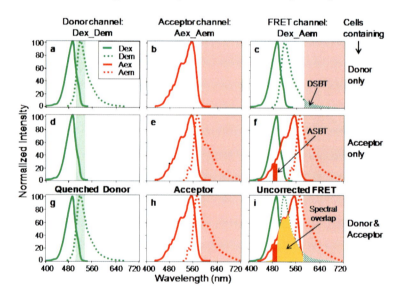

Figure 10.2 Excitation and emission spectra of donor–acceptor AF488–AF555 FRET pair.

Then, those images are processed by currently available algorithm-based software, such as PFRET (23;41), riFRET (42), PixFRET (43), and FRET analyzer (44), or other algorithms such as the FRET application of the Zeiss Physiology AxioVision module and similar modules in Leica, Olympus, and Nikon systems. The processing of FRET images by the SBT removal algorithms results in a corrected FRET signal, representing the actual energy transfer levels, which can be used to calculate $E\%$. Some of the referenced methods were compared by Berney et al. (45). We use our proprietary PFRET software—processed FRET (PFRET) microscopy (23;41), which requires donor and acceptor reference specimens containing only donor- or acceptor-labeled molecules, respectively, at identical imaging conditions, as described above. The PFRET algorithm produces processed (or corrected) FRET images (PFRET), which represent the actual energy transfer levels, in a pixel-by-pixel manner, depending on the uncorrected FRET image (uFRET; Dex_Aem) as shown in Eq. 10.6. In Eq. 10.4, $E\%$ is an expression of the energy transfer as a percentage of the unquenched donor (D), which can also be calculated as described in Eq. 10.7, where qD represents the quenched donor (Dex_Dem) (5;23;25;36;41;46).

$$\text{PFRET} = \text{uFRET} - \text{DSBT} - \text{ASBT} \tag{10.6}$$

$$D = qD + \text{PFRET} \times \gamma \tag{10.7}$$

The factor γ in Eq. 10.7, which is a coefficient that is used for relating intensities measured in the donor channel to the corresponding intensities collected in the acceptor channel, plays a crucial role in recording precise $E\%$ values and distances between fluorophores. Determination of the γ coefficient involves the quantum yields of the donor and acceptor and also the detector gains and spectral sensitivities (23). Since the excitation efficiencies, quantum yields of the fluorophore molecules, and the detection efficiencies remain constant throughout the experiments, the γ factor does not affect the relative comparison of $E\%$ measurements used in FRET quantitative analysis (5;36;40). Therefore, for simplicity, in our FRET analysis we used $\gamma = 1$, as described previously (5;25;36;40;46). Nevertheless, the filter-based FRET assays carried out with different imaging parameters will by definition have distinct γ factors. Other practical considerations for filter-based FRET microscopy are potential autofluorescence, the rare "back-bleedthrough" (the acceptor wavelength absorbed by the donor), and the low signal-to noise ratio, to be judged, corrected, or ameliorated during the optimization phase (23).

Recently, a quantitative FRET imaging assay has been demonstrated in epithelial cells (30) by taking advantage of a well-recognized biological model system, that is, the Transferrin-Receptor (TFR)-Transferrin (Tfn), which is required for iron uptake into the cells (47;48). AF488 (donor) or AF555 (acceptor) are conjugated to iron-bound Tfn molecules to track the intracellular endocytic trafficking of TFR–Tfn complexes (Fig. 10.3). Receptor–ligand complexes are endocytosed via clathrin-coated pits (CCPs) under specific experimental conditions and can be tracked spatially and temporally in live cells using confocal microscopy. Each apo-Tfn molecule binds two Fe^{3+} (asterisk) in the extracellular media to become holo-Tfn (Tfn). Two molecules of Tfn bind to each TFR homodimer. In Fig. 10.3, one Tfn–AF488 (donor, green circle) and one Tfn–AF555 (acceptor, red circle) bind a TFR homodimer at the plasma membrane. Other possibilities include TFR homodimers bound to two Tfn–AF488 or two Tfn–AF555 molecules, which will affect FRET efficiency. This assay evaluates the ability of Tfn–AF488 donor conjugates to transfer energy to Tfn–AF555 acceptor molecules upon binding to TFR homodimer and subsequent internalization into the endocytic pathway (Fig. 10.3) (5;25;40). Subsequent formation of higher-order clusters of TFR–Tfn during endocytic trafficking cannot be excluded (40). Images from cells cointernalized with Tfn–AF488 and Tfn–AF555 were collected using a Zeiss LSM510 Meta microscope and processed using the PFRET algorithm (23) to remove DSBT and ASBT and generate corrected FRET (PFRET) images. In Fig. 10.4, pseudocolor images depict the pixel-by-pixel distribution of Acceptor, qDonor, uFRET, PFRET, which represents the actual energy transfer levels, and $E\%$ levels. A significant level of SBT is removed from the uFRET image upon PFRET algorithm implementation to generate PFRET images. The PFRET algorithm provides a tool to adequately remove SBT from uFRET, taking into consideration intracellular variability of fluorophore expression levels. Furthermore, $E\%$ images clearly confirm the existence of energy transfer from Tfn–AF488 (donor) to Tfn–AF555 (acceptor) in intracellular punctate endocytic structures. qD, A, and $E\%$ images provide important information about the morphology pattern of these parameters; as expected, the typical endocytic pattern of irregular and punctate structures is clearly detected across all images (Fig. 10.4).

Organic Dye Donor–Acceptor FRET Pair | 271

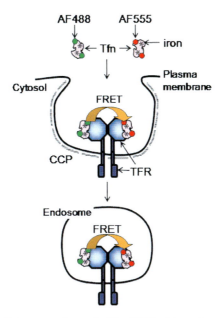

Figure 10.3 Cellular uptake of Tfn-AF88 (donor) and Tfn-AF555 (acceptor).

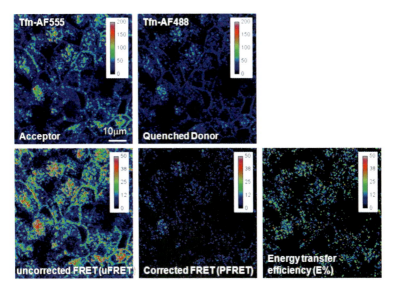

Figure 10.4 FRET processing and analysis of TFR–Tfn complexes in intracellular endosomal compartments.

In summary, the main problem with AF488–AF555 as a FRET pair relates to the presence of significant SBT that has to be removed using reference images containing AF488 or AF555 alone, followed by processing using an SBT correction algorithm. On average, PFRET-based SBT correction in the Tfn–AF488 and Tfn–AF555 FRET-based cellular assay reaches up to ~70%, of which 70% corresponds to ASBT and the remaining 30% to DSBT (data not shown). As shown in Fig. 10.10f, subjecting Tfn–AF488 and Tfn–AF555 cellular images to the PFRET SBT correction algorithm results in an average $E\%$ ≈ 19%, whereas processing of those images by the PFRET ASBT-only correction algorithm results in an average $E\%$ ≈ 28.9% and by the PFRET DSBT-only algorithm results in an average $E\%$ ≈ 38.2%. These results confirm the definite requirement for SBT correction when using this FRET pair since DSBT- or ASBT-only corrections lead to significant differences in $E\%$ values.

One of the main advantages of the AF488–AF555 FRET pair has to do with the availability of the donor (488 nm) and acceptor (543 nm) excitation wavelengths as well as donor and acceptor emission filters in the majority of the optical instruments used for FRET imaging. As shown in Fig. 10.1, this FRET pair can probe for a wide range of nanoscale distances, making it one of the best pairs for FRET imaging and analysis in cellular systems.

10.5 FP Donor–Acceptor FRET Pair: mTFP-mKO2

Target proteins fused to donor or acceptor FPs, such as CFP-YFP, are normally expressed in a wide variety of cultured cells to measure protein–protein interactions using intermolecular FRET (7). Alternatively, FPs may be used to construct biosensors, where donor and acceptor FPs are fused to the same target protein to measure changes in signaling events such as calcium binding (49;50) and activation of guanosine-5′-triphosphate (GTP)-binding proteins (51;52) or kinases (53) using intramolecular FRET. It is important to mention that FRET between FPs can occur at a reduced level in fixed cells (54) and, thus, should be performed predominantly in live cells. The large number of FP varieties, beyond the standard CFP/GFP/YFP/RFP, such as Cerulean, Teal, Venus, mOrange, mCherry, mTFP, and mKO2, plus other "fruit-named" fusion proteins, have provided an enormous flexibility in the development of FRET-based assays

(55;56) for the study of biological processes in live cells. mTFP (donor) and mKO2 (acceptor) are new variants of FPs that show a high quantum yield (0.85 and 0.62, respectively) and nearly identical brightness. The spectra of mTFP and mKO2 FPs are available from Allele Biotechnology and MBL International, respectively. In Fig. 10.5, the orange area indicates the spectral overlap between the mTFP (donor) emission (dotted blue line) and the mKO2 (acceptor) excitation spectra (solid red line). Furthermore, the excellent photostability of mTFP and mKO2 together with their large spectral overlap (Fig. 10.5), which is greater than 50%, resulting in R_0 values ~5.29 nm (Fig. 10.1), makes them an attractive pair for live-cell FRET microscopy to detect varying separation distances by measuring FRET efficiencies. In summary, the photophysical properties of mTFP and mKO2 suggest they are a good FRET pair.

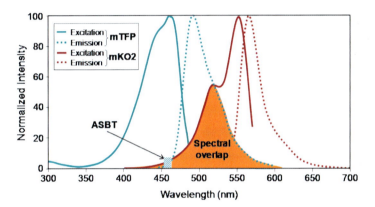

Figure 10.5 Excitation and emission spectra of mTFP and mKO2 FRET pair.

10.5.1 Spectral FRET Microscopy

Recent advances in spectral microscopy have allowed the development of spectral FRET microscopy to improve the accuracy of FRET nanoscale measurements of FP donor–acceptor pairs. In spectral imaging microscopy, a specimen is scanned at a single focal plane to produce a series of x-y images at discrete wavelength bands, normally called a λ-stack. This approach provides high spectral resolution and thus improves the accuracy of measurements of the contaminations by DSBT, ASBT, and background noise (27;37). The spectral signatures for the individual fluorophores are obtained from λ-stacks collected from reference specimens containing only donor or acceptor FPs. λ-stacks from specimens containing both donor and acceptor FPs are then subjected to spectral linear unmixing using reference spectra, mentioned above, to separate the contributions of individual donor and acceptor emission signals in each pixel of an acquired image. Therefore, spectral imaging provides one of the most accurate methods for identification of the DSBT contributions (26;27;57). This is especially important for the use of FPs in FRET microscopy, since due to the FP's broad emission profile, DSBT is normally very significant in filter-based FRET of FP pairs. Therefore, spectral FRET microscopy has expanded significantly the number of FP pairs that can be used in FRET-based cellular assays. However, since the spectra of ASBT and FRET signals are identical, the spectral FRET microscopy method still requires an algorithm and acceptor-only labeled specimens to separate ASBT from FRET signals.

For our spectral FRET study of mTFP-5aa-mKO2 expressed in live cells, we used the Leica TCS SP5 X imaging system, equipped with a 25 mW Argon laser, a white light laser (WLL), a fast resonant point scanner (8000 Hz), and a Leica 63X/1.4NA oil objective (37). Cells expressing mTFP-5aa-mKO2 fusion proteins were subjected to spectral FRET microscopy, and several λ-stacks were acquired using a donor excitation wavelength (458 nm Argon laser line) and an acceptor excitation wavelength (540 nm generated by the WLL) and then decomposed and reassigned to the donor (mTFP) and the acceptor (mKO2) emission channels through spectral linear unmixing. In Fig. 10.6a, a representative image is shown from λ-stacks acquired from a cell expressing mTFP-5aa-mKO2 upon excitation with the donor excitation wavelength (458 nm Argon laser) (left panel). The emission spectrum (right panel) was plotted from the region of interest shown in the left panel (red square). The emission signals at the 565 nm peak clearly indicate FRET signals, although the spectrum still contains both DSBT and ASBT. The same cell was also excited at the acceptor excitation wavelength (540 nm generated by the white light laser) (left panel), and the respective acceptor emission spectrum is shown in the right panel (Fig. 10.6b). In panel (c), a cell expressing only mKO2 was excited at both the donor and the acceptor excitation wavelengths. Acceptor excitation produced the mKO2 reference spectrum used for spectral linear unmixing showing its typical emission peak at 565 nm (solid line), whereas at

FP Donor–Acceptor FRET Pair | 275

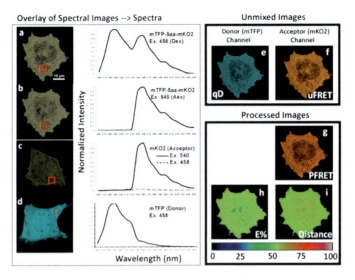

Figure 10.6 Image acquisition and processing of the mTFP-5aa-mKO2 construct expressed in live cells in spectral FRET microscopy.

donor excitation (dashed line) there is virtually no signal, indicating that ASBT is trivial. In panel (d), a representative mTFP-expressing cell was excited only at the donor excitation wavelength to generate the mTFP reference spectra used for spectral linear unmixing, which shows its typical emission peak at 492 nm and virtually no emission at 565 nm, indicating reduced DSBT. The reference spectra used for unmixing were obtained from cells expressing only either mKO2 or mTFP. After the unmixing process, the DSBT is separated from the FRET signal; however, the ASBT and FRET signals are still combined in the acceptor emission spectrum. To determine the contribution of the ASBT to the FRET signal, the λ-stacks of the mTFP-5aa-mKO2 cells excited by the acceptor excitation wavelength and the λ-stacks of the mKO2-alone cells excited by both the donor and the acceptor excitation wavelengths were acquired. The qD (Fig. 10.6e) and uFRET (Fig. 10.6f) images were obtained from unmixing the λ-stack of the mTFP-5aa-mKO2 cell excited by the donor excitation and together with the acceptor image (data not shown) obtained from unmixing the λ-stack of the same cell excited by the acceptor excitation were processed by our spectral FRET algorithm to remove the ASBT and generate the PFRET image (Fig. 10.6g) (27). The $E\%$ levels per pixel are shown as an image (Fig. 10.6h), where the value at each pixel is calculated based on Eqs. 10.6 and 10.7 using the corresponding pixel values in the PFRET and qD images. Given the average energy transfer efficiency (E = 0.5476) of mTFP-5aa-mKO2 and the Förster distance of mTFP-mKO2 (R_0 = 5.29 nm) to Eq. 10.1, the average distance between mTFP and mKO2 was estimated to be 5.12 nm, resulting in the distance image (Fig. 10.6i). Figure 10.6 was modified from Figs. 10.3 and 10.4 in (37). In addition, the SBT correction (SBT%) for this FRET pair was small, <5%, which is consistent with our expectations based on the mTFP and mKO2 excitation and emission spectra (Fig. 10.6). Since in spectral FRET microscopy the DSBT is removed through spectral linear unmixing, the reduced ASBT (due to the low absorption of mKO2 at the donor excitation wavelength for mTFP) makes this pair very valuable for spectral FRET microscopy.

10.5.2 FLIM-FRET Microscopy

We also characterized the mTFP-5aa-mKO2 fusion proteins expressed in live cells using time-correlated single-photon counting (TCSPC) FLIM-FRET microscopy. The experiments were carried out on a Biorad Radiance 2100 confocal/multiphoton imaging system, which is equipped with a Becker & Hickl SPC-150 board, a 10W Verdi pumped tunable (Coherent Mira 900) mode-locked ultrafast (78 MHz) pulsed (<150 femtoseconds) laser tuned to 870 nm, and a Nikon 60X/1.2NA water immersion infrared (IR) objective (37). In Fig. 10.7, the change in the donor (mTFP) lifetime in the absence and presence of the acceptor (mKO2) is measured from cells expressing mTFP alone or mTFP-5aa-mKO2, respectively. The instrument response function (IRF) of the FLIM system was measured through acquiring the second-harmonic-generation (SHG) signals emitted from urea crystals, and its utility was verified using a fluorescence lifetime standard—Cresyl Violet dissolved in ethanol (Fig. 10.7a) (37).

The FLIM-FRET E% is calculated from the quenched and unquenched mTFP lifetimes based on Eq. 10.5. To determine the unquenched lifetime (τ_D) of mTFP, we acquired data sets from the cells that were only labeled with mTFP. These data sets were then analyzed using the Becker & Hickl SPCImage software based on a single-exponential decay model (37). Given the input of the measured IRF, the fitting routine yielded an average unquenched mTFP lifetime of 2.684 ns. Better approximation of the data was

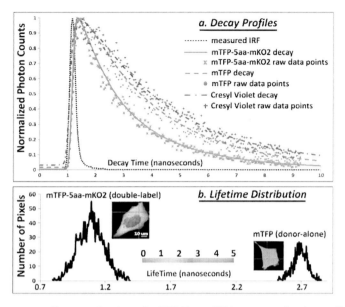

Figure 10.7 Characterization of mTFP-5aa-mKO2 expressed in live cells in FLIM-FRET microscopy.

not observed from the double exponential fitting of these data sets. Using the SPCImage software and the same measured IRF, we performed both single and double exponential fittings on the data sets acquired from the mTFP-5aa-mKO2-expressed cells. The monoexponential model was rejected because the biexponential model yielded a much better approximation of the data in terms of $\chi 2$, residuals, and visually comparing the fitting curves with the raw data. The obtained average quenched lifetime (τ_{DA}) of mTFP was 1.111 ns. Figure 10.7a demonstrates the decay rate differences between the mTFP without mKO2 and the mTFP in the presence of mKO2. In Fig. 10.7b, the lifetime distributions of representative cells labeled with mTFP-5aa-mKO2 (left; quenched) and of representative cells labeled with mTFP-only (right; unquenched) are shown with a nanosecond time scale. A wider distribution of the quenched mTFP lifetime was observed compared to the unquenched mTFP, since distance and dipole moment variations existed between the mTFP- and mKO2-folded proteins in cells even though they were tethered by a five-amino-acid linker. Figure 10.7 was modified from Fig. 10.5 in (37). Given the quenched (τ_{DA} = 1.111 ns) and unquenched (τ_D

= 2.684 ns) lifetimes of mTFP to Eq. 10.5, we calculated the FLIM-FRET E% resulting in an average FRET efficiency of approximately 58.63%, which is close to that (54.76%) obtained from spectral FRET microscopy imaging.

10.6 QD–Organic Dye FRET Pairs: QD566–AF568 and QD580–AF594

Semiconductor nanocrystals, widely known as QDs, are a powerful fluorescent tool in biological research (58–61). QDs contain a semiconductor core, for example, Cd and Se, which is surrounded by a semiconductor shell (ZnS) to improve their optical properties. The wide range of QD fluorescence emission peaks stems from their tunable core sizes (1–10 nm) (Fig. 10.8A). QDs show nearly Gaussian emission peaks, without the "shoulders" displayed by the emission spectra of most organic dyes and FPs. QDs show broad absorption patterns and can be efficiently excited far from their emission spectra to avoid background scattering. The combination of narrow emission spectra, minimal overlap between multiple emission spectra, and broad absorption spectra make QDs very useful for fluorescence multiplexing and high-throughput screening. Furthermore, the photostability as well as quantum yields of QDs are greater than those of organic dyes at similar wavelengths (62). Although QDs show long lifetimes that allow for enhanced sensitivity and signal-to-noise ratio, their complex decay behavior leads to a very complex and challenging analysis of time-resolved lifetime fluorescence measurements. When compared to standard organic fluorophore dyes and FPs, QDs show few disadvantages, such as their bulky size (Fig. 10.8A) and their "blinking" behavior, in which dark periods of no emission interrupt longer periods of fluorescence (62). Core-shell QDs can be manipulated in several different ways to generate stable water-stabilized QDs, which show improved brightness, aqueous solubility, colloidal stability, and functionality (Fig. 10.8A). Water-stabilized QDs have a wide range of applications in fluorescence bioimaging due to their ability to retain high quantum yields and luminosities over long lifetimes. One method to produce water-stabilized QDs is ligand exchange, in which hydrophobic capping surface ligands are replaced by hydrophilic bifunctional ligands (66;67). Another approach includes covering the hydrophobic surface groups with block copolymers or phospholipid micelles (68–70). The main commercial sources of QDs use this strategy in which core-shell QDs are encapsulated into a micelle of amphiphilic polymers. Both encapsulated and DHLA-PEG-ligand exchange QDs show good dispersion in water, stability in a wide range of pH, and a low nonspecific binding to cellular compartments (67;71;72). Currently, both types of QDs can be easily conjugated to proteins, oligonucleotides, and other biomolecules using straightforward bioconjugation methods and have been used routinely in cellular targeting, sensing, and imaging (73). QD–protein conjugates can be used as probes toward specific biological mechanisms in immunoassays and live-cell imaging, as well as in a variety of other fluorescence-based detection assays, including FRET (Fig. 10.8).

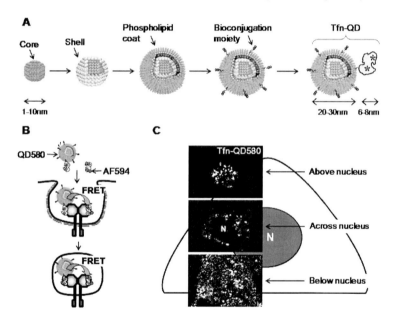

Figure 10.8 Generation and characterization of Tfn–QD conjugates.

10.6.1 Application of QDs as Donor Molecules in FRET Pairs

In acceptor-sensitized intensity-based FRET microscopy, the spectral characteristics of QDs, such as their increased photostability, tunable and narrow emission spectra, and broad excitation spectra, make them extremely good candidates for donor fluorophores. Recently, QDs have been shown to act as strong FRET donors to dye acceptors in a variety of biological applications (35;74;75). Steady-state and time-resolved fluorescence measurements have also demonstrated the efficient nonradiative energy transfer between QD donors and organic dye acceptors, in a manner consistent with the Förster theory of dipole–dipole interactions (35;66;74;76). However, because of their broad absorption spectra and long excited-state lifetimes, QDs do not perform adequately as acceptors in FRET reactions (35;74;75).

Here we will address several properties of QDs that should be considered when planning to use them as FRET donors in live cells. QDs are multivalent nanoparticles, allowing each QD molecule to be conjugated to multiple acceptor molecules. This can have a significant advantage in the assembly of biosensor machineries, where a central QD donor acts as a scaffold for the binding of multiple biosensor molecules that can be labeled directly or indirectly with acceptor dye fluorophores (35;74;75). The ability to multifunctionalize QDs can also be used to induce aggregation as shown for the amyloid aggregation of the protein α-synuclein using streptavidin QDs loaded with biotinylated α-synuclein in vitro and in cells (77). Furthermore, low donor/acceptor ratios due to the attachment of multiple acceptors to multivalent QDs will result in improved FRET efficiencies and highly sensitive QD-mediated FRET reactions. On the other hand, the development of monovalent QDs is important for the development of FRET-based approaches to follow interactions between membrane-bound receptors (78–80), since they may be affected by cross-linking/oligomerization.

We have described above the need for SBT correction when using traditional organic fluorophores as donor and acceptor molecules. In contrast, QDs offer unique advantages in this respect with their broad absorption spectra and the ability to apply excitation wavelengths that minimize, in particular, ASBT, making them great donor molecules when paired with organic dyes for FRET reactions (30) and FPs (70), improving the signal-to-noise ratio of the FRET signal and simplifying FRET processing and analysis; SBT is significantly reduced, while preserving the strong spectral overlap necessary for FRET.

While QDs provide some clear advantages, we have described some potential drawbacks: the large size of QDs, which may result in strong geometrical exclusion and reduce FRET detection sensitivity (36); the heterogeneity in the orientation of QD–protein conjugates, which may lead to a mixed functional behavior, resulting in nonoptimal performance of QD–protein conjugates as FRET donors (66;76); and the QD "blinking" behavior, which may reduce the ability of QD donors to transfer energy to transiently proximally located acceptor molecules. QD-mediated nonspecific adsorption to various cellular components may reduce the ability of QD–protein conjugates to interact with their respective acceptor-labeled target molecules, leading to reduced signal-to-noise ratios of FRET signal and significantly lower sensitivity of QD-based biosensors and binding assays (81). This again is a demonstration of the importance of the judicious selection of fluorophore pairs to be tailored to specific research goals.

10.6.2 Filter-Based and Spectral FRET Confocal Microscopy of QD566–AF568 and QD580–AF594

While the majority of QD–protein conjugates have been characterized in in vitro assays, here we will focus on QD–AF FRET-based assays that have been performed in cultured cells. The microscopy instruments used to visualize the FRET reaction in cells may limit the selection of QD–dye pairs, although the use of spectral imaging may alleviate such concerns. Considering that the size of QDs is within the range of FRET measurable distances, it is important to discuss their ability to actually transfer energy to AF acceptor molecules (Fig. 10.8B). An interesting hypothesis is that the coat, enveloping the QD core, displays a nonuniform diameter shell, which may explain the ability of QDs to act as FRET donors to AF acceptors at distances that may be close to the R_0 values of these QD–AF FRET pairs.

Here we will describe experiments showing that Tfn–QD conjugates can track the cellular uptake of TFR–Tfn complexes as well as act as donor molecules and transfer energy to Tfn–AF acceptor molecules upon binding to TFR homodimers and subsequent internalization into cells (30). In Fig. 10.8B,C, the cellular uptake and FRET reactions of Tfn–QD580 (donor) and Tfn–AF594 (acceptor) conjugates are shown, as described in Fig. 10.3 and in Refs. 5, 25, 36, 40, and 46. In Fig. 10.8C, Tfn–QD580 was internalized into Madin–

Darby canine kidney (MDCK)-pTR cells for one hour at 37°C. After washing, cells were subjected to fixation and imaging using a Zeiss LSM510 Meta confocal microscope. Vertical sections were collected every 0.5 µm, and image representatives show the typical punctuate distribution of TFR–Tfn complexes in endocytic structures located above, across, and below the nucleus.

The TFR–Tfn cell-based FRET assay showcases the capabilities of quantitative FRET analysis concerning the trafficking of membrane-bound receptor–ligand complexes (5;25;40) and the advantage of QDs as donors in FRET events to significantly reduce the overall SBT signal, while preserving the strong spectral overlap necessary for FRET. Whereas ASBT can be lowered by exciting the QDs at wavelengths that minimize the acceptor excitation, DBST is reduced since QDs display narrow emission peaks that minimize the donor emission bleedthrough into the acceptor channel upon donor excitation. This was tested with two different QD–AF FRET pairs using either filter-based or spectral confocal FRET microscopy, followed by the PFRET SBT correction processing algorithm. As shown in Fig. 10.9, there is a significant spectral overlap between the emission spectra of QD566–Tfn and the absorption spectra of AF568–Tfn (Fig. 10.9A) as well as that of QD580–Tfn and AF594–Tfn (Fig. 10.9B), suggesting that both QD566–AF568 and QD580–AF594 make good FRET pairs. In agreement, R_0 = 6 nm for both these FRET pairs, suggesting that both QD566 and QD580 can act as strong donors for their respective AF acceptor partners (Fig. 10.1).

These different confocal imaging approaches were used to test whether FRET occurs between these two donor–acceptor pairs after uptake into cells; differences in SBT levels were also tested. In one approach, filter-based FRET images were collected from cells internalized with either Tfn–QD566 or Tfn–AF568 and cells cointernalized with both (5;25;30;40). In another approach, λ-stacks were collected from cells internalized with either Tfn–QD580 or Tfn–AF594 and cells cointernalized with both Tfn conjugates, subjected to linear unmixing and processed by spectral FRET microscopy (27;30). In both cases, the images were processed using the PFRET algorithm (36;41;46). In Fig. 10.10, the pseudocolor image depicts acceptor (panel a; AF594–Tfn), qD (panel b; QD580–Tfn), uFRET (panel c), PFRET (panel d), and $E\%$ (panel e) in a pixel-by-pixel manner, as described above (25;30;40). The typical endocytic morphology of irregular, punctuate, and peripherally localized structures and a

QD–Organic Dye FRET Pairs | 283

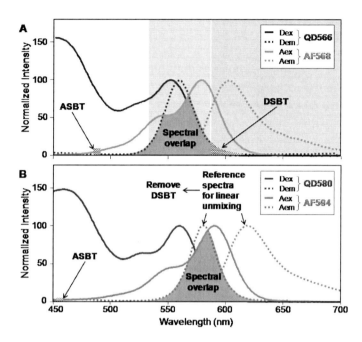

Figure 10.9 Excitation and emission spectra of QD566–AF568 and QD580–AF594 FRET pairs.

centrally located nucleus is detected across all FRET images (Fig. 10.10). To determine the level of SBT contamination due to ASBT or DSBT in both QD–AF FRET pairs, the PFRET algorithm was used to process uFRET images in the presence of only the acceptor or donor reference images, respectively. For the QD566–AF568 pair using filter-based confocal FRET imaging, the ASBT levels are significantly higher than that of DSBT (~70% ASBT vs. 30% DSBT), suggesting that ASBT is the predominant correction in the overall SBT. In contrast, for the QD580–AF594 using spectral confocal FRET imaging, ASBT and DSBT levels are similar but strongly reduced (9% ASBT vs. 11% DSBT). E% measurements indicate that the FRET behavior of TFR–Tfn complexes occurs similarly in a manner independent of the QD–AF FRET pair used (30). Furthermore, as shown in Fig. 10.10f, E% averages after removal of both ASBT and DSBT (solid) or of only ASBT (narrow diagonal lines) or DSBT (wide diagonal lines) are plotted, indicating that only the QD580–AF594 FRET pair analyzed by spectral FRET imaging shows similar E%

levels independently of whether ASBT or DSBT or both are removed by the PFRET correction algorithm. In summary, a careful selection of QD–AF FRET pairs allows for a significant reduction of the ASBT since QDs, due to their broad excitation spectra, can be excited at wavelengths that minimize acceptor excitation. Spectral imaging, which uses a linear unmixing algorithm to separate the donor and acceptor emission spectra, can then be used as the sole methodology necessary to remove the DSBT.

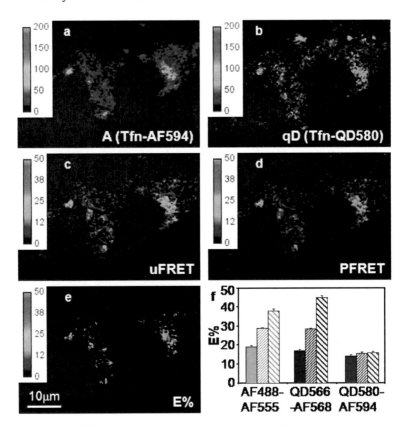

Figure 10.10 FRET imaging of QD–AF FRET pairs in cells.

Future research in QDs will have a positive impact in the application of FRET microscopy to the study of complex biological processes in cells and tissues. For example, long-range FRET detection up to 13 nm may be achieved within supramolecular complexes containing QD, FPs, and organic dyes (83). Furthermore, the characterization of

QD–dye FRET in the far-red region will have significant benefits for biological imaging, since it minimizes autofluorescence, increases the depth of tissue penetration and facilitates multiphoton absorption/excitation (84). Finally, the spectral characteristics of QDs make them very useful for the development of multiplex FRET, in which use multiple distinct dye acceptors are conjugation with the same QD donor (85;86). Multiplex FRET approaches would track one or more QD–protein donor conjugates as they interact with multiple target acceptor molecule(s). The development of nonblinking QDs may lead to straightforward methods for single-molecule FRET correlated with single-molecule particle tracking. Therefore, these novel QD-based FRET approaches will significantly expand our ability to track the temporal dynamics of protein–protein interactions in live cells.

10.7 Conclusions and Outlook

The significance of the research described here lies in the development of platform technology that will impact FRET-based imaging approaches designed to probe nanoscale molecular mechanisms in live cells. We have shown that the selective choice of FRET pairs and imaging conditions may avoid the need to process FRET images to remove SBT. For example, spectral FRET microscopy of mTFP–mKO2 and the QD580–AF594 FRET pairs show significantly reduced SBT levels. Such a development would allow the collection of real-time FRET images and data, which together with the increased photostability of available fluorophores, would make real-time live-cell FRET a reality. In the future, new development in fluorophore and FRET microscopy research will draw upon the unique biophysical properties of different fluorophores to eliminate much of the processing analysis required in determining $E\%$, strengthen the quantitative data obtained in FRET imaging, make live-cell FRET imaging faster, brighter, and more quantitative, and allow for application of FRET-based approaches to tissue biology in vivo and ex vivo.

Acknowledgments

We would like to thank Evident Technologies, Inc., for providing the QDs used in this study. This work was supported by an award

(AP) from the National Center for Research Resources, the National Institutes of Health (RR025616).

References

1. N. R. Gascoigne, et al., *Curr. Top. Microbiol. Immunol.*, **334**, 31–46 (2009).
2. H. Wallrabe, A. Periasamy, *Curr. Opin. Biotechnol.*, **16**(1), 19–27 (2005).
3. J. Gandia, C. Lluis, S. Ferre, R. Franco, F. Ciruela, *Bioessays*, **30**(1), 82–89 (2008).
4. R. M. Clegg, *J. Biotechnol.*, **82**(3), 177–179 (2002).
5. A. Periasamy, H. Wallrabe, Y. Chen, M. Barroso, *Methods Cell Biol.*, **89**, 569–598 (2008).
6. S. S. Vogel, C. Thaler, S. V. Koushik, *Sci. STKE*, **2006**(331), re2 (2006).
7. D. W. Piston, G. J. Kremers, *Trends Biochem. Sci.*, **32**(9), 407–414 (2007).
8. R. B. Sekar, A. Periasamy, *J. Cell Biol.*, **160**(5), 629–633 (2003).
9. A. Periasamy, R. N. Day, *Molecular Imaging: FRET Microscopy and Spectroscopy*, New York: Oxford University Press, 2005.
10. L. Stryer, *Annu. Rev. Biochem.* **47**, 819–846 (1978).
11. J. R. Lakowicz, I. Gryczynski, Z. Gryczynski, J. D. Dattelbaum, *Anal. Biochem.*, **267**(2), 397–405 (1999).
12. T. Forster, "Delocalized excitation and excitation transfer," in O. Sinanoglu, ed., *Modern Quantum Chemistry Part III: Action of Light and Organic Crystals*, New York: Academic Press, 1965, 93–137.
13. R. M. Clegg, "Fluorescence resonance energy transfer," in X. F. Wang, B. Herman, eds., *Fluorescence Imaging Spectroscopy and Microscopy*, New York: John Wiley & Sons, 1996, 179–251.
14. J. R. Lakowicz, *Principles of Fluorescence Spectroscopy*, 2nd ed., New York: Kluwer Academic/Plenum, 1999.
15. A. K. Kenworthy, *Methods*, **24**(3), 289–296 (2001).
16. A. K. Kenworthy, N. Petranova, Edidin M, *Mol. Biol. Cell*, **11**(5), 1645–1655 (2000), PMCID:PMC14873.
17. M. K. Raarup, A. W. Fjorback, S. M. Jensen, H. K. Muller, M. M. Kjaergaard, H. Poulsen, O. Wiborg, J. R. Nyengaard, *J. Biomed. Opt.*, **14**(3), 034039 (2009).

18. L. Tron, J. Szollosi, S. Damjanovich, S. H. Helliwell, D. J. Arndt-Jovin, T. M. Jovin, *Biophys. J.*, **45**(5), 939–946 (1984).
19. L. Matyus, *J. Photochem. Photobiol. B*, **12**(4), 323–337 (1992).
20. Z. Kam, T. Volberg, B. Geiger, *J. Cell Sci.*, **108**(Pt 3), 1051–1062 (1995).
21. G. W. Gordon, G. Berry, X. H. Liang, B. Levine, B. Herman, *Biophys. J.*, **74**(5), 2702–2713 (1998), PMCID:PMC1299610.
22. A. Hoppe, K. Christensen, J. A. Swanson, *Biophys. J.*, **83**(6), 3652–3664 (2002).
23. Y. Chen, M. Elangovan, A. Periasamy, "FRET data analysis: the algorithm," in A. Periasamy, R. N. Day, eds., *Molecular Imaging: FRET Microscopy and Spectroscopy*, New York: Oxford University Press, 2005, 126–145.
24. Y. Chen, A. Periasamy, *J. Fluoresc.*, **16**(1), 95–104 (2006)
25. H. Wallrabe, Y. Chen, A. Periasamy, M. Barroso, *Microsc. Res. Tech.*, **69**(3), 196–206 (2006).
26. C. Thaler, S. V. Koushik, P. S. Blank, S. S. Vogel, *Biophys. J.*, **89**(4), 2736–2749 (2005).
27. Y. Chen, J. P. Mauldin, R. N. Day, A. Periasamy, *J. Microsc.*, **228**(Pt 2), 139–152 (2007), PMCID:PMC2874973.
28. D. Megias, R. Marrero, P. B. Martinez Del, M. A. Garcia, J. J. Bravo-Cordero, A. Garcia-Grande, A. Santos, M. C. Montoya, *Microsc. Res. Tech.*, **72**(1), 1–11 (2009).
29. V. Raicu, M. R. Stoneman, R. Fung, M. Melnichuk, D. B. Jansma, L. F. Pisterzi, S. Rath, M. Fox, J. W. Wells, D. K. Saldin, *Nat. Photon.*, **3**(2), 107–113 (2009).
30. N. McGrath, M. Barroso, *J. Biomed.Opt.*, **13**(3), 031210 (2008).
31. D. M. Grant, W. Zhang, E. J. McGhee, T. D. Bunney, C. B. Talbot, S. Kumar, I. Munro, C. Dunsby, M. A. Neil, M. Katan, et al., *Biophys. J.*, **95**(10), L69–L71 (2008).
32. A. Periasamy, R. M. Clegg, *FLIM Microscopy in Biology and Medicine*, London: CRC Press, 2009.
33. S. V. Koushik, S. S. Vogel, *J. Biomed. Opt.*, **13**(3), 031204 (2008).
34. G. H. Patterson, D. W. Piston, B. G. Barisas, *Anal. Biochem.*, **284**(2), 438–440 (2000).
35. I. L. Medintz, H. Mattoussi, *Phys.Chem.Chem.Phys.*, **11**(1), 17–45 (2009).
36. H. Wallrabe, M. Elangovan, A. Burchard, A. Periasamy, M. Barroso, *Biophys. J.*, **85**(1), 559–571 (2003). PMCID:PMC1303111.

37. Y. Sun, C. F. Booker, S. Kumari, R. N. Day, M. Davidson, A. Periasamy, *J. Biomed. Opt.*, **14**(5), 054009 (2009).
38. P. Konig, G. Krasteva, C. Tag, I. R. Konig, C. Arens, W. Kummer, *Lab. Invest.*, **86**(8), 853–864 (2006).
39. M. Keese, R. J. Magdeburg, T. Herzog, T. Hasenberg, M. Offterdinger, R. Pepperkok, J. W. Sturm, P. I. Bastiaens, *J. Biol. Chem.*, **280**(30), 27826–27831 (2005).
40. H. Wallrabe, G. Bonamy, Periasamy, M. Barroso, *Mol. Cell Biol.*, **18**, 2226–2243 (2007). PMCID:PMC1877110.
41. M. Elangovan, H. Wallrabe, Y. Chen, R. N. Day, M. Barroso, A. Periasamy, *Methods*, **29**(1), 58–73 (2003).
42. J. Roszik, D. Lisboa, J. Szollosi, G. Vereb, A. *Cytometry*, **75**(9), 761–767 (2009).
43. J. N. Feige, D. Sage, W. Wahli, B. Desvergne, L. Gelman, *Microsc. Res. Tech.*, **68**(1), 51–58 (2005).
44. M. Hachet-Haas, N. Converset, O. Marchal, H. Matthes, S. Gioria, J. L. Galzi, S. Lecat, *Microsc. Res. Tech.*, **69**(12), 941–956 (2006).
45. C. Berney, G. Danuser, *Biophys.J.*, **84**(6), 3992–4010 (2003). PMCID:PMC1302980.
46. H. Wallrabe, M. Stanley, A. Periasamy, M. Barroso, *J. Biomed. Opt.*, **8**(3), 339–346 (2003).
47. Y. Cheng, O. Zak, P. Aisen, S. C. Harrison, T. Walz, *Cell*, **116**(4), 565–576 (2004).
48. C. M. Lawrence, S. Ray, M. Babyonyshev, R. Galluser, D. W. Borhani, S. C. Harrison, *Science*, **286**(5440), 779–782 (1999).
49. T. Nagai, S. Yamada, T. Tominaga, M. Ichikawa, A. Miyawaki, *Proc. Natl. Acad. Sci. U S A*, **101**(29), 10554–9 (2004).
50. K. Takao, K. Okamoto, T. Nakagawa, R. L. Neve, T. Nagai, A. Miyawaki, T. Hashikawa, S. Kobayashi, Y. Hayashi, *J. Neurosci.*, **25**(12), 3107–3112 (2005).
51. L. Hodgson, O. Pertz, K. M. Hahn, *Methods Cell Biol.*, **85**, 63–81 (2008).
52. K. Aoki, M. Matsuda, *Nat. Prot.*, **4**(11), 1623–1631 (2009).
53. K. J. Herbst, Q. Ni, J. Zhang, *IUBMB Life*, **61**(9), 902–908 (2009).
54. M. Anikovsky, L. Dale, S. Ferguson, N. Petersen, *Biophys. J.*, **95**(3), 1349–1359 (2008)
55. L. Albertazzi, D. Arosio, L. Marchetti, F. Ricci, F. Beltram, *Photochem. Photobiol.*, **85**(1), 287–297 (2009).

56. A. W. Nguyen, P. S. Daugherty, *Nat. Biotechnol.*, **23**(3), 355–360 (2005).
57. T. Zimmermann, J. Rietdorf, A. Girod, V. Georget, R. Pepperkok, *FEBS Lett.*, **531**(2), 245–249 (2002).
58. X. Gao, S. Nie, *Trends Biotechnol.*, **21**(9), 371–373 (2003).
59. T. M. Jovin, *Nat. Biotechnol.*, **21**(1), 32–33 (2003).
60. W. C. Chan, D. J. Maxwell, X. Gao, R. E. Bailey, M. Han, S. Nie, *Curr. Opin. Biotechnol.*, **13**(1), 40–46 (2002).
61. W. W. Yu, E. Chang, R. Drezek, V. L. Colvin, *Biochem. Biophys. Res. Commun.*, **348**(3), 781–786 (2006).
62. U. Resch-Genger, M. Grabolle, S. Cavaliere-Jaricot, R. Nitschke, T. Nann, *Nat. Methods*, **5**(9), 763–775 (2008).
63. X. Wang, X. Ren, K. Kahen, M. A. Hahn, M. Rajeswaran, S. cagnano-Zacher, J. Silcox, G. E. Cragg, A. L. Efros, T. D., *Nature*, **459**(7247), 686–689 (2009).
64. A. M. Smith, S. Nie, *Nat. Biotechnol.*, **27**(8), 732–733 (2009).
65. W. C. Law, K. T. Yong, I. Roy, H. Ding, R. Hu, W. Zhao, P. N. Prasad, *Small*, **5**(11), 1302–1310 (2009).
66. I. L. Medintz, A. R. Clapp, H. Mattoussi, E. R. Goldman, B. Fisher, J. M. Mauro, *Nat. Mater.*, **2**(9), 630–638 (2003).
67. K. Susumu, B. C. Mei, H. Mattoussi, *Nat. Prot.*, **4**(3), 424–436 (2009).
68. D. S. Lidke, K. A. Lidke, B. Rieger, T. M. Jovin, D. J. rndt-Jovin, *J. Cell Biol.*, **170**(4), 619–626 (2005).
69. L. Groc, M. Heine, L. Cognet, K. Brickley, F. A. Stephenson, B. Lounis, D. Choquet, *Nat. Neurosci.*, **7**(7), 695–696 (2004).
70. B. Dubertret, P. Skourides, D. J. Norris, V. Noireaux, A. H. Brivanlou, A. Libchaber, *Science*, **298**(5599), 1759–1762 (2002).
71. O. Carion, B. Mahler, T. Pons, B. Dubertret, *Nat. Prot.*, **2**(10), 2383–2390 (2007).
72. W. Liu, M. Howarth, A. B. Greytak, Y. Zheng, D. G. Nocera, A. Y. Ting, M. G. Bawendi, *J. Am. Chem. Soc.*, **130**(4), 1274–1284 (2008).
73. M. F. Frasco, N. Chaniotakis, *Anal. Bioanal. Chem.*, **396**(1), 229–240 (2010).
74. W. R. Algar, U. J. Krull, *Anal. Bioanal. Chem.*, **391**(5), 1609–1618 (2008).
75. W. Zhong, *Anal. Bioanal. Chem.*, **394**(1), 47–59 (2009)

76. I. L. Medintz, J. H. Konnert, A. R. Clapp, I. Stanish, M. E. Twigg, H. Mattoussi, J. M. Mauro, J. R. Deschamps, *Proc. Natl. Acad. Sci. U S A*, **101**(26), 9612–9617 (2004).

77. M. J. Roberti, M. Morgan, G. Menendez, L. I. Pietrasanta, T. M. Jovin, E. A. Jares-Erijman, *J. Am. Chem. Soc.*, **131**(23), 8102–8107 (2009).

78. M. Howarth, W. Liu, S. Puthenveetil, Y. Zheng, L. F. Marshall, M. M. Schmidt, K. D. Wittrup, M. G. Bawendi, A. Y. Ting, *Nat. Methods*, **5**(5), 397–399 (2008).

79. N. L. Andrews, K. A. Lidke, J. R. Pfeiffer, A. R. Burns, B. S. Wilson, J. M. Oliver, D. S. Lidke, *Nat. Cell Biol.*, **10**(8), 955–963 (2008).

80. S. J. Clarke, C. A. Hollmann, F. A. Aldaye, J. L. Nadeau, *Bioconjug. Chem.*, **19**(2), 562–568 (2008).

81. I. Yildiz, X. Gao, T. K. Harris, F. M. Raymo, *J. Biomed. Biotechnol.*, **2007**, 18081 (2007).

82. M. Barroso, E. S. Sztul, *J. Cell Biol.*, **124**(1–2), 83–100 (1994).

83. H. Lu, O. Schops, U. Woggon, C. M. Niemeyer, *J. Am. Chem. Soc.*, **130**(14), 4815–4827 (2008).

84. E. Z. Chong, D. R. Matthews, H. D. Summers, K. L. Njoh, R. J. Errington, P. J. Smith, *J. Biomed. Biotechnol.*, **2007**(7), 54169 (2007).

85. A. R. Clapp, I. L. Medintz, H. T. Uyeda, B. R. Fisher BR, E. R. ER, M. G. Bawendi, H. Mattoussi, *J. Am. Chem. Soc.*, **127**(51), 18212–18221 (2005).

86. M. Suzuki, Y. Husimi, H. Komatsu, K. Suzuki, K. T. Douglas, *J. Am. Chem. Soc.*, **130**(17), 5720–5725 (2008).

Chapter 11

Cancer Detection and Biosensing Applications with Quantum Dots

Ken-Tye Yong
School of Electrical and Electronic Engineering, Nanyang Technological University, Singapore 639798, Singapore
ktyong@ntu.edu.sg

Cancer is the second-leading cause of death, with millions of newly diagnosed cases worldwide every year. Because this disease is usually detected in an advanced stage, when the primary tumor has metastasized and invaded other organs, and is beyond surgical intervention, the yearly cancer death rate has not changed significantly with recently developed drugs and surgical techniques. Therefore, to advance cancer prevention and therapy, it is imperative to explore biofriendly and target-specific probes capable of early diagnosis and treatment of this disease. Quantum dots (QDs) possess unique optical properties that make them powerful candidates as probes or anticancer drug carriers for cancer research and bioimaging applications. The surface of QDs can be modified with functional biomaterials to selectively interact with molecular markers to map out the cancerous area. The flexibility and versatility of QDs may

---

Luminescence: The Instrumental Key to the Future of Nanotechnology
Edited by Adam M. Gilmore
Copyright © 2014 Pan Stanford Publishing Pte. Ltd.
ISBN 978-981-4241-95-3 (Hardcover), 978-981-4267-72-4 (eBook)
www.panstanford.com

provide the keys to answer important cancer biology questions for improving cancer-related mortality and overall survival. In this chapter, we review the design of biocompatible QDs and how we exploit their unique characteristics for cancer detection and biosensing applications.

11.1 Introduction

Human cancers were once considered to be an incurable disease; however, the advancement of nanotechnology and biotechnology has changed the perception of this incurable disease.[1,2] Today, many cancer patients will survive from their illness with proper treatment when they are diagnosed at an early stage.[3] Advances in diagnosis and therapy of cancer are largely responsible for the significantly reduction of the yearly cancer death rate. This was observed from the five-year survival rate for all cancers diagnosed between 1996 and 2002.[3] Despite these advances, cancer remains the second leading cause of death in the U.S., exceeded by only heart disease, and accounts for one in four deaths.

The development of particles in the nanometer scale has become an important field in chemistry research as methods of manipulation and characterization of such structures have dramatically expanded over the past several years.[4] This interest is also driven by advances in applications ranging from medical imaging to treatment of human diseases.[5-8] The preparation of nanoparticles dates back to the 19th century, with Michael Faraday first reporting the discovery of a monodispersed gold nanoparticle colloid.[9] For the last two decades, various synthesis methods have been explored to make nanoparticles with tunable size and novel optical properties.[10] To date, numerous recipes are available in the literature for synthesizing functional inorganic isotropic and anisotropic nanoparticles.[11] To name a few, solution-phase synthesis approaches include coprecipitation[12], hot colloidal synthesis[13,14], sol-gel processing[15], microemulsions[16], hydrothermal method[17], solvothermal method[18], templated syntheses[19], and biomimetic syntheses[20,21], all standard techniques in making nanoparticles with uniformity sizes. Among the types of nanoparticles, semiconductor nanoparticles (also known as quantum dots [QDs]) represent the most widely and intensively studied nanomaterials in the last decade.[22] This is because semiconductor nanoparticles offered tailorable electronic optical properties and the flexibility of surface functionalization for biological applications.[23,24]

When the diameter of the semiconductor nanoparticles becomes smaller than the bulk exciton radius, the energy levels in these nanometer-sized crystals become quantized and the transitions are locked into discrete energy states, as opposed to the band structures in bulk semiconductors.[25,26] QDs are luminescent nanoparticles with sizes ranging from 2 nm to 10 nm that contain hundreds to thousands of atoms of group II–VI elements (e.g., CdSe, CdTe, CdS, ZnSe, ZnTe, and ZnS) or group III–V elements (e.g., InP, InAs) or group IV–VI elements (e.g., PbSe, PbTe, and PbS) or group IV elements (e.g., Si, Ge). QDs have significant advantages over traditional organic dyes as luminescence markers for biology and biomedical applications.[27] For example, different colors of QDs can be simultaneously excited with a single light source, with minimal spectral overlap, providing significant advantages for multiplex and dynamic imaging of molecular targets.[28,29] Also, QDs have been shown to remain brightly stable emissive even after long periods of excitation, whereas organic dyes are photobleached quickly. In addition, QDs can be made to emit in a range of wavelengths by changing their size, shape, and composition, whereas the new structure of organic dyes must be designed to shift their emission toward desirable wavelengths.[30] This flexibility in optical tuning allows the QDs to emit in the near-infrared (NIR) region, which is optimal for deep tissue imaging. These advantages of QDs have led to their development as a new generation of optical probes for diagnostic imaging and sensing.[31-33]

This chapter presents the current research on the use of QDs for cancer detection, sensing, and therapy. The main objectives of this chapter are not only to highlight the development of QDs for cancer detections but also to explore the challenges and opportunities we face while utilizing them for human disease treatments. Understanding the underlying interaction behavior between functionalized QDs and the biological environment will provide us with helpful guidelines in designing novel and powerful "tools" based on nanoscale materials for specific cancer applications. In this chapter, we give an overview of the basic methods adopted for QD preparation in Section 11.2. Section 11.3 summarizes commonly used QDs in biological applications. Section 11.4 briefly discusses the synthesis of water-dispersible QDs. Section 11.5 focuses on the functionalization of water-dispersible QDs with biomolecules. Section 11.6 presents bioconjugated QDs for in vitro cancer studies. Section 11.7 discusses the potential of QDs as ultrasensitive probes for in vivo cancer imaging. Section 11.8 considers the risks and benefits of QDs for biomedical applications. Finally, Section 11.9 briefly discusses the conclusion and future outlook of QDs for cancer applications.

The authors wish to stress that there have been more studies on biocompatible QDs in the literature, with promises for human cancer applications; however, due to the page limits in this chapter, the author is forced to pick and choose the relevant reports that are most closely related to the subject discussed here.

11.2 Preparation of Quantum Dots with the Hot Colloidal Synthesis Method

The hot colloidal synthesis technique is still the most effective method to prepare highly crystalline and monodisperse QDs. The synthesis of QDs requires a nucleation burst event followed by a slower growth on the nucleated nuclei.[34] This can be achieved by injecting required chemical precursors into a hot reaction mixture containing surfactants and high-boiling-point solvents.[35] With the introduction of the precursors into the hot colloidal reaction mixture, it will raise the precursor concentration above the nucleation threshold and result in a supersaturation of nanocrystals in the solution.[36] Most importantly, the consumption of the precursors during the growth of the nanocrystals must not go beyond the rate of the precursor introduced into the mixture, because this will hinder the formation of new nuclei.[37] By systematic tailoring the synthesis parameters, such as temperature, precursor concentrations, growth time, and the choice of surfactants, one can manipulate the size and crystallinity of the QDs.[38] In general, the concentration of the surfactants is commonly three to four times higher than that of the precursors because high surfactant concentrations will provide better passivation to the QD surface and lead to the formation of monodisperse QDs. The surfactants are attached to the surfaces of the nanocrystals, creating a hydrophobic organic moieties layer that protects the nanocrystals from forming aggregates in solution.[34] Ligands that bind more strongly to the surface of the nanocrystals will generate greater steric hindrance that slows the growth of the nanocrystals and allows one to obtain a desirable nanocrystal size upon the addition of appropriate surfactants. An alternative method to narrow the size distribution of the particles involves the use of a binary or even tertiary surfactant mixture in which the surfactants attach tightly to the nanocrystal surface, thereby slowing their growth in the solution.[34] When the nanocrystal reaches the desired size, addition of organic solvents such as chloroform or toluene into the hot colloidal mixture will cool the solution and prevent the growth of the nanocrystals. A semipolar solvent such as ethanol, methanol, or butanol is introduced into the colloidal nanocrystal solution to clean and collect the nanocrystals. These alcohol solvents have an unfavorable interaction with the surfactant molecules anchored on the nanocrystal surface and will destabilize the nanocrystals' dispersion and cause them to agglomerate. Centrifugation of the aggregate suspension allows the surfactant solution and unreacted precursors to be removed, and the precipitate of the nanocrystals can be obtained.

The above method is generally used in the production of the "core" of QDs. These "core" nanocrystals must be coated with a biocompatible layer before they can be safely applied in biological applications.[28] Methods for coating a semiconductor nanocrystal core with a second semiconductor shell are well reported, and different kinds of core-shell structures have been prepared[39] (CdSe/ZnS, CdSe/ZnSe, CdTe/ZnS, CdHgTe/ZnS, and InP/ZnS core-shell nanocrystals, to name a few). There are common guidelines for the preparation of high-quality core-shell nanocrystals:[34] (i) the synthesized nanocrystal cores must remain their structure and crystallinity when the second semiconductor material is deposited; (ii) the lattice mismatch between the two semiconductor materials must be minimized so that a uniform epitaxial growth of the shell can occur around the core, resulting in low density of defects in the shell; and (iii) the coating of semiconductor nanoparticle cores must be chosen in a way that they will not break down under the deposition reaction or else this will cause the loss of particles or generate foreign nanocrystal species. Generally, nanocrystal cores are prepared first and then redispersed in a low-boiling-point organic solvent (e.g., hexane or chloroform). The colloidal nanocrystal dispersion is then mixed with high-boiling-point surfactants and heated to the desired reaction temperature, and at the same time the inorganic shell precursors are slowly introduced into the reaction mixture. This process will allow the shell materials to heterogeneously nucleate on the nanocrystal cores. It is important to note that as long as the rate of the inorganic shell precursor addition does not go beyond the rate of its deposition onto the cores, the precursor concentration never reaches the threshold for homogeneous nucleation of foreign nanocrsytals.[34]

11.3 Types of Quantum Dots Available for Biomedical and Cancer Applications

We here briefly discuss the types of QDs utilized in biomedical imaging and therapy. Figure 11.1 shows commonly used QDs and specific biological applications where they are suited. The emission band of these QDs can be tuned from the visible to NIR region by manipulating their size, shape, composition, and structure. Each type of QDs has its own advantages and disadvantages, and the discussion below is generated to provide useful guidelines for researchers to engineer QD probes meeting their own specific needs in cancer-relevant applications.

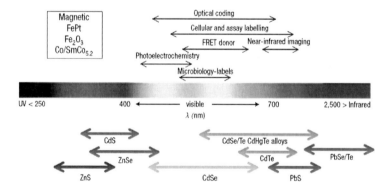

Figure 11.1 Commonly used QD core materials and their emission wavelength. Representative areas of biological interest are also presented corresponding to the pertinent emission. (Reprinted by permission from Macmillan Publishers Ltd.: *Nature Materials* (Ref. 114, http://www.nature.com), Copyright 2005.)

11.3.1 CdSe/ZnS Core-Shell Quantum Dots

The synthesis of high-quality CdSe/ZnS QDs by hot colloidal chemistry is well established.[40,41] It usually involves injection of organometallic precursors into a surfactant solution (e.g., trioctylphosphine oxide [TOPO], carboxylic acid, or phosphonic acids) at high temperatures (typically >250°C), and this yields highly monodisperse, spherical CdSe nanoparticles.[42] The CdSe QDs are size tunable and highly luminescent. The emission peak window of the QDs is generally ranged between 500 nm and 650 nm.[43] A higher-bandgap ZnS material as the shell can passivate the surface trap sites, which drastically enhances the emission efficiency of the QDs.[44] This is usually done by the addition of zinc and sulfur precursors into solutions containing CdSe cores dispersed in TOPO at high temperature. CdSe/ZnS QDs have been extensively used for in vitro and in vivo studies since the release of two *Science* papers from Nie's and Alivisatos' group in the year 1998.[45,46] They have demonstrated that biomolecule-conjugated QDs can be used as ultrabright, photostable, and biocompatible contrast agents for targeted cell imaging and sensing.[47,48]

11.3.2 CdTe/ZnS Core-Shell Quantum Dots

CdTe QDs have found numerous applications in electronics, sensors, and solar energy applications. As seen in Fig. 11.1, the CdTe QD emission wavelength can be systematically tuned from 500 nm to 800 nm.[49] CdTe QDs are generally prepared using the aqueous-phase synthesis approach.[50] However, the use of "naked" CdTe QDs for biological applications is still very limited to date. This is because the CdTe QDs are more toxic than the core-shell CdSe/ZnS, CdSe/CdS/ZnS, and InP/ZnS QDs. It was argued that the long-term contact of the functionalized CdTe QDs with biological fluids can result in partial desorption of the hydrophilic moieties, thus exposing the bare nanocrystal surface to the biological system, with high chances of releasing toxic cadmium ions into the surrounding. As a result, "naked" CdTe QDs are not widely used for biological applications. To overcome this challenge, some groups have proposed approaches to protect the CdTe QDs from further degradation. Tsay et al. reported the use of the two-step synthesis method to fabricate CdTe/ZnS QDs.[51] Basically, CdTe cores were synthesized in water first, and then they were transferred into organic solvents using the phase-transferring technique. Subsequently, the CdTe cores were coated with a ZnS shell in organic phase. This hybrid approach affords a facile growth of shells around cores, producing ultrasmall and highly luminescent NIR QDs.

11.3.3 InP/ZnS Core-Shell Quantum Dots

InP QDs as group III–V semiconductor nanocrystals are well known for their structural robustness.[52] This result from the greater degree of covalent bonding in these materials compared to those made up of group II–VI elements (e.g., CdSe, CdTe, CdS, ZnSe, ZnTe, ZnS). This structural robustness leads to enhanced optical stability and, most importantly, reduced toxicity due to lower chances of releasing constituent heavy metal ions. It was reported that the optical properties of the core-shell structures strongly depended on the nature of the shell around the InP[4]. By coating the InP core with a ZnS, ZnSe, or CdSe shell, the emission wavelength can be systematically tuned from 500 nm to 650 nm.

Peng's group has been actively producing high-quality InP QDs for potential biological applications.[53] The group first reported the synthesis of InP QDs without any size sorting in a noncoordinating solvent by strictly controlling the reaction parameters.[54] This method is much faster, less expensive, and "greener" than any other methods in the synthesis of InP QDs. They have discovered that fatty acids with well-defined chain lengths as ligands, careful selection of appropriate noncoordinating solvents, and a thorough degassing process are important factors in forming high-quality InP QDs. Later, they have improved the synthesis procedures to form InP/ZnS QDs[53] in high yield and quality. The sizes of InP QDs they produced ranged from 1 nm to 8 nm at very low temperature (<190°C) as compared to the current recipes available for making CdSe/ZnS QDs at (>350°C). The low temperature was enabled by the activation of In carboxylate precursors with fatty amines. By simply varying the reaction parameters, the absorption peak position can be systematically tuned from 390 nm to 720 nm without prolonged heating. This new synthetic scheme further enabled the direct synthesis of InP/ZnS core-shell nanocrystals in a one-pot fashion. The resulting InP/ZnS core-shell nanocrystals can emit from about 450 nm to 750 nm, with a quantum yield as high as over 40%. The bright core-shell nanocrystals were stable in air and could be dispersed in water with hydrophilic thiols ligands. The thiol ligand replacement did not quench the emission, and the nanocrystals were stable under ambient conditions for at least three months.

Recently, our group has reported a facile method in the synthesis of hydrophilic InP/ZnS QDs and used them as luminescence probes for live-cancer-cell imaging.[55] Efficient two-photon excitation and two-photon imaging are possible with these QDs.

11.3.4 PbS Quantum Dots

The bulk phase of the PbS semiconductor has a bandgap energy of 0.41 eV at a temperature of 300 K. Since this gap is larger than that of other QDs, such as CdSe, CdTe, PbSe, and PbTe, it is not necessary to make ultrasmall PbS QDs to adjust the effective gap across the 1,000–1,500 nm spectral region.[56] Sargent's group has developed biocompatible PbS QDs for biological applications.[57] The group has shown that the PbS QDs can be prepared through a one-pot synthesis method. The method is simple, with the use of nontoxic reagents, at low reaction temperatures.[58] The photoluminescence (PL) properties can be maintained in biological fluids, and the tuning of QD size can be achieved without the need of size-selective precipitation. They have demonstrated that the QDs emission can be tailored from 800 nm to 1,400 nm in the NIR spectral range. Moreover, the width of the emission peak is narrower than that for the PbS nanocrystals obtained by organometallic methods.[59,60] Even though the group has demonstrated the biocompatibility of the PbS QDs, no studies are conducted on using these functionalized nanoparticles for targeted bioimaging. More recently, Hyun et al. have shown a simple procedure to make water-dispersible PbS QDs.[61] The QD surface was functionalized with mercapto ligands that allow target-specific labeling of the cancer cells. For the first time, NIR luminescence imaging of human colon cancer cells is demonstrated with these water-dispersible PbS QDs.

11.3.5 Type II CdTe/CdSe Core-Shell Quantum Dots

So far, we have been discussed the preparation and characterization of type I semiconductor nanocrystals, where the band offsets are engineered in such a way that the conduction band of the semiconductor shell material is having a higher energy than that of the core and the valence band of the semiconductor shell material is having a lower energy than that of the core. In this scenario, both electrons and holes are trapped in the QD core. As for type II semiconductor QDs, the QDs are now designed to have both the valence and conduction bands in the core either higher or lower than in the semiconductor shell material.[62] As a result, one carrier is mostly confined to the core, while the other is mostly confined to the shell. It was reported that type II QDs are expected to have many novel properties that are fundamentally different from the type I QDs because of the spatial separations of carriers. Type II QDs can be made to emit in the NIR wavelengths that would otherwise not be available with type I QDs. Bawendi's group has reported the preparation of type II CdTe/CdSe QDs by a hot colloidal synthesis method.[63,64] They demonstrated that functionalized CdTe/CdSe QDs can be used as ultrasmall optical probes for sentinel lymph node mapping.[63]

11.3.6 Silicon Quantum Dots

A great alternative probe to replace current heavy metal–based QDs in cancer detection would be silicon QDs. Silicon QDs are expected to be far less toxic than group II–VI QDs. However, they have not been widely used in biomedical applications because they are difficult to synthesize in the aqueous phase and therefore are not compatible with biological fluids.[65] Silicon QDs are generally made dispersible in organic solvents such as chloroform and hexane because hydrophobic moieties (e.g., styrene and octene) can be simply attached to the QDs surface.[66] However, silicon QDs suffered from their poor quantum yield and colloidal instability in biological buffers. The use of red-emitting silicon QDs grafted with poly(acrylic acid) for fixed-cell labeling was reported.[67] However, these silicon QDs cannot maintained colloidal and spectral stability in biological fluids over a long period of time. Our group have prepared silicon QDs well dispersed in water with the help of poly(ethylene glycol) (PEG)ylated phospholipid micelles, in which the optical properties of silicon nanocrystals are maintained for over a few months.[68] In this approach, the silicon QDs were initially synthesized by laser-driven pyrolysis of silane and then followed by $HF-HNO_3$ etching.[69] The combination of both top-down and solution-phase synthesis methods allow the ease of scale-up production of silicon QDs. We have used styrene, octadecene, or ethyl undecylenate to functionalize the silicon QD surfaces, rendering them dispersible in organic solvents such as chloroform, hexane, and toluene. PEGylated phospholipid micelles were then used to enable transferring silicon QDs into biological buffers, creating a hydrophilic shell with PEG groups on their surface. More importantly, the PEG groups can be modified with functional groups such carboxyl, amine, and maleimide for a further bioconjucation process. All these promising results will aid the development of silicon QDs as valuable optical probes for biomedical diagnostics.

11.3.7 Other Types of Quantum Dots

In addition to CdSe/ZnS, CdTe/ZnS, PbS, InP/ZnS, and Si QDs, other types of nanocrystals such as InAs[70], ZnSe[71], ZnTe[72], CdHgTe[73,74], CuInSe[75], and CuInS[76,77] have been prepared through the hot colloidal synthesis technique. These QDs are less common for biomedical applications, either due to the fact that their emission wavelengths are in the ultraviolet (UV) range, which is not suitable for imaging studies, or due to the ease of degradation of these QDs in biological fluids. It is also difficult to produce these QDs with the desired quantum yield. Nevertheless, these particles may be further engineered for targeted imaging applications. For example, Pradhan et al. have reported the preparation of Mn-doped ZnSe QDs using the hot colloidal synthesis method.[78,79] The PL quantum yield of QDs can be as high as 50% by simply controlling the formation of small-sized MnSe nanoclusters as the core and the diffused interface between the nanocluster core and the ZnSe-overcoating layers. Also, Zhang et al. has demonstrated the preparation of Mn-doped Si nanoparticles using a low-temperature solution-phase synthesis technique.[80] They have shown that by doping 5% Mn in the silicon particles, the nanoparticles exhibited green emission with a slightly lower quantum yield compared to the undoped ones (pure silicon). Santra et al. have prepared ~3.1 nm Mn-doped CdS/ZnS core-shell QDs with fluorescent, radio-opacity, and paramagnetic properties for labeling brain blood vessels in small animals.[81] Qian et al. reported the fabrication of CdHgTe/CdS QDs with the aqueous-phase synthesis method and used them as probes for small-animal imaging.[82] These reports opened new routes in tailoring alternative QDs to emit in the visible spectrum region for biomedical applications.

11.3.8 CdSe/CdS/ZnS Quantum Rods

Many physical properties of semiconductor nanoparticles depend strongly on their shape structure.[83-87] Thus, the developments of well-controlled synthetic methods to manipulate the nanoparticle shape and elucidate the mechanisms by which the size and shape of the nanocrystals can be controlled are important issues in nanomaterial chemistry.[88-91] CdSe-based quantum rods (QRs) show advantages over corresponding spherical QDs in biological and medicinal applications.[92] For example, color control is achievable by tuning the rod diameter that governs the bandgap energy of CdSe QRs, and the Stokes shift is strongly dependent on the aspect ratio (length/diameter) of the rod.[93,94] These unique characteristics of QRs as biological markers may provide an avenue for further improvements in ultrasensitive bioimaging strategies. Alivisatos' and our groups have independently reported a general synthetic method to produce functionalized CdSe/CdS/ZnS QRs as ultrasensitive contrast agents for in vitro cancer imaging.[95,96] More specifically, Alivisatos' group has reported the use of QRs for breast cancer cell labeling. Water-dispersible QRs were prepared using the surface silanization technique. They have discovered that QRs are brighter single-molecule probes compared to QDs. On the other hand, our group has shown that QRs functionalized with a lysine shell can be used as sensitive contrast agents for two-photon fluorescence imaging of cancer cells, with reduced photodamage compared to UV-excited imaging. More recently, we demonstrated for the first time that peptide-labeled QRs were capable of active tumor targeting and imaging in live animals.[97]

11.4 Preparation of Water-Dispersible Quantum Dots

QDs are mostly prepared in the organic phase, and they are only dispersible in solvents such as toluene, hexane, and chloroform. A step to prepare water-dispersible QDs is vital for them to be useful in biomedical applications.[98,99] Two general methods have been commonly used to modify the hydrophobic surface of QDs to make them dispersible in biological fluids—the ligand exchange process and encapsulation of QDs with biocompatible materials (see Fig. 11.2).[100–102] In the case of the ligand exchange process (Fig. 11.2a–c), a hydrophobic moiety-functionalized QD solution is mixed with an excessive amount of heterobifunctional surfactants such as mercaptoacetic acid, mercaptosucinnic acid, mercaptopropionic

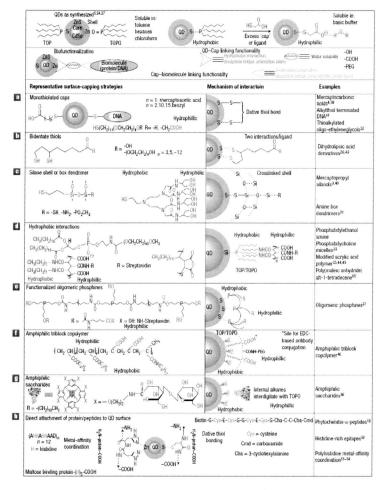

Figure 11.2 Schematic of generic QD solubilization and biofunctionalization (a–h) (reprinted by permission from Macmillan Publishers Ltd.: *Nature Materials* (Ref. 114, http://www.nature.com), Copyright 2005).

acid, and animoethanethiol, allowing functional groups (e.g., thiol or amine group) anchoring on the QD surface and the other functional groups (e.g., carboxyl and hydroxyl groups) protruding out from the QDs surface to favorably interact with water. During the ligand exchange process, the hydrophobic capping agents on the QD surface are "ripped off" by the hetrobifunctional surfactant through a mass-driven process and the bifunctional ligand binding to the QD surface will provide them with water dispersibility.[102] However, this approach comes with some disadvantages.[103] These include (i) instability optical properties of QDs, (ii) aggregation of functionalized QDs in biological fluids, and (iii) complex steps required to remove excess ligands from the QD solution. These challenges can be overcome by covering the hydrophobic QDs with amphiphilic biocompatible polymers. The encapsulation method (Fig. 11.2d–g) allows QDs to be dispersed in a biological buffer solution and remain stable for months due to the strong hydrophobic interaction between the hydrophobic moieties on the QD surface and the hydrophobic segment of the polymer. No matter what method is used to suspend the QDs in aqueous buffers, they should be purified from residual ligands and excess amphiphiles before use in biological assays by ultracentrifugation, dialysis, or filtration. Also, when choosing a water solubilization method, it should be noted that many biological and physical properties of the QDs may be affected by the surface coating and the overall physical dimensions of the QDs are dependent on the coating thickness. Typically the QDs are much larger when coated with amphiphiles compared to those coated with a monolayer of a ligand.

11.5 Preparation of Bioconjugated Quantum Dots

Water-dispersible QDs can be functionalized with a variety of biomolecules through well-known conjugation techniques due to their large surface-area-to-volume ratio that gives multiple surface attachment points for conjugation purpose.[104–107] First, QDs functionalized with carboxylic acid groups on their surface can be reacted with amines by using simple 1-ethyl-3-(3-dimethylaminopropyl)-carbodiimide (EDC) coupling chemistry.[102] The EDC approach for conjugation can be sometimes troublesome because it takes time to optimize the concentration ratio of QDs to targeting ligands to produce monodispersed bioconjugates. In an alternative approach, the amine groups on the nanocrystal surface can be reacted with active esters or converted to maleimide groups for conjugating with thiolated peptides, cysteine-tagged proteins, or thiolated antibodies.[108] Third, the hydrophobic coating of QDs can be directly displaced with thiolated peptides or proteins to provide a direct linking of targeting ligands to the nanocrystals surface.[109] Finally, nanocrystal conjugation can also be fabricated using a noncovalent self-assembly method with engineered proteins. It has been discovered that decorating QDs with PEG molecules can significantly reduce the uptake of QDs in the liver, spleen, and bone marrow.[109] Recently, QDs coated with PEG molecules were applied in noninvasive whole-body luminescence imaging. The PEGylated QDs were observed in the spleen, bone marrow, and lymph nodes of mice without observing any ill effects, illustrating their biocompatibility for in vivo studies. Many surface coatings (e.g., PEG, lysine, dihydrolipoic acid, PEGylated phospholipids, and cysteine) have been employed in making colloidal stable QD formulation. QDs terminated with a carboxyl group are reported to be rapidly taken up by the reticuloendothelial system (RES) system. PEGylated QDs terminated with anime groups have circulation half-lives depending on the molecular weight of PEG. Also, it was reported that when a "pure" PEG coating was applied on the QDs (without functional groups), the distribution of QDs in the whole body would depend on the length of the PEG molecule.[110] The biodistribution of QDs may vary slightly in the small-animal body, depending on the types of surface coatings used. Amino-PEG-, carboxy-PEG-, and mPEG-700-coated QDs are all deposited in the liver, spleen, and lymph nodes. Most importantly, the QD formulation tested in these studies was all excreted from the body within a few days.

11.6 Bioconjugated Quantum Dots and Quantum Rods for in vitro Cancer Imaging and Sensing

QDs have made significant progress in cellular labeling, imaging[111–113], and sensing.[27,114–116] Wu et al. reported the use of QDs conjugated with immunoglobulin G (IgG) and streptavidin to label the breast cancer marker human epidermal growth factor receptor 2 (Her2) on the surface of fixed and live cancer cells (see Fig. 11.3), to stain actin and microtubule fibers in the cytoplasm, and to detect nuclear antigens inside the nucleus.[117] Weng et al. fabricated multifunctional QD-conjugated immunoliposomes for breast cancer in vitro imaging.[118] Duan et al. reported a new class of cell-penetrating QDs with multivalent and endosome-disrupting (endosomolytic) surface coatings such as PEG-grafted polyethylenimine (PEI).[119] These QDs are able to penetrate cancer cell membranes and disrupt endosomal organelles in living cells. Jiang et al. demonstrated the use of lysine-coated QDs for fixed cancer cell imaging.[120] Our group has applied functionalized QRs for multiplex imaging of pancreatic cancer cells.[121] Bioconjugated QDs will be used as probes to label specific cellular components. The entry of bioconjugated QDs into fixed cancer cells is usually either through antibody–antigen interaction or through chemically creating pores.[122] For live-cell labeling with functionalized QDs, the process is more complex and other parameters such as cell viability, media, temperature, and population of cells should be considered for each designed set of experiments. So far, bioconjugated QDs (e.g., CdSe/ZnS, CdTe/ZnS, and CdTe/CdSe) with different emitting colors have been employed to label membrane-bound proteins, cytoplasmic compartments, subcellular components, and nucleus of live cancer cells.[117,123] For targeted live-cancer-cell imaging, the major hurdle is the transport of QDs to cross the cell membrane lipid layers.[120] There are a few ways to overcome this challenge.[124] It was reported that the positively charge QDs can be used to facilitate the crossing of the membrane lipid barrier since the surface of the cells is known to be slightly negatively charged.[125] Negatively charged QDs are generally terminated with carboxyl groups, and these particles can be conjugated with biomolecules (e.g., transferrin and folic acid) and delivered to the cells by a receptor-mediated process since many cancer cell lines are known to be overexpressed with certain receptors (e.g., transferrin and folate receptor) on their surface.[126] In addition to surface-charged QDs, there are other routes for targeted delivery of QDs to live cells, such as nonspecific pinocytosis, microinjection, and peptide-induced transport.[127] More recently, it was discovered that millions of QDs can be spontaneously delivered to the nucleus of a single cell without affecting the cell viability and functionality. The potential ability of QDs to monitor migration and differentiate single cell in real time is useful to answer important biology questions such as embryogenesis, cancer metastasis, and stem cell therapeutics.[128] The real advantage of using QDs for in vitro imaging comes from their resistance to photobleaching, which enhances the resolution and contrast of the images during the acquisition time.[129] With the unlimited emitting colors of functionalized QDs in the market, all can be excited with a single wavelength source, demonstrating the potential of QDs for various microscopy applications.[130]

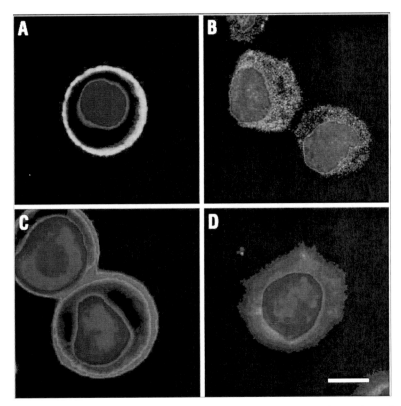

Figure 11.3 The QD–IgG probes successfully labeled Her2 on the surface of human breast cancer cells, SK-BR-3. Her2 was clearly labeled with (a) QD 535–IgG and (c) QD 630–IgG. When cells were incubated with normal mouse IgG and QD–IgG, there were no detectable or very weak nonspecific signals on the cell surface (b, d). (Reprinted by permission from Macmillan Publishers Ltd.: *Nature Biotechnology* (Ref. 131, http://www.nature.com), Copyright 2003.)

11.7 Multifunctional Quantum Dots and Quantum Rods for in vivo Cancer Targeting and Imaging

QDs have significantly improved over the organic dyes used for in vivo imaging because they have a nonspecific in vivo distribution after functionalization with biomolecules and rapid renal clearance.[131] They have excellent stability at high concentrations and a long circulation time in vivo for more than a few days.[132]

Two approaches have been used to deliver the QD formulation to cancerous areas, passive targeting and active targeting.[133] Passive targeting delivery relies on two unique properties of the tumor microenvironment[2]. One is the presence of leaky vasculatures in the tumor, and this makes them highly permeable to macromolecules in comparison with normal tissue.[134] Two is the presence of a dysfunctional lymphatic drainage system in the tumor, which improves the fluid retention in the tumor interstitial space.[135] Due to these factors, the nanoparticle concentration (e.g., micelles, liposome particles) found in the tumor matrix can be as high as 100 to 150 times than that in the normal tissue. This specific deposition of nanoparticles in the tumor environment is well known as the enhanced permeability and retention (EPR) effect that occurs as nanoparticles extravasate out of tumor microvasculature, leading to an accumulation of particles in the tumor interstitium.[134,136] The nanoparticle extravasation process depends on the size of open interendothelial gap junctions and transendothelial channels.[137] The pore cutoff size of these transport pathways is reported to be in the range of 400 nm to 600 nm. It was reported that the cutoff size of liposomes extravasation into tumors in vivo was around 400 nm. Thus, nanoparticles with sizes less than 200 nm will be quite effective for extravasating the tumor microenvironment. The hydrodynamic size of functionalized QDs is generally between 5 nm and 60 nm, suggesting that they are suitable for tumor targeting and imaging using the EPR effect. On the other hand, active targeting is achieved by both local and systemic administration of QDs, with targeting molecules conjugated on the QD surface that can specifically recognize and bind to receptors that are overexpressed on the cancer cells or tissue surface.[2,138,139] In the case of targeted delivery, the nanoparticles can be delivered to cancer cells, while minimizing ill effects to normal tissue.[138,140] This method is particularly useful for the surgeons to differentiate the primary tumors that have not metastasized to other major organs.[141]

The use of engineered QDs as multifunctional probes with an emission wavelength ranging from 500 nm to 900 nm for noninvasive tumor targeting and multiplex imaging in live animals[142] has been reported. Gao et al. demonstrated the use of QDs for tumor imaging in nude mice[143] (see Fig. 11.4). These QDs were carefully encapsulated with a block copolymer to improve intravascular circulation and colloidal stability. Further linking the QDs with biomolecules such as anti-prostate-specific membrane antigen (PSMA), the QD bioconjugates allowed specific targeting of human prostate cancer xenografts. They have shown that polymer beads doped with CdSe/ZnS QDs emitting different colors can be visualized in nude mice at the same time. Recently, Stroh et al. demonstrated that QDs can be used for staining tumor blood vessels in vivo.[144] The authors suggested that the highly luminescent QDs (e.g., CdSe/ZnS, CdSe/CdS/ZnS, and CdTe/ZnS) will provide the capability to differentiate the tumor vasculature and tumor matrix. Akerman et al. demonstrated targeting of breast cancer xenografts using bioconjugated CdSe/ZnS QDs.[145] However, when QDs were conjugated with endotheliocyte-targeting molecules, the conjugates were localized in the tumor vasculature. Chen et al. employed CdSe/ZnS QDs linked with an alpha-fetoprotein monoclonal antibody for targeted imaging of a human hepatocellular carcinoma xenograft growing on nude mice.[146] Also, the authors have investigated the cytotoxicity, hemodynamics, and distribution of the QDs in tissue. We have applied cyclic arginine-glycine-aspartic acid (RGD)-peptide-labeled CdSe/CdS/ZnS QRs for tumor targeting and imaging in live animals (see Fig. 11.5).[97] The targeted luminescent probes involve encapsulating hydrophobic CdSe/CdS/ZnS QRs with PEGylated phospholipids, followed by conjugation of these PEGylated phospholipids to ligands that specifically target the tumor vasculature. In vivo optical imaging studies in nude mice bearing pancreatic cancer xenografts, both subcutaneous and orthotopic, indicate that the QR probes accumulate at tumor sites via the cyclic RGD peptides binding to the $\alpha_V\beta_3$ integrins overexpressed in the tumor vasculature, following systemic injection. More importantly, cytotoxicity studies indicated the absence of any toxic effect in the cellular and tissue levels arising from functionalized QRs.

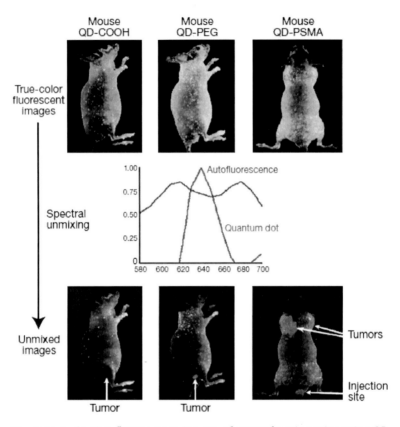

Figure 11.4 In vivo fluorescence images of tumor-bearing mice using QD probes with three different surface modifications: carboxylic acid groups (left), PEG groups (middle), and PEG-PSMA Ab conjugates (right). (Reprinted by permission from Macmillan Publishers Ltd.: *Nature Biotechnology* (Ref. 164, http://www.nature.com), Copyright 2004.)

The use of QDs that emit in the NIR spectrum will be an excellent choice for ultrasensitive in vivo cancer imaging.[147] By manipulating the composition and structure of QDs, these NIR QDs can be made to display emission peaks ranging from 700 nm to 950 nm.[148] This emission window is well separated from the autofluorescence (400 to 600 nm) region.[133] Bawendi's group has used CdTe/CdSe QDs for optically tracing lymph nodes by NIR luminescence imaging. The authors have reported that upon injection of the QDs into the paw of a mouse, the exact location of the sentinel lymph node could be determined. These important findings could simplify the surgical procedure in removing lymph nodes for pathology evaluation.

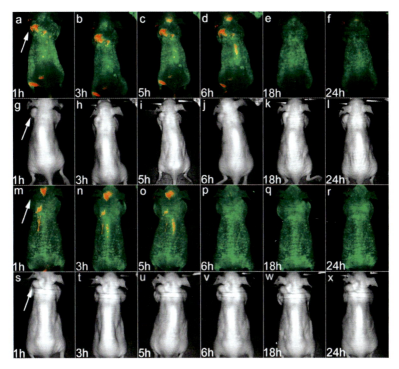

Figure 11.5 Time-dependent in vivo luminescence imaging of Panc-1 tumor-bearing mice (left shoulder, indicated by white arrows) injected with 1 mg of cRGD-peptide-conjugated QRs (a–f) and nonconjugated QRs (m–r), respectively. All images were acquired under the same experimental conditions. The autofluorescence from tumor-bearing mice is coded in green, and the unmixed QR signal is coded in red. Prominent uptake in the liver, spleen, and lymph nodes was also visible. TEM images in panels g–i and panels s–x correspond to the luminescence images in panels a–f and panels. (Reprinted with permission from Ref. 111, Copyright 2009 American Chemical Society.)

Besides serving as powerful optical targeted probes, QDs have been recently engineered as "transport cargo" vehicles for traceable drug delivery applications.[149] Langer's group is the pioneer in developing functionalized CdSe/ZnS QDs for cancer therapy.[150] For example, the group has fabricated QDs functionalized with both an aptamer and doxorubicin for targeted cancer imaging, therapy, and sensing. The developed QD system is capable of targeting and imaging prostate cancer cells that express the PSMA protein. The intercalation of doxorubicin in the double-stranded stem of the aptamer results in bioconjugated QDs with reversible self-quenching properties based on a bi-fluorescence resonance energy transfer (FRET) mechanism. This multifunctional QD system can deliver doxorubicin to the targeted prostate cancer cells, and the authors subsequently sense the delivery of doxorubicin by activating the luminescence of QDs. This work can be further developed for in vivo cancer imaging and therapy applications in the near future.

11.8 The Risk and Benefits of Using Functionalized Quantum Dots for Biomedical Health Care

Though QDs are promising for cancer imaging and therapy, concerns about using them for in vivo applications are constantly raised due to their potential toxicity, especially with heavy metal–based QDs.[151-157] Scientists also have doubts about whether the QDs formulation can be excreted from the body after they have performed their "programmed" tasks (e.g., imaging, drug delivery).[158-160] Before addressing all these issues, the author would like to start the discussion with doxorubicin formulation, a chemotherapy drug available in most hospitals and clinics in the world for many human cancer therapies.[161] Doxorubicin is an anthracyclines that is usually used to treat a large variety of cancers.[162] Doxorubicin was first used for treating both hematological and solid tumors in the 1970s.[163] The side effects include nausea, vomiting, heart arrhythmias, neutropenia, alopecia, and overall stress to the mind and body.[164] To date, the clinically recommended doxorubicin dose for cancer patients usually ranges from 60 mg/m^2 to 75 mg/m^2 for every three weeks or 20 mg/m^2 for a one-week dose.[165] The major acute toxicity of doxorubicin is bone marrow suppression. In addition, congestive heart failure will occur if cumulative drug doses exceeded 500 mg/m^2. Though doxorubicin is known for its deadly toxicity and side effects to the body, many countries are continuing to use it as a final resource to treat and cure cancer patients.[166]

Many patients were successfully treated and cured with appropriate dosage of doxorubicin formulations. It is worth mentioning that it takes more than 40 years to evaluate, test, examine, investigate, and re-engineer the doxorubicin formulations to be safely used and treat cancer patients today because many problems were encountered during their usage period.[167] Similarly, QD formulations are currently experiencing the same history as that of doxorubicin for both in vitro and in vivo cancer imaging and therapy. It will of course take a considerable amount of time before QDs can be realized in the clinical trials, because statistical meaningful data is needed to engineer the safe and optimized dosage of QD formulations for personalized cancer patients' diagnoses and treatment. However, it is very fortunate that QDs as diagnostic and therapeutic probes have been receiving great attention for the last few years[168,169], which will eventually speed up the development of safe QD formulation for early cancer imaging and therapy.

11.9 Conclusions and Outlook

This chapter has described the use of QDs as multifunctional probes for biomedical detection. These techniques are limited to laboratory settings, and the clinical use of QDs awaits unambiguous proof of beneficial effects for a given application. This presentation is aimed to provide a base for researchers to develop novel functional QD platforms as new building blocks for advancing approaches in disease diagnosis. In future, we envision that the versatile and rich surface chemistry of QDs combined with a better understanding of the influence of these nanocrystals on cells and bodies will pave the way for the use of QDs in sensing and treatment of cancers and other human diseases.

Acknowledgments

This work was supported by the AACR-Pancreatic Cancer Action Network Fellowship for Pancreatic Cancer Research, a startup grant from Nanyang Technological University (No.M58040022), and tier 1 academic research funds from Nanyang Technological University (No.M52040149).

References

1. P. Mayer-Kuckuk, D. Banerjee, M. A. Hayat, "Strategies for imaging biology in cancer and other diseases," in *Cancer Imaging*, San Diego: Academic Press, 2008, 3–14.
2. F. X. Gu, R. Karnik, A. Z. Wang, F. Alexis, E. Levy-Nissenbaum, S. Hong, R. S. Langer, O. C. Farokhzad, *Nano Today*, 14–21 (2007).
3. F. Alexis, J.-W. Rhee, J. P. Richie, A. F. Radovic-Moreno, R. Langer, O. C. Farokhzad, *Urol. Oncol.: Semin. Ori. Invest.*, 74–85.
4. P. N. Prasad, *Nanophotonics*, New York: Wiley-Interscience, 2004.
5. P. N. Prasad, *Biophotonics*, New York: Wiley-Interscience, 2004.
6. H. M. E. Azzazy, M. M. H. Mansour, S. C. Kazmierczak, *Clin. Biochem.*, 917–927 (2007).
7. I. Brigger, C. Dubernet, P. Couvreur, *Adv. Drug Delivery Rev.*, 631–651 (2002).
8. L. Brannon-Peppas, J. O. Blanchette, *Adv. Drug Delivery Rev.*, 1649–1659 (2004).
9. K.-T. Yong, Y. Sahoo, M. T. Swihart, P. N. Prasad, *Colloids Surf., A*, 89–105 (2006).
10. W.-T. Liu, *J. Biosci. Bioeng.*, 1–7 (2006).
11. Y. Yin, A. P. Alivisatos, *Nature*, 664–670 (2005).
12. K. Tao, H. Dou, K. Sun, *Chem. Mater.*, 5273–5278 (2006).
13. K.-T. Yong, Y. Sahoo, M. T. Swihart, P. N. Prasad, *Adv. Mater.*, 1978–1982 (2006).
14. H. Yu, J. Li, R. A. Loomis, P. C. Gibbons, L. W. Wang, W. E. Buhro, *J. Am. Chem. Soc.*, 16168–16169 (2003).
15. Y.-L. Shi, T. Asefa, *Langmuir*, 9455–9462 (2007).
16. M. M. Husein, E. Rodil, J. H. Vera, *Langmuir*, 2264–2272 (2006).
17. T.-J. Zhu, X. Chen, Y.-Q. Cao, X.-B. Zhao, *J. Phys. Chem. C*, 8085–8091 (2009).
18. X. Chen, S. S. Mao, *Chem. Rev.*, 2891–2959 (2007).
19. X. Wen, Y. Xie, C. L. Choi, K. C. Wan, X.-Y. Li, S. Yang, *Langmuir*, 4729–4737 (2005).
20. Z. Zheng, Huang, Ma, Zhang, Liu, Z. LiuLiu, K. W. Wong, W. M. Lau, *Cryst. Growth Des.*, 1912–1917 (2007).
21. B. L. Cushing, V. L. Kolesnichenko, C. J. O'Connor, *Chem. Rev.*, 3893–3946 (2004).

22. H. Arya, Z. Kaul, R. Wadhwa, K. Taira, T. Hirano, S. C. Kaul, *Biochem. Biophys. Res. Commun.*, 1173–1177 (2005).
23. R. E. Bailey, A. M. Smith, S. Nie, *Phys. E: Low-dimens. Syst., Nanostruct.* 1–12 (2004).
24. D. K. Chatterjee, Y. Zhang, *Sci. Technol. Adv. Mater.*, 131–133 (2007).
25. A. P. Alivisatos, *J. Phys. Chem.*, 13226–13239 (1996).
26. J. M. Pietryga, R. D. Schaller, D. Werder, M. H. Stewart, V. I. Klimov, J. A. Hollingsworth, *J. Am. Chem. Soc.*, 11752–11753 (2004).
27. A. Zajac, D. Song, W. Qian, T. Zhukov, *Colloids Surf., B*, 309–314 (2007).
28. M. Danek, K. F. Jensen, C. B. Murray, M. G. Bawendi, *Chem. Mater.*, 173–180 (1996).
29. W. C. W. Chan, D. J. Maxwell, X. Gao, R. E. Bailey, M. Han, S. Nie, *Curr. Opin. Biotechnol.*, 40–46 (2002).
30. X. Michalet, F. F. Pinaud, L. A. Bentolila, J. M. Tsay, S. Doose, J. J. Li, G. Sundaresan, A. M. Wu, S. S. Gambhir, S. Weiss, *Science*, 538–544 (2005).
31. J. M. Costa-Fernández, R. Pereiro, A. Sanz-Medel, *Trends Anal. Chem.*, 207–218 (2006).
32. F. Pinaud, X. Michalet, L. A. Bentolila, J. M. Tsay, S. Doose, J. J. Li, G. Iyer, S. Weiss, *Biomaterials*, 1679–1687 (2006).
33. J. K. Jaiswal, S. M. Simon, *Trends Cell Biol.*, 497–504 (2004).
34. C. Burda, X. Chen, R. Narayanan, M. A. El-Sayed, *Chem. Rev.*, 1025–1102 (2005).
35. S. Kim, M. G. Bawendi, *J. Am. Chem. Soc.*, 14652–14653 (2003).
36. Z. A. Peng, X. Peng, *J. Am. Chem. Soc.*, 183–184 (2001).
37. Z. A. Peng, X. Peng, *J. Am. Chem. Soc.*, 3343–3353 (2002).
38. L. Qu, Z. A. Peng, X. Peng, *Nano Lett.*, 333–337 (2001).
39. D. V. Talapin, I. Mekis, S. Gotzinger, A. Kornowski, O. Benson, H. Weller, *J. Phys. Chem. B*, 18826–18831 (2004).
40. C. B. Murray, C. R. Kagan, M. G. Bawendi, *Science*, 1335–1338 (1995).
41. P. Sharma, S. Brown, G. Walter, S. Santra, B. Moudgil, *Adv. Colloid Interface Sci.*, 471–485 (2006).
42. C. B. Murray, D. J. Norris, M. G. Bawendi, *J. Am. Chem. Soc.*, 8706–8715 (1993).
43. X. Peng, M. C. Schlamp, A. V. Kadavanich, A. P. Alivisatos, *J. Am. Chem. Soc.*, 7019–7029 (1997).

44. X. Peng, J. Wickham, A. P. Alivisatos, *J. Am. Chem. Soc.*, 5343–5344 (1998).
45. M. Bruchez, Jr., M. Moronne, P. Gin, S. Weiss, A. P. Alivisatos, *Science*, 2013–2016 (1998).
46. W. C. W. Chan, S. Nie, *Science*, 2016–2018 (1998).
47. A. Fu, C. M. Micheel, J. Cha, H. Chang, H. Yang, A. P. Alivisatos, *J. Am. Chem. Soc.*, 10832–10833 (2004).
48. P. M. A. Farias, B. S. Santos, A. Fontes, A. A. S. Vieira, D. C. N. Silva, A. G. Castro-Neto, C. R. Chaves, A. H. G. B. Da Cunha, D. Scordo, J. C. O. F. Amaral, V. Moura-Neto, *Appl. Surf. Sci.*, 691–693 (2008).
49. C. Wang, Q. Ma, W. Dou, S. Kanwal, G. Wang, P. Yuan, X. Su, *Talanta*, 1358–1364 (2009).
50. A. L. Rogach, T. Franzl, T. A. Klar, J. Feldmann, N. Gaponik, V. Lesnyak, A. Shavel, A. Eychmuller, Y. P. Rakovich, J. F. Donegan, *J. Phys. Chem. C*, 14628–14637 (2007).
51. J. M. Tsay, M. Pflughoefft, L. A. Bentolila, S. Weiss, *J. Am. Chem. Soc.*, 1926–1927 (2004).
52. E. P. A. M. Bakkers, M. A. Verheijen, *J. Am. Chem. Soc.*, 3440–3441 (2003).
53. R. Xie, D. Battaglia, X. Peng, *J. Am. Chem. Soc.*, 15432–15433 (2007).
54. D. Battaglia, X. Peng, *Nano Lett.*, 1027–1030 (2002).
55. D. J. Bharali, D. W. Lucey, H. Jayakumar, H. E. Pudavar, P. N. Prasad, *J. Am. Chem. Soc.*, 11364–71 (2005).
56. J. S. Steckel, B. K. H. Yen, D. C. Oertel, M. G. Bawendi, *J. Am. Chem. Soc.*, 13032–13033 (2006).
57. S. Hinds, S. Myrskog, L. Levina, G. Koleilat, J. Yang, S. O. Kelley, E. H. Sargent, *J. Am. Chem. Soc.*, 7218–7219 (2007).
58. W. Lin, K. Fritz, G. Guerin, G. R. Bardajee, S. Hinds, V. Sukhovatkin, E. H. Sargent, G. D. Scholes, M. A. Winnik, *Langmuir*, 8215–8219 (2008).
59. D. A. R. Barkhouse, A. G. Pattantyus-Abraham, L. Levina, E. H. Sargent, *ACS Nano*, 2356–2362 (2008).
60. E. Istrate, S. Hoogland, V. Sukhovatkin, L. Levina, S. Myrskog, P. W. E. Smith, E. H. Sargent, *J. Phys. Chem. B*, 2757–2760 (2008).
61. B.-R. Hyun, Chen, D. A. Rey, F. W. Wise, C. A. Batt, *J. Phys. Chem. B*, 5726–5730 (2007).
62. S. Kim, B. Fisher, H. J. Eisler, M. Bawendi, *J. Am. Chem. Soc.*, 11466–11467 (2003).

63. S. Kim, Y. T. Lim, E. G. Soltesz, A. M. De Grand, J. Lee, A. Nakayama, J. A. Parker, T. Mihaljevic, R. G. Laurence, D. M. Dor, L. H. Cohn, M. G. Bawendi, J. V. Frangioni, *Nat. Biotech.*, 93–97 (2004).
64. S. W. Kim, S. Kim, J. B. Tracy, A. Jasanoff, M. G. Bawendi, *J. Am. Chem. Soc.*, 4556–4557 (2005).
65. F. Hua, M. T. Swihart, E. Ruckenstein, *Langmuir*, 6054–6062 (2005).
66. X. Li, Y. He, S. S. Talukdar, M. T. Swihart, *Langmuir*, 8490–8496 (2003).
67. Z. F. Li, E. Ruckenstein, *Nano Lett.*, 1463–1467 (2004).
68. F. Erogbogbo, K.-T. Yong, I. Roy, G. Xu, P. N. Prasad, M. T. Swihart, *ACS Nano*, 873–878 (2008).
69. X. Li, Y. He, M. T. Swihart, *Langmuir*, 4720–4727 (2004).
70. F. Wang, H. Yu, S. Jeong, J. M. Pietryga, J. A. Hollingsworth, P. C. Gibbons, W. E. Buhro, *ACS Nano*, 1903–1913 (2008).
71. S. Schlecht, S. Tan, M. Yosef, R. Dersch, J. H. Wendorff, Z. Jia, A. Schaper, *Chem. Mater.*, 809–814 (2005).
72. S. H. Lee, Y. J. Kim, J. Park, *Chem. Mater.*, 4670–4675 (2007).
73. B. Tang, F. Yang, Y. Lin, L. Zhuo, J. Ge, L. Cao, *Chem. Mater.*, 1212–1214 (2007).
74. H. Sun, H. Zhang, J. Ju, J. Zhang, G. Qian, C. Wang, B. Yang, Z. Y. Wang, *Chem. Mater.*, 6764–6769 (2008).
75. P. M. Allen, M. G. Bawendi, *J. Am. Chem. Soc.*, 9240–9241 (2008).
76. R. Xie, M. Rutherford, X. Peng, *J. Am. Chem. Soc.*, 5691–5697 (2009).
77. H. Zhong, Y. Zhou, M. Ye, Y. He, J. Ye, C. He, C. Yang, Y. Li, *Chem. Mater.*, 6434–6443 (2008).
78. N. Pradhan, D. Goorskey, J. Thessing, X. Peng, *J. Am. Chem. Soc.*, 17586–17587 (2005).
79. Y. Zhang, C. Gan, J. Muhammad, D. Battaglia, X. Peng, M. Xiao, *J. Phys. Chem. C*, 20200–20205 (2008).
80. X. Zhang, M. Brynda, R. D. Britt, E. C. Carroll, D. S. Larsen, A. Y. Louie, S. M. Kauzlarich, *J. Am. Chem. Soc.*, 10668–10669 (2007).
81. S. Santra, H. Yang, P. H. Holloway, J. T. Stanley, R. A. Mericle, *J. Am. Chem. Soc.*, 1656–1657 (2005).
82. H. Qian, C. Dong, J. Peng, X. Qiu, Y. Xu, J. Ren, *J. Phys. Chem. C*, 16852–16857 (2007).
83. C. J. Barrelet, Y. Wu, D. C. Bell, C. M. Lieber, *J. Am. Chem. Soc.*, 11498–11499 (2003).
84. M. J. Bierman, Y. K. A. Lau, S. Jin, *Nano Lett.*, 2907–2912 (2007).

85. K. T. Yong, Y. Sahoo, K. R. Choudhury, M. T. Swihart, J. R. Minter, P. N. Prasad, *Chem. Mater.*, 5965–5972 (2006).
86. K. T. Yong, Y. Sahoo, H. Zeng, M. T. Swihart, J. R. Minter, P. N. Prasad, *Chem. Mater.*, 4108–4110 (2007).
87. K.-T. Yong, Y. Sahoo, M. Swihart, P. Schneeberger, P. Prasad, *Top. Catal.*, 49–60 (2008).
88. S. Acharya, I. Patla, J. Kost, S. Efrima, Y. Golan, *J. Am. Chem. Soc.*, 9294–9295 (2006).
89. L. Manna, D. J. Milliron, A. Meisel, E. C. Scher, A. P. Alivisatos, *Nat. Mater.*, 382–385 (2003).
90. K. T. Yong, Y. Sahoo, K. R. Choudhury, M. T. Swihart, J. R. Minter, P. N. Prasad, *Nano Lett.*, 709–714 (2006).
91. K. T. Yong, Y. Sahoo, M. T. Swihart, P. N. Prasad, *J. Phys. Chem. C*, 2447–2458 (2007).
92. X. Peng, L. Manna, W. Yang, J. Wickham, E. Scher, A. Kadavanich, A. P. Alivisatos, *Nature*, 59–61 (2000).
93. T. Mokari, U. Banin, *Chem. Mater.*, 3955–3960 (2003).
94. Z. A. Peng, X. Peng, *J. Am. Chem. Soc.*, 1389–1395 (2001).
95. A. Fu, W. Gu, B. Boussert, K. Koski, D. Gerion, L. Manna, M. LeGros, C. A. Larabell, A. P. Alivisatos, *Nano Lett.*, 179–182 (2007).
96. K. T. Yong, J. Qian, I. Roy, H. H. Lee, E. J. Bergey, K. M. Tramposch, S. He, M. T. Swihart, A. Maitra, P. N. Prasad, *Nano Lett.*, 761–765 (2007).
97. K.-T. Yong, R. Hu, I. Roy, H. Ding, L. A. Vathy, E. J. Bergey, M. Mizuma, A. Maitra, P. N. Prasad, *ACS Appl. Mater. Interfaces*, 710–719 (2009).
98. P. K. Bae, K. N. Kim, S. J. Lee, H. J. Chang, C. K. Lee, J. K. Park, *Biomaterials*, 836–842 (2009).
99. D. Maysinger, J. Lovric, A. Eisenberg, R. Savic, *Eur. J. Pharm. Biopharm.*, 270–281 (2007).
100. I. L. Medintz, H. T. Uyeda, E. R. Goldman, H. Mattoussi, *Nat. Mater.*, 435–446 (2005).
101. Y. Xing, M.-k. So, A. L. Koh, R. Sinclair, J. Rao, *Biochem. Biophys. Res. Commun.*, 388–394 (2008).
102. A. M. Smith, H. Duan, A. M. Mohs, S. Nie, *Adv. Drug Delivery Rev.*, 1226–1240 (2008).
103. J. R. Krogmeier, H. Kang, M. L. Clarke, P. Yim, J. Hwang, *Opt. Commun.*, 1781–1788 (2008).
104. D. Gerion, F. Pinaud, S. C. Williams, W. J. Parak, D. Zanchet, S. Weiss, A. P. Alivisatos, *J. Phys. Chem. B*, 8861–8871 (2001).

105. A. F. E. Hezinger, J. Teßmar, A. Göpferich, *Eur. J. Pharm. Biopharm.*, 138–152 (2008).
106. K. K. Jain, *Trends in Biotechnology*, 143–145 (2006).
107. J. D. Smith, G. W. Fisher, A. S. Waggoner, P. G. Campbell, *Microvasc. Res.*, 75–83 (2007).
108. X. Gao, L. Yang, J. A. Petros, F. F. Marshall, J. W. Simons, S. Nie, *Curr. Opin. Biotechnol.*, 63–72 (2005).
109. Y. Xing, Q. Chaudry, C. Shen, K. Y. Kong, H. E. Zhau, L. W. Chung, J. A. Petros, R. M. O'Regan, M. V. Yezhelyev, J. W. Simons, M. D. Wang, S. Nie, *Nat. Protoc.*, 1152–1165 (2007).
110. B. Ballou, B. C. Lagerholm, L. A. Ernst, M. P. Bruchez, A. S. Waggoner, *Bioconjugate Chem.*, 79–86 (2003).
111. Y. Higuchi, M. Oka, S. Kawakami, M. Hashida, *J. Controlled Release*, 131–136 (2008).
112. W. A. Hild, M. Breunig, A. Goepferich, *Eur. J. Pharm. Biopharm.*, 153–168 (2008).
113. Z. Li, K. Wang, W. Tan, J. Li, Z. Fu, C. Ma, H. Li, X. He, J. Liu, *Anal. Biochem.*, 169–174 (2006).
114. D. Gerion, F. Chen, B. Kannan, A. Fu, W. J. Parak, D. J. Chen, A. Majumdar, A. P. Alivisatos, *Anal. Chem.*, 4766–4772 (2003).
115. T. Zhukov, M. Dubiec, P. Zhukov, N. Korsunska, S. Ostapenko, J. Zhang, A. Cantor, M. Tockman, *Lung Cancer*, S84–S84 (2005).
116. A. M. Smith, X. Gao, S. Nie, *Photochem. Photobiol.*, 377–385 (2004).
117. X. Wu, H. Liu, J. Liu, K. N. Haley, J. A. Treadway, J. P. Larson, N. Ge, F. Peale, M. P. Bruchez, *Nat. Biotech.*, 41–46 (2003).
118. K. C. Weng, C. O. Noble, B. Papahadjopoulos-Sternberg, F. F. Chen, D. C. Drummond, D. B. Kirpotin, D. Wang, Y. K. Hom, B. Hann, J. W. Park, *Nano Lett.*, 2851–2857 (2008).
119. H. Duan, S. Nie, *J. Am. Chem. Soc.*, 3333–3338 (2007).
120. W. Jiang, S. Mardyani, H. Fischer, C. W. Chan, *Chem. Mater.*, 872–878 (2006).
121. K.-T. Yong, I. Roy, H. E. Pudavar, E. J. Bergey, K. M. Tramposch, M. T. Swihart, P. N. Prasad, *Adv. Mater.*, 1412–1417 (2008).
122. E. B. Garon, L. Marcu, Q. Luong, O. Tcherniantchouk, G. M. Crooks, H. P. Koeffler, *Leukemia Res.*, 643–651 (2007).
123. K. Susumu, B. C. Mei, H. Mattoussi, *Nat. Protoc.*, 424–436 (2009).
124. J. M. Klostranec, W. C. W. Chan, *Adv. Mater.*, 1953–1964 (2006).

125. D. Li, G. Li, W. Guo, P. Li, E. Wang, J. Wang, *Biomaterials*, 2776–2782 (2008).
126. G. Xu, K.-T. Yong, I. Roy, S. D. Mahajan, H. Ding, S. A. Schwartz, P. N. Prasad, *Bioconjugate Chem.*, 1179–1185 (2008).
127. A. M. Derfus, W. C. W. Chan, S. N. Bhatia, *Adv. Mater.*, 961–966 (2004).
128. S. Courty, C. Bouzigues, C. Luccardini, M. Ehrensperger, S. Bonneau, M. Dahan, I. James, "Tracking individual proteins in living cells using single quantum dot imaging," in *Methods in Enzymology*, Vol. 414, Academic Press, 2006, 211–228.
129. T. Jamieson, R. Bakhshi, D. Petrova, R. Pocock, M. Imani, A. M. Seifalian, *Biomaterials*, 4717–4732 (2007).
130. S.-C. Hsieh, F.-F. Wang, C.-S. Lin, Y.-J. Chen, S.-C. Hung, Y.-J. Wang, *Biomaterials*, 1656–1664 (2006).
131. B. Dubertret, P. Skourides, D. J. Norris, V. Noireaux, A. H. Brivanlou, A. Libchaber, *Science*, 1759–1762 (2002).
132. X. Gao, J. Chen, J. Chen, B. Wu, H. Chen, X. Jiang, *Bioconjugate Chem.*, 2189–2195 (2008).
133. M. V. Yezhelyev, X. Gao, Y. Xing, A. Al-Hajj, S. Nie, R. M. O'Regan, *Lancet Oncol.*, 657–667 (2006).
134. J. Cheng, B. A. Teply, I. Sherifi, J. Sung, G. Luther, F. X. Gu, E. Levy-Nissenbaum, A. F. Radovic-Moreno, R. Langer, O. C. Farokhzad, *Biomaterials*, 869–876 (2007).
135. O. C. Farokhzad, R. Langer, *ACS Nano*, 16–20 (2009).
136. O. C. Farokhzad, R. Langer, *Adv. Drug Delivery Rev.*, 1456–1459 (2006).
137. A. Potineni, D. M. Lynn, R. Langer, M. M. Amiji, *J. Controlled Release*, 223–234 (2003).
138. S. A. Townsend, G. D. Evrony, F. X. Gu, M. P. Schulz, R. H. Brown Jr, R. Langer, *Biomaterials*, 5176–5184 (2007).
139. J. D. Byrne, T. Betancourt, L. Brannon-Peppas, *Adv. Drug Delivery Rev.*, 1615–1626 (2008).
140. Y. Lu, E. Sega, C. P. Leamon, P. S. Low, *Adv. Drug Delivery Rev.*, 1161–1176 (2004).
141. V. Bagalkot, O. C. Farokhzad, R. Langer, S. Jon, *Nanomed.: Nanotechnol., Biol. Med.*, 352–352 (2007).
142. H. Kobayashi, Y. Hama, Y. Koyama, T. Barrett, C. A. S. Regino, Y. Urano, P. L. Choyke, *Nano Lett.*, 1711–1716 (2007).
143. X. Gao, Y. Cui, R. M. Levenson, L. W. K. Chung, S. Nie, *Nat. Biotech.*, 969–976 (2004).

144. M. Stroh, J. P. Zimmer, D. G. Duda, T. S. Levchenko, K. S. Cohen, E. B. Brown, D. T. Scadden, V. P. Torchilin, M. G. Bawendi, D. Fukumura, R. K. Jain, *Nat. Med.*, 678–682 (2005).

145. M. E. Akerman, W. C. W. Chan, P. Laakkonen, S. N. Bhatia, E. Ruoslahti, *Proc. Natl. Acad. Sci. U S A*, 12617–12621 (2002).

146. L.-D. Chen, J. Liu, X.-F. Yu, M. He, X.-F. Pei, Z.-Y. Tang, Q.-Q. Wang, D.-W. Pang, Y. Li, *Biomaterials*, 4170–4176 (2008).

147. R. E. Bailey, S. Nie, *J. Am. Chem. Soc.*, 7100–7106 (2003).

148. W. Jiang, A. Singhal, B. Y. S. Kim, J. Zheng, J. T. Rutka, C. Wang, W. C. W. Chan, *J. Assoc. Lab. Autom.*, 6–12 (2008).

149. P. Juzenas, W. Chen, Y.-P. Sun, M. A. N. Coelho, R. Generalov, N. Generalova, I. L. Christensen, *Adv. Drug Delivery Rev.*, 1600–1614 (2008).

150. V. Bagalkot, L. Zhang, E. Levy-Nissenbaum, S. Jon, P. W. Kantoff, R. Langer, O. C. Farokhzad, *Nano Lett.*, 3065–3070 (2007).

151. Z. Chen, H. Chen, H. Meng, G. Xing, X. Gao, B. Sun, X. Shi, H. Yuan, C. Zhang, R. Liu, F. Zhao, Y. Zhao, X. Fang, *Toxicol. Appl. Pharmacol.*, 364–371 (2008).

152. S. J. Cho, D. Maysinger, M. Jain, B. Roder, S. Hackbarth, F. M. Winnik, *Langmuir*, 1974–1980 (2007).

153. A. M. Derfus, W. C. W. Chan, S. N. Bhatia, *Nano Lett.*, 11–18 (2004).

154. J. M. Tsay, X. Michalet, *Chem. Biol.*, 1159–1161 (2005).

155. K. Yamamoto, A. Hoshino, H. Mohamed, "Quantum dot modification and cytotoxicity," in *Handbook of Self Assembled Semiconductor Nanostructures for Novel Devices in Photonics and Electronics*, Amsterdam: Elsevier, 2008, 799–809.

156. L. W. Zhang, W. W. Yu, V. L. Colvin, N. A. Monteiro-Riviere, *Toxicol. Appl. Pharmacol.*, 200–211 (2008).

157. Y. Su, Y. He, H. Lu, L. Sai, Q. Li, W. Li, L. Wang, P. Shen, Q. Huang, C. Fan, *Biomaterials*, 19–25 (2009).

158. W. Liu, H. S. Choi, J. P. Zimmer, E. Tanaka, J. V. Frangioni, M. Bawendi, *J. Am. Chem. Soc.*, 14530–14531 (2007).

159. H. Soo Choi, W. Liu, P. Misra, E. Tanaka, J. P. Zimmer, B. Itty Ipe, M. G. Bawendi, J. V. Frangioni, *Nat. Biotech.*, 1165–1170 (2007).

160. H. C. Fischer, L. Liu, K. S. Pang, W. C. W. Chan, *Adv. Funct. Mater.*, 1299–1305 (2006).

161. A. Gabizon, D. Goren, A. T. Horowitz, D. Tzemach, A. Lossos, T. Siegal, *Adv. Drug Delivery Rev.*, 337–344 (1997).

162. B. Ceh, M. Winterhalter, P. M. Frederik, J. J. Vallner, D. D. Lasic, *Adv. Drug Delivery Rev.*, 165–177 (1997).

163. D. N. Waterhouse, P. G. Tardi, L. D. Mayer, M. B. Bally, "A comparison of liposomal formulations of doxorubicin with drug administered in free form: changing toxicity profiles," in *Drug Safety*, Vol. 24, ADIS International, 2001, 903–920.

164. S. A. Abraham, D. N. Waterhouse, L. D. Mayer, P. R. Cullis, T. D. Madden, M. B. Bally, D. Nejat, "The liposomal formulation of doxorubicin," in *Methods in Enzymology*, Vol. 391, Academic Press, 2005, 71–97.

165. D. N. Waterhouse, T. Denyssevych, N. Hudon, S. Chia, K. A. Gelmon, M. B. Bally, *Pharm. Res.*, 915–922 (2005).

166. D. N. Waterhouse, K. A. Gelmon, D. Masin, M. B. Bally, *J. Exp. Ther. Oncol.*, 261–271 (2003).

167. Robert M. O'Bryan, James K. Luce, Robert W. Talley, Jeffrey A. Gottlieb, Laurence H. Baker, Gianni Bonadonna, *Cancer*, 1–8 (1973).

168. W. E. Bawarski, E. Chidlowsky, D. J. Bharali, S. A. Mousa, *Nanomed.: Nanotechnol., Biol. Med.*, 273–282 (2008).

169. K.-T. Yong, I. Roy, M. T. Swihart, P. N. Prasad, *J. Mater. Chem.*, 4655–4672 (2009).

Chapter 12

Zinc Oxide Nanoparticles in Biosensing Applications

Linda Y. L. Wu
*Surface Technology Group, Singapore Institute of Manufacturing Technology,
71 Nanyang Drive, Singapore 638075, Singapore*
ylwu@simtech.a-star.edu.sg

For the visualization of biologically relevant molecules and activities inside living cells and early detection of diseases, semiconductor quantum dots offer high-quality optical imaging and various particle-cell immobilization advantages than organic markers. But the toxic ligands used in most of the synthesis processes and the cytotoxicity due to the release of heavy metals are the major issues. ZnO is a good candidate to overcome these issues due to its biocompatibility and excellent optical properties. However, the required conditions for bioimaging are not simultaneously satisfied by ZnO due to one of the following problems: high temperature or too fast chemical reaction, leading to surface defects, resulting in poor optical properties; no suitable surface capping, leading to particle agglomeration in water or uncontrolled particle shape; toxic ligands used in the process, leading to post-treatment and potential contamination/toxicity

Luminescence: The Instrumental Key to the Future of Nanotechnology
Edited by Adam M. Gilmore
Copyright © 2014 Pan Stanford Publishing Pte. Ltd.
ISBN 978-981-4241-95-3 (Hardcover), 978-981-4267-72-4 (eBook)
www.panstanford.com

of the final colloidal solution; and no suitable doping, leading to only ultraviolet (UV) emission, not detectable by common confocal microscopy. In this chapter, we report the synthesis and surface modification of biofriendly ZnO-based colloids, which have been tested both in vitro and in vivo in human tumor cells and rat models, providing significant advantages on quantum confinement effects, superior optical properties, nontoxicity, and a unique dual-color imaging feature.

12.1 Introduction

Advances in molecular medicines require the optimum detection of individual biomolecules, cell components, and other biological entities. Traditional fluorescent dyes (e.g., fluorescein, ethidium, methyl coumain, rhodamine, etc.)[1] have a number of physical and chemical limitations that include low photostability, narrow absorption bands, broad emission spectra, and differences in chemical properties of the dyes, making multiple, parallel assays impractical.[2] Fluorescent semiconductor (groups II–VI) nanocrystals, often referred to as quantum dots (QDs), have unique optical and electrical properties, such as size- and composition-tunable fluorescence emission from visible to infrared wavelengths, large absorption coefficients across a wide spectral range, and a very high level of brightness and photostability.[3] The major issue of Cd-containing QDs is the potential cytotoxicity; therefore, alternative materials that do not contain cadmium and are more biocompatible are required.[4] ZnO is a versatile semiconductor with a wide bandgap (~3.37eV) and an extremely large exciton-binding energy (60 meV), which makes the exciton state stable at room temperature and above. Its unique optical properties and biocompatibility are huge advantages as a better candidate for bioimaging than metals and chalcogenide (S, Se, Te) nanoparticles. However, research on ZnO for bioapplications considerably lags behind the other candidates because the synthesis is not as well developed and the doping and surface modification of ZnO are less well understood. A few methods for obtaining ZnO nanocrystals in aqueous solution at low temperature have been reported, but they all involve strong alkaline media[5-8] or annealing[9], and, therefore, are not suitable for bioapplications. To make ZnO QDs for in vivo and in vitro bioapplications, the synthesis process has to meet several requirements: (1) only contains biocompatible materials, (2) uses a suitable surface-capping agent to ensure nanosized ZnO particles in a stable colloidal solution, (3) provides chemical functional groups on the ZnO surface that will eventually bind to biomolecules, and (4) ensures photostability and efficient fluorescence of the ZnO nanocrystals. A biofriendly synthesis method for ZnO using the buffer tris(hydroxymethyl)aminomethane was reported[10], but no actual biotest was conducted. A surface modification method using an organosilane cross-linker was reported[11], but only an organic dye was tested as an indication for further biomolecular attachment. Recently, we reported the preliminary study of bioimaging on plant cells by ZnO.[12,13]

Since ZnO intrinsically emits in the UV wavelength, doping with suitable elements is an effective method to adjust their electrical, optical, and magnetic properties, which is crucial for its practical applications. After bandgap modification, ZnO could provide both UV and blue-violet emissions, which fill up the missing spectra range of current QDs. Doping of ZnO by Mg, Ni, Cu, In, and Al has been reported to improve the electrical conductivity.[14–17] Doping of ZnO by Mn, Ni, Cu, and Co has been reported to improve ferromagnetic properties.[18–21] Some other dopants such as Li, Na, and K were also reported[22–23], but due to their small atomic radii, these elements occupy the interstitial sites, rather than substitution sites, inducing strain and increasing the formation of native defects (vacancies). For bioimaging applications, the preferred dopants should have similar atomic radii with Zn and can reduce the bandgap of ZnO to enhance photoluminescence (PL) emission in visible wavelengths. We need to avoid surface defects, have a controllable bandgap, and have strong and stable PL emission. It is an added value if one marker can label both the nucleus and the cytoplasm of a cell simultaneously. This eliminates the complexity of two-step labeling using two different markers and the need of two excitation sources in the case of fluorescent dyes. The chemical synthesis and surface modification of ZnO nanocrystals have been reported by us recently.[24,25] Many different types of surface-capping agents have been studied, including 3-aminopropyl trimethoxysilane (Am), mercaptosuccinic acid (Ms), 3-mercaptopropyl trimethoxysilane (Mp), and polyvinylpyrrolidone (Pv); two types of aminosilanes, aminoethyl aminopropyl trimethoxysilane (Z60) and aminoethyl aminopropylsilane triol homopolymer water solution (Z61); and titania (TiO$_2$) and silica (SiO$_2$) through the sol-gel route. Bandgap modification of ZnO has been achieved by doping ZnO nanocrystals with Co, Cu, and Ni cations in combination with surface capping. The applications of these nanoparticles on bioimaging of both human cells and plant systems were tested and proven. Due to the small size, the double amino surface functional groups, and the high PL emission intensity, ZnO nanoparticles showed dual-colored images with blue emission at the nucleus and turquoise emission at the cytoplasm simultaneously. Cytotoxicity was tested on human osteosarcoma cells, which proved the nontoxicity of the Z60- and Z61-capped ZnO nanocrystals. The maximum inhibitory nanoparticle loadings corresponding to 50% cell viability (IC$_{50}$) of the synthesized ZnO were compared to commercial CdSe/ZnS QDs. The quantum yields (QYs) of our nanocrystals varied from 79% to 95%, which are higher than most of the current Cd-based QDs (typically 25–30%).[26]

12.2 Particle Size Control through Chemical Synthesis and Surface Modifications

To visualize the biological details in the living cells and the nucleus, the particle size of the biomarker must be below a few nanometers. Therefore, the control of particle size in an aqueous system is crucial. Both the synthesis parameters and the surface-capping method are important control factors for particle size. In the current study, ZnO colloidal solutions were synthesized by refluxing zinc acetate dehydrate and Zn(Ac)$_2$·2H$_2$O methanol in a molar ratio of 0.03:4 at 66–67°C for six to seven hours. Nickel, cobalt, and copper dopants were added with different molar ratios (x = 0.05, 0.1, 0.15, 0.2) during the synthesis. A capping agent was added after cooling the reactants in ice water. The reaction time and temperature, fast cooling, and proper surface modification by the above-described capping agents are the key parameters in controlling the particle size.

Figure 12.1 shows the field-emission scanning electron microscope (FESEM) images of (a) Z60-capped 5%Ni-doped ZnO particles precipitated on an AAO template, the particle size smaller than 10 nm, (b) uncapped 5%Ni-doped ZnO particles in different sizes from submicron to 1.5 µm, (c) uncapped Cu-doped ZnO particles, and (d) uncapped pure ZnO particles in irregular shape and different sizes. These images indicate that capping is very effective in controlling the particle size; without capping, the particles grow to a larger size with a wider size distribution. It is noted that Ni-doped ZnO grew to a cabbage-like shape with perfect circular outlines, but uncapped pure ZnO grew to irregular shapes. This means that the dopant changes the lattice structure and facial energies of the ZnO, leading to preferential shapes. Similar shape and surface structures were observed in Co-doped ZnO particles (not shown here), while Cu-doped particles only grew into spherical particles without surface structures, as shown in Fig. 12.1c. Figure 12.2 shows the particle size distribution curve of a typical effectively capped ZnO colloid synthesized by this study, measured by Zetasizer (Malvern Nano ZS) using the 5%Co-ZnO-Z60 colloidal sample (7 days after synthesis). The size is between 1 to 2 nm with 97% particles between 1.5 to 2 nm.

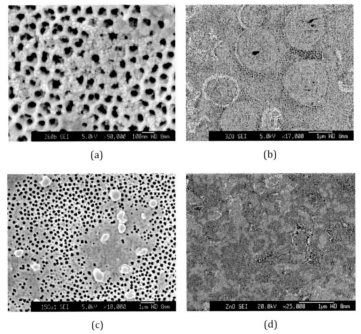

Figure 12.1 FESEM image of (a) Z60-capped 5%Ni-doped ZnO particles precipitated on an AAO template; the particle size is smaller than 10 nm; (b) uncapped 5%Ni-doped ZnO particles in different sizes from submicron to 1.5 µm; (c) uncapped 15%Cu-doped ZnO particles; and (d) uncapped pure ZnO particles in different sizes and irregular shapes.

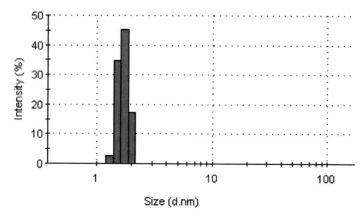

Figure 12.2 Particle size distribution curve of a typical ZnO colloid (effectively capped) synthesized by this study, showing a narrow size distribution by effective capping. (This curve was obtained from 5%Co-ZnO-Z60 colloid on the 7th day after synthesis.)

The stability of the colloid was monitored by measuring the particle size on daily intervals after the colloid was synthesized. It was found that Z60-capped particles are more stable than Z61- and TiO$_2$-capped particles. Co- and Ni-doped ZnO nanocrystals showed a similar particle size and size-growing trend, and Z60-capped 5%Co- and 5%Ni-doped ZnO colloids were stable for more than one year if they were kept in the refrigerator (about 4°C). Uncapped ZnO particles grew to about 600 nm in size in 30 days. This confirms the effectiveness of using capping agents to control the particle size by providing terminal groups with surface charges on the particle surfaces.

12.3 Bandgap Modification for Visible Emission

For easy observation of bioimages, bandgap modification of ZnO for visible emission is essential as the intrinsic emission of pure ZnO is in the UV wavelength range. This can be obtained by doping ZnO with suitable dopants such as Co, Ni, and Cu. The mechanism for bandgap modification is to insert lower energy levels in the wide bandgap of pure ZnO, inducing PL emission at lower energy levels. The PL spectra and QYs of the doped ZnO are discussed below.

12.3.1 Photoluminescence Spectra of Pure and Doped ZnO

PL spectra of doped and capped ZnO nanocrystals were measured under UV (325 nm) excitation and compared in Fig. 12.3a,b,c. As seen in Fig. 12.3a, pure ZnO without a capping layer emits at 377 nm with high PL intensity, which corresponds to the intrinsic bandgap of bulk ZnO. A general trend for Ni-doped ZnO is that the higher the Ni%, the lower the PL intensity. This means nickel, as a dopant for ZnO, inhibits the PL brightness, due to the nonradiative nature of the Ni^+ sites. Both the Z61 and Z60 capping agents caused a PL peak red shift of 20 nm, and while the TiO_2 capping agent does not change the PL peak position, it lowers the PL intensity. For Cu-doped ZnO, as seen in Fig. 12.3b, the overall PL intensity is lower than for Ni-doped ZnO. This indicates that copper, as a dopant for ZnO, caused more PL deficiency than nickel. The general trend is the same as Ni-doped ZnO: higher dopant lowers the PL intensity, and Z60-capped ZnO particles are lower than Z61-capped ZnO particles. A major benefit of Cu doping is the larger PL peak red shift of 40 nm from the 15%Cu-Z61 and 20%Cu-Z61 samples, and the intensity of the 15%Cu-Z61 is still moderate. The PL intensity of ZnO decreases by Ni and Cu doping and was also reported by other researchers[19], and we found that PL intensity could be increased by a larger particle size and more particle precipitation due to the higher energy absorption. For Co-doped ZnO, as seen in Fig. 12.3c, the overall intensity is similar to Ni-doped ZnO and higher than Cu-doped ZnO. It was found that the Co-doped ZnO particles were very stable even without a capping agent. So the colloids of 10%Co and 20%Co without capping were also tested. Their PL intensities were high, but no peak shifting was observed. Therefore, few dopants with different capping agents were further studied. It is seen from Fig. 12.3c that 5%Co-Z60 causes a 40 nm red shift of PL emission, and 5%Co-Z61 causes a 30 nm red shift. Since the confocal microscope used for the detection of bioimages of plant cells is equipped with an excitation source of 488 nm, we also measured the PL spectra of several doped ZnO particles under excitation at 488 nm, as shown in Fig. 12.3d. It is well known that the PL emission peak will shift to a longer wavelength if excitation is at a longer wavelength. From our results, we found that the PL spectrum did not simply shift; it also changed its shape and the peak position relative to the others. Although the PL intensity is lower than for those excited at 325 nm, all the nanocrystals exhibited green emissions (510–570 nm) under the Leica True Scanner SP2 Confocal Laser Scanning Microscope

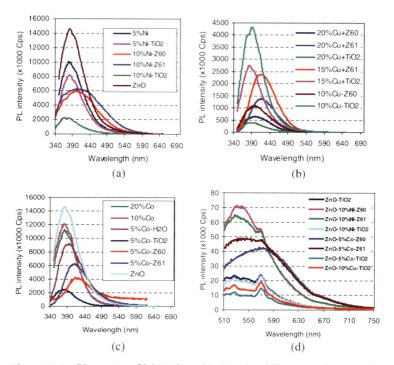

Figure 12.3 PL spectra of (a) Ni-doped ZnO with a different surface capping, (b) Cu-doped ZnO with a different surface capping, and (c) Co-doped ZnO with different surface capping agents. (a, b, and c) are under excitation at 325 nm. (d) PL spectra of Ni-, Cu-, and Co-doped ZnO under excitation at 488 nm.

12.3.2 Quantum Yield of Pure and Doped ZnO Colloids

The fluorescence QY is defined as the ratio of photons absorbed to photons emitted through fluorescence. A reliable method for recording the QY is the comparative method suggested by Williams et al.[27], which involves the use of well-characterized standard samples with known QY values. A detailed description of the method and a list of standard samples with their literature QYs are given by Jobin Yvon Horiba[28], who is the manufacturer of the spectrofluorometer. Four standard solutions were made according to the original literature: fluorescein in 0.01 M NaOH[29] in concentrations of 3 × 10⁻⁵ and 7 × 10⁻⁵ mol/L respectively, and anthracene in pure ethanol[30] in concentrations of 3 × 10⁻³ and 7 × 10⁻³ mol/L respectively. The absorbance values at the excitation wavelength of all the solutions were recorded by the ultraviolet-visible-near-infrared (UV-vis-NIR) scanning spectrophotometer. PL emission spectra of all the solutions were measured by the Fluorolog-3 spectrofluorometer under an excitation wavelength of 350 nm. The integrated PL intensity is the area under the PL curve in the wavelength range from 370 nm to 600 nm. The equation to calculate the QY of a test sample relative to a standard sample is presented as:[28]

$$\Phi_X = \Phi_{ST} \left(\frac{\text{Grad}_X}{\text{Grad}_{ST}} \right) \left(\frac{\eta_X^2}{\eta_{ST}^2} \right)$$

where the subscripts ST and X denote standard and test, respectively, Φ is the fluorescence QY, Grad is the gradient from the plot of the integrated PL intensity versus the absorbance, and η is the refractive index of the solvent. Using this equation, the fluorescein and anthracene standard solutions were first cross-calibrated. Then the QYs of the test samples were calculated using either fluorescein or anthracene as the standard. All the data is tabulated in Table 12.1.

Table 12.1 QY results of standard and test samples and the data used for the calculations

Compound/solvent	Gradient	Refractive index	Measured QY	Literature QY
Fluorescein/0.01 M NaOH	3.7758	1.34	0.7854	0.79
Anthracene/ethanol	1.2532	1.361	0.2715	0.27
ZnO/methanol	4.5786	1.328	0.9417	—
ZnO-TiO₂/methanol	3.8752	1.328	0.7963	—
Fluorescein/0.01 M NaOH	3.8212	1.34	0.789	0.79
Anthracene/ethanol	1.1926	1.361	0.254	0.27
5%Co with ZnO/methanol	3.8418	1.328	0.779	—
10%Co with ZnO/methanol	4.5928	1.328	0.931	—

* The standard deviation of the test data is below 2%.

The QYs of the pure ZnO and TiO$_2$-capped ZnO were measured as being 94% and 79%, respectively.[13,26] Using the same method and the two commercial fluorescence materials, fluorescein and anthracene, as the reference, the QYs of the 5%Co- and 10%Co-doped ZnO were obtained as being 79% and 95%, respectively. The variation of the QY result in the two separate tests on fluorescein and anthracene materials is below 2%, indicating that the comparative method is trustful. Since the QYs of SiO$_2$-coated CdSe-ZnS and uncoated CdSe-ZnS core-shell QDs in water solutions were reported as being 17% and 10%, respectively[31], our QY values are much higher. This further confirmed that Co-doped ZnO is the best choice for its excellent optical property, small and stable particle size in colloidal solution, and dual-color bioimaging capability (shown in the next section).

12.4 Bioimaging Using ZnO Nanocrystals

12.4.1 In vitro Bioimaging on Human and Animal Cells

Bioimaging tests were conducted on two human cell lines (osteosarcoma MG-63 cells and histiocytic lymphoma U-937 monocyte cells both from ATCC) and mung bean (*Vigna radiata*) plant cells. The MG-63 cells were cultivated in a glass-based culture dish for one day. Then a nanocrystal colloidal solution was added. Bioimaging was carried out within four hours by the Nikon C1Si Spectral Imaging Confocal Laser Scanning Microscope System (Nikon C1Si). Figure 12.4a shows the image of the MG-63 cells labeled by ZnO-Z61. It is seen from the image that the nucleus shows a blue color and the cytoplasm shows a turquoise (blue-green) color. The monocyte U-937 cells were grown in a culture dish, and a small amount was transferred onto glass slides. After nanoparticles were added, a drop of 2% paraformaldehyde in a phosphate buffered saline (PBS) buffer solution (pH 7.4) was added to fix the cells. Bioimaging was carried out within four hours using the Nikon C1Si. Figure 12.4b shows the image of one U-937 cell labeled by 5%Co-ZnO-Z60. The same dual-color emission can be seen. At the right side and bottom of the main image, the two side views of the cell are shown, which further confirmed the location of the blue color emission at the nucleus and the turquoise color at the cytoplasm.

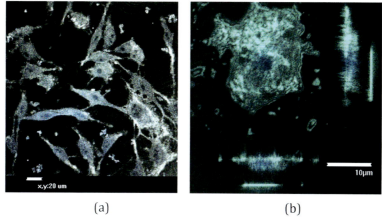

(a)   (b)

Figure 12.4 Confocal images of labeled human cells. (a) Image of MG-63 cells labeled by ZnO-Z61 and (b) image of one U-937 cell labeled by 5%Co-ZnO-Z60.

The detailed structures of the whole cell are clearly labeled by our nanocrystals. Although further investigation is required to confirm the principle of the dual-color emission, we believe it is due to the combination of the quantum confinement effect and the chemical interactions between the functional groups of the nanoparticles with the cells. Since the particle size ranges from below 1 nm to 2 nm, it is in the quantum size range. In this size range, smaller particle emits in a shorter wavelength (or blue shift). The smaller particles in our QDs can penetrate into the nucleus, while the relatively larger particles stay in the cytoplasm part; therefore, the emission color in the nucleus is blue-shifted compared to the color in the cytoplasm. This forms the dual-color emission.

Meanwhile, since the chemical status of the nucleus and the cytoplasm of the cell are different, the dual-color emission could be due to the different chemical interactions in the nucleus and the cytoplasm of the cell, resulting from the addition of the functionalized ZnO nanoparticles. In a separate experiment, 10%Cu-ZnO-Z60 in a slightly larger size and low PL emission was added to MG-63 cells. Only one color emission at the cytoplasm is seen. From these results we can conclude that three important factors contribute to the dual-color imaging feature: (1) particle size in the quantum size range with a narrow size distribution, (2) high PL intensity and a suitable emission wavelength range, and (3) suitable chemical functions on the nanoparticle surface, for example, double amino surface functional groups, which ensure the adsorption of nanoparticles by the whole cell.

Comparing with the single amino terminal groups and inorganic TiO_2 and SiO_2 capping layers reported by us earlier[12], the current capping agents (Z60 and Z61) have better performance in bioimaging due to the better-controlled PL emissions and higher adsorption by cells.

12.4.2 Bioimaging on a Plant System

Bioimaging on a plant system was conducted by adding ZnO colloids to the growing mung bean (*Vigna radiata*) seedlings and measuring the fluorescence emissions of the cross sections of the seedlings after 24 hours using the CLSM to detect the second-order PL emission from the cells. Figure 12.5a shows the images labeled by 5%Co-ZnO-Z60, and Fig. 12.5b shows the images labeled by 5%Cu-ZnO-TiO_2 at the cross section of the shoot–root junctions with all color emissions at the vascular cylinder area. It is seen that all the cell structures are labeled by 5%Co-ZnO-Z60, while only cell walls are selectively labeled by 5%Cu-ZnO-TiO_2. It was found that the plant had different absorption behaviors toward different nanocrystals. This may be due to the different chemical nature and cytotoxicities of the differently doped and capped nanocrystals. With less cytotoxicity, more nanocrystals can be absorbed and transported by plant, whereby more emission can be obtained on the cross sections.

12.4.3 In vivo Bioimaging in a Rat Model

In vivo tests were performed in rats using two QDs (ZnO-5%Cu-Z60 and ZnO-5%Co-Z60). six-eight-week-old male rats were weighed, labeled, and injected intraperitoneally with the QDs at a concentration of 50 µM. After 1 or 24 hours, the rats were given 3 mL anesthesia consisting of ketamine/diazepam, and fluorescence at the rats' external bodies was analyzed using an Ocean Optics USB 2000 fiber optic spectrofluorometer. The rats were then dissected to measure the fluorescence spectrum from four organs: liver, kidney, lung, and spleen. Five data points were taken from each organ, and an average was computed. Tissue specimens from the liver, kidney, lung, and spleen were taken for fixation and embedding in paraffin.

Bioimaging Using ZnO Nanocrystals | **335**

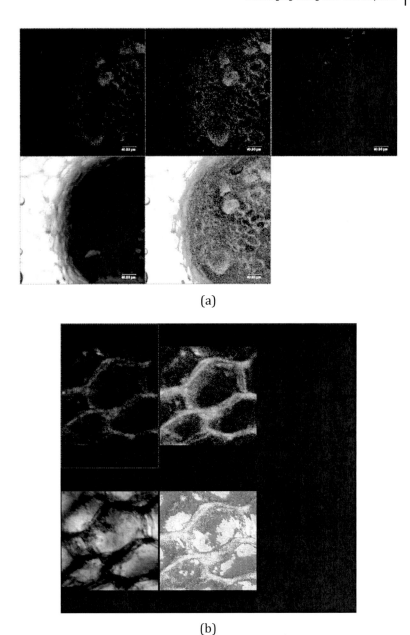

Figure 12.5 Fluorescence emissions on cross sections of seedlings at the shoot–root junctions of (a) a 5%Co-ZnO-Z60-added seedling and (b) a 5%Cu-ZnO-TiO$_2$-added seedlings, taken 24 hours after adding a QD colloidal solution, when excited at 488 nm.

An automatic tissue processor, Leica TP1020, was used to process the fixed tissue specimens. The specimens were then sectioned to a 6 µm thickness using a Leica RM2135 rotary microtome and placed onto microscope slides for taking the fluorescence images using the CLSM. Figure 12.6 shows the emission spectra from the lung at 1 and 24 hours after injection. High emission at 1 hour and much lower emission at 24 hours were found. This implies that the nanoparticles have transported to either deep tissue or other organs within 24 hours.

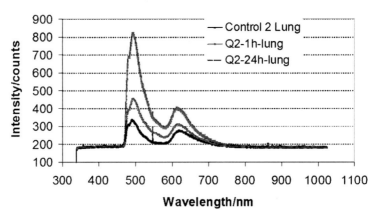

Figure 12.6 Fluorescence spectra from the lung of a rat after 1 and 24 hours of injection with ZnO-5%Co-Z60 (50 µM) QDs.

Figure 12.7 shows the confocal images of the rat's (a) kidney and (b) lung, which were taken after 1 hour of injection of the ZnO colloids. No negative effect was observed on the rats within 24 hours after injection of QDs, which confirmed the suitability of the ZnO QDs for in vivo test on live animals. The ZnO nanocrystals can be easily further attached with proteins or antibodies for specific cancer detection, as described and reviewed by Biju et al.[26]

12.5 Cytotoxicity Tests

Cytotoxicity of QDs is always a concern for in vivo bioapplications. For plant cells, we observed the seedling growth rate after adding

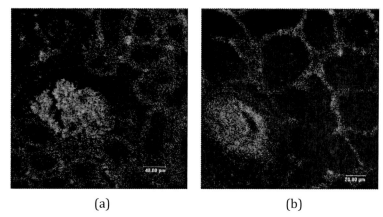

Figure 12.7 Fluorescence images of the kidney (a) and lung (b) of a rat after 1 hour of injection with ZnO-5%Co-Z60 (50 μM) QDs.

nanoparticles at the roots and growing them for another two days, and we found that Co-doped ZnO was absorbed by the plant and did not affect the growth of the seedling, while Ni-doped and Cu-doped ZnO slowed down the growth of the seedlings. We then tested the bioimaging for these slow-growing seedlings on their cross sections and detected the same fluorescence emission from the cell walls, which confirmed the intake of the nanocrystals by the plant cells. This is only possible when the nanoparticles are small enough to be transported from the root to the whole sprout. For human cells, we tested the cytotoxicity of the ZnO, 5%Co-, 5%Cu-, and 10%Ni-doped ZnO nanoparticles using an established method with methylthiazolyldiphenyl-tetrazolium (MTT) cell proliferation assay[32] on MG-63 cells. The maximum inhibitory nanoparticle loading corresponding to 50% cell viability (IC_{50}) was measured for each type of nanoparticles. We found that the maximum nanoparticle loadings for TiO_2-capped nanoparticles (ZnO, 5%Cu-ZnO, 10%Ni-ZnO) were about 500 μg/mL, which may be due to the acidic catalyst that was added to the TiO_2 solution. All other capped pure ZnO and doped ZnO nanoparticles were nontoxic; nanoparticle loading higher than 1,000 μg/mL did not cause any cell proliferation. The reported maximum inhibitory nanoparticle loading corresponding to 50% cell viability (IC_{50}) for Cd-containing QDs was 100 μg/mL, which was obtained from silica-coated CdSe QDs.[31] The nanoparticle loading of normal MAA-coated CdSe was only 7 μg/mL. This means that our nanoparticle loading is the highest of the reported values so far.[31-36]

To have a fair comparison between our ZnO nanocrystals and the commercial CdSe/ZnS QDs, both the selected ZnO nanocrystals and a commercial QD were tested in the same batch of an MTT test on MG63 cells. The commercial QD was first treated by ligand exchange and dispersed in water according to published procedures[37] and then added to live cells in different concentrations. Figure 12.8 shows the cell viability after different concentrations of the ZnO and commercial QDs were added separately. It is seen that the cell viability dropped to 50% at a 300 μg/mL concentration with commercial CdSe QD added, while the cells with 5%Co-ZnO-Z60 and ZnO-Z60 nanocrystals were in good conditions at QD concentrations of 500 μg/mL and above. This shows that the cytotoxicity of commercial CdSe-based QDs is relatively higher than the (doped) ZnO QDs synthesized in this research. For typical in vitro cell imaging studies, 4 to 100 μg/mL was used.[22,38] For in vivo applications, relatively higher concentrations of QDs from 100 to 250 μg/mL were delivered to mice through intravenous injection.[32,39] Therefore, a QD concentration above 500 μg/mL is more than enough for both in vitro and in vivo studies. From these comparisons, we can conclude that our nanoparticles are nontoxic and much safer to be used as in vivo biomarkers.

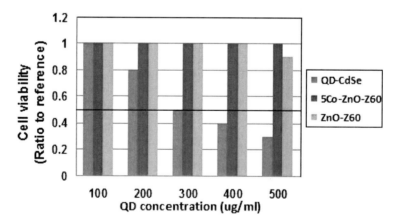

Figure 12.8 Cell viability results comparing synthesized ZnO QDs to a commercial CdSe/ZnS QD product on MG-63 cells.

12.6 Conclusions and Outlook

Doped ZnO nanocrystals have shown great potential in biosensing applications due to the bio-friendly synthesis process, controllable material purity, particle size and surface functions, superior optical properties, and unique bioimaging capability. The major advantage over other semiconductor QDs is the nontoxicity, which has been proven by both in vitro and in vivo biomedical tests. These results have provided a new platform for substitution of the conventional biomarkers and overcoming the current issues of QDs in toxicity by heavy metals, poor clearance of nanoparticles (>5 nm) from the body, and possible deterioration of their surface coatings. The advancement in semiconductor synthesis and experimental and theoretical understandings of their optical properties at ensemble and single-particle levels will accelerate their practical applications in different fields from optical devices to biomedical detection and imaging.

Acknowledgments

This work was supported by the Agency for Science, Technology and Research (A*STAR) of Singapore. I thank my collaborators in Nanyang Technological University: Prof. Alfred Tok, Dr. Fu Sheng, and Prof. L C Kwek for the invaluable discussions, thank Ms. L. L. Ma and Ms. G. J. Loh for their help in cell culture and in vivo tests, respectively, and thank the Nikon Imaging Center of Singapore for the in vitro bioimaging tests.

References

1. J. C. Politz, *Trends Cell Biol.*, **9**(7), 284–287 (1999).
2. M. Bruchez, Jr., M. Moronne, *Science*, **281**(5385), 2013 (1998).
3. A. P. Alivisatos, *Science*, **271**, 933–937 (1996).
4. C. J. Lin, T. Liedl, R. A. Sperling, M. T. Fernandez, R. Pereiro, A. Medel, W. H. Chang, W. J. Parak, *J. Mater. Chem.*, **17**, 1343–1346 (2007).
5. M. Ristic, S. Music, M. Ivanda, S. Popovic, *J. Alloys Compd.*, **397**(1–2), L1–L4 (2005).
6. C. F. Jin, X. Yuan, W. W. Ge, J. M. Hong, X. Q. Xin, *Nanotechnology*, **14**(6), 667–669 (2003).

7. Z. Wang, X. F. Qian, J. Yin, Z. K. Zhu, *Langmuir*, **20**(8), 3441–3448 (2004).
8. Z. R. Tian, J. A. Voigt, J. Liu, B. McKenzie, M. J. McDermott, M. A. Rodriguez, H. Konishi, H. Xu, *Nat. Mater.*, **2**(12), 821–826 (2003).
9. R. Hoffmann, T. Fuchs, T. P. Niesen, J. Bill, F. Aldinger, *Surf. Interface Anal.*, **34**(1), 708–711 (2002).
10. L. P. Bauermann, J. Bill, F. J. Aldinger, *Phys. Chem. B*, **110**, 5182–5185 (2006).
11. J. W. Soares, D. M. Steeves, D. Ziegler, B. S. DeCristofano, *Proc. SPIE*, **6370**, 637011 (2006).
12. Y. L. Wu, C. S. Lim, S. Fu, A. I. Y. Tok, H. M. Lau, F. Y. C. Boey, X. T. Zeng, *Nanotechnology*, **18**, 215604 (2007).
13. Y. L. Wu, S. Fu, A. I. Y. Tok, X. T. Zeng, C. S. Lim, L. C. Kwek, F. C. Y. Boey, *Nanotechnology*, **19**, 345605 (2008).
14. G. H. Ning, X. Zhao, J. Li, *Opt. Mater.*, **27**, 1–5 (2004).
15. A. E. J. J. Gonzalez, *Solid State Chem.*, **128**, 176–180 (1997).
16. Y. W. Chen, Y. C. Liu, S. X. Lu, C. S. Xu, C. L. Shao, C. Wang, J. Y. Zhang, Y. M. Lu, D. Z. Shen, X. W. Fan, *J. Chem. Phys.*, **123**, 134701 (2005).
17. G. Westin, M. Wijk, A. Pohl, *J. Sol-Gel Sci. Technol.*, **31**, 283–286 (2004).
18. C. J. Cong, L. Liao, J. C. Li, L. X. Fan, K. L. Zhang, *Nanotechnology*, **16**, 981–984 (2005).
19. O. P. Perez, A. P. Palomino, R. Singhal, P. M. Voyles, Y. Zhu, W. Jia, M. S. Tomar, *Nanotechnology*, **18**, 315606 (2007).
20. Y. Liao, T. Huang, M. Lin, K. Yu, H. C. Hsu, T. Lee, C. Lee, J. C. A. Huang, *J. Magn. Magn. Mater.*, **310**, e818–e820 (2007).
21. C. J. Cong, J. H. Hong, Q. Y. Liu, L. Liao, K. L. Zhang, *Solid State Commun.*, **138**, 511–515 (2006).
22. M. S. Tokumoto, S. H. Pulcinelli, C. V. Santilli, V. J. Briois, *Phys. Chem. B*, **107**, 568–574 (2003).
23. Y. Inubushi, R. Takami, M. Iwasaki, H. Tada, S. Ito, *J. Colloid Interface Sci.*, **200**, 220–227 (1998).
24. Y. L. Wu, A. I. Y. Tok, F. Y. C. Boey, X. T. Zeng, X. H. Zhang, *Appl. Surf. Sci.*, **253**, 5473–5479 (2007).
25. Y. L. Wu, A. I. Y. Tok, F. Y. C. Boey, X. T. Zeng, X. H. Zhang, *IEEE Trans. Nanotechnol.*, **6**(5), 497–503 (2007).
26. V. Biju, T. Itoh, A. Anas, A. Sujith, M. Ishikawa, *Anal. Bioanal. Chem.*, **391**, 2469 (2008).

27. A. T. R. Williams, S. A. Winfield, J. N. Miller, *Analyst*, **108**, 1067 (1983).
28. A guide to recording fluorescence quantum yields (http://www.jyhoriba.co.uk/)
29. J. Q. Umberger, V. K. Lamer, *J. Am. Chem. Soc.*, **67**, 1099 (1945).
30. W. H. Melhuish, *J. Phys. Chem.*, **65**, 229 (1961).
31. S. T. Selvan, T. T. Tan, J. Y. Ying, *Adv. Mater.*, **17**, 1620 (2005).
32. A. M. Derfus, W. C. W. Chan, S. N. Bhatia, *Nano Lett.*, **4**, 11 (2004).
33. A. Hoshino, K. Fujioka, M. Suga, Y. F. Sasaki, T. Ohta, M. Yasuhara, K. Suzuki, K. Yamamoto, *Nano Lett.*, **4**(11), 2163–2169 (2004).
34. C. Kirchner, T. Liedl, S. Kudera, T. Pellegrino, A. M. Javier, H. E. Gaub, S. Stolzle, N. Fertig, W. J. Parak, *Nano Lett.*, **5**(2), 331–338 (2005).
35. S. J. Clarke, C. A. Hollmann, Z. Zhang, D. Suffern, S. E. Bradforth, N. M. Dimitrijevic, W. G. Minarik, J. L. Nadeau, *Nat. Mater.*, **5**, 409–417 (2006).
36. J. K. Jaiswal, H. Mattoussi, J. M. Mauro, S. M. Simon, *Nat. Biotechnol.*, **21**, 47–51 (2003).
37. Y. Chen, Z. Rosenzweig, *Nano Lett.*, **2**(11), 1299–1302 (2002).
38. X. Wu, H. Liu, J. Liu, K. N. Haley, J. A. Treadway, J. P. Larson, N. Ge, F. Peale, M. P. Bruchez, *Nat. Biotechnol.*, **21**, 41–46 (2003).
39. D. R. Larson, W. R. Zipfel, R. M. Williams, S. W. Clark, M. P. Bruchez, F. W. Wise, W. W. Webb, *Science*, **300**, 1434–1436 (2003).

Chapter 13

Use of QDOT Photoluminescence for Codification and Authentication Purposes

Shoude Chang

National Research Council Canada, 1200 Montreal Road, Ottawa, Canada
shoude.chang@nrc.ca

This chapter addresses the information-coding and retrieval technology by using photoluminescent (PL) semiconductor quantum dots (QDOTs) synthesized via wet-chemistry approaches. QDOTs with different wavelengths and intensities span a 2D coding space. Secure information is coded in such a group of QDOTs, which is then applied to various substrates, such as ink, paint, and labels. When an exciting light beam shines at these information carriers, their emitting spectral features, that is, wavelengths and intensities, provide the encoded information. As the wavelength of the photoluminescence can be precisely controlled when designing the code, these QDOTs are able to be made to emit at Fraunhofer line positions, namely, black lines in the solar spectrum; thus, the retrieval system can even extract useful information in sunshine-covered areas. Furthermore, multiphoton excitation (MPE) technology enables the retrieval

---

Luminescence: The Instrumental Key to the Future of Nanotechnology
Edited by Adam M. Gilmore
Copyright © 2014 Pan Stanford Publishing Pte. Ltd.
ISBN 978-981-4241-95-3 (Hardcover), 978-981-4267-72-4 (eBook)
www.panstanford.com

system to perform multilayer information extraction, which greatly increases the dimensions of coding space. Anticipated applications include security, military, and law enforcement, for example, coding and retrieval information from military helmets, vehicles, and even fingernails. In addition, QDOT-based security information can be easily destroyed by preset expiration in the presence of timing agents, which results in the expiration of this information carrier.

13.1 Introduction

The most important issues in security technologies are information storage and retrieval. The prevailing technology used nowadays for object identification is the use of a one-dimensional (1D) or two-dimensional (2D) barcode. The visibility of the printed pattern of a barcode is vulnerable to counterfeiting; therefore, a barcode is definitely not qualified in many security applications. In addition, as the barcode reader has to scan the 1D bar sequence or to register the 2D dot image, these procedures make the system bulky and complicated. Regarding high-level security, a hidden information carrier, which is invisible to the human eye and tiny in size, is mandatory; moreover, the invariance under the changes of position and rotation of the information encoded in the carrier is critical to simplify the information retrieval procedure with enhanced reliability.

Multiplexed spectral coding technology that makes use of multiple wavelengths and multiple intensities as the coding space meets the security and invariant requirements to a certain degree. Organic dyes and metal complexes are commonly used as fluorescence-sensing materials in various applications. Basically, they seem to be suitable in the multiplexed spectral coding technology; however, their intrinsic optical properties limit them to be the ideal candidates. For example, different dye molecules require different excitation wavelengths; furthermore, it is different to retrieve information from a mixture of these fluorescent molecules due to their broad emission bandwidth and asymmetry. The limitation may also be related to the certain reaction of different dye molecules or the immiscibility of some of the dye molecules in a common matrix material. From a technical point of view, an ideal set of luminescent substances should have the following properties in order to meet the prerequisite for multiplexed encoding and decoding:

(1) A single light source for all materials to emit at different wavelengths;
(2) Each emission independent of the excitation wavelength;
(3) Stable emission with narrow bandwidth from each luminescent material;
(4) No chemical interaction among the fluorescent molecules;
(5) No emission from matrix materials; and
(6) Good miscibility of all luminescent materials in the selected matrix materials.

Recently, colloidal photoluminescent (PL) semiconductor quantum dots (QDOTs) have demonstrated many of the above-mentioned characteristics. For example, QDOTs exhibit emission with narrower bandwidth, broader absorption, and better photostability compared to traditional luminescent materials. Furthermore, they are excellent with multiphoton excitation due to large multiphoton action cross sections; accordingly, they are magnificent candidates for multilayer information extraction. Moreover, their bandgap emission and absorption peak positions can be easily and accurately tuned via the control of their size, structure, and composition. Due to their intrinsic optical properties, QDOTs are suitable in the multiplexed optical coding technology with potential in security and defense.[1-5] Basically, QDOTs are ultrasmall nanocrystals and are spherical in shape, usually in the range of 1–10 nm. Elements made up of QDOTs are often from group IIB and group VIA in the periodic table. QDOTs can be binary compounds, such as cadmium selenide (CdSe) and zinc sulfide (ZnS); ternary compounds, such as CdTeSe and ZnCdS; or layered structures, such as core-shell CdSe/ZnS and CdTe/CdSe/CdTe.[6-12] When these nanocrystals dispersed in a transparent solution are excited, they provide coding information based on their emission position and intensity; such a solution with the secure information encoded is the so-called info-ink. A mixture of various QDOTs with different emission positions can be dedicatedly designed to feature a special code with a set of data consisting of the emission positions and intensity. Such coding information is hidden in a fluorescence spectrum; accordingly a spectroscopic device is needed rather than a scanner or a camera to decode the encoded information. It is necessary to point out that such QDOT-based information carriers are miniature in size and invisible to human eyes, demonstrating their potential in security applications.

Colloidal PL QDOTs as info-inks for biolabeling were proposed about 10 years ago[13–15]; meanwhile, significant advances in solution chemistry synthesis expand such applications of QDOT-based info-inks to many areas, including security.[11,16–21]

Nie et al. proposed the idea of using colloidal QDOTs in multiplex bioimaging and biolabeling[22]; such a multiplex detection only involved single-photon excitation in solution. Bawendi et al in the Massachusetts Institute of Technology (MIT) reported the design of using QDOTs for inventory control[16,17]; however, no practical approaches were provided about their information retrieval system. Also, we demonstrated a QDOT-based information retrieval system with patents granted.[18,19] In this chapter, we describe, in detail, how to prepare/encode a QDOT-based info-ink and how to process and retrieve information from a fluorescent signal emitted from QDOTs. Furthermore, we will discuss how to expand the coding capability and the possibility of detecting a QDOT signal from sunshine-covered areas. Applications, from anticounterfeiting to friend/enemy discrimination using QDOT-based technologies, are proposed, described, and discussed.

13.2 QDOTs Used as Information Carriers

Various PL QDOTs with different emission wavelengths provide a great number of combinations of wavelength and intensity. For example, an encoder using a 6-wavelength and 10-intensity scheme has a theoretical coding capacity of about one million discrimination codes. The coding capacity can be even expanded by utilizing a third parameter, such as a 1D sequence or a 2D array of QDOTs. To be able to use QDOTs for the spectral coding of nonbiological objects such as banknotes, passports, certificates, and other valuable documents, paintable or printable QDOT/polymer/solvent info-inks are needed. The info-inks consisting of different QDOTs, polymers, solvents, and additives can be applied onto various surfaces for coding purposes. A hybrid optic-electronic-digital system is used to extract the data from the resulting emission spectra. A detailed description is given in the following sections.

After an info-ink is applied to a target surface and dried, the polymer becomes a matrix material, in which the mixture of different QDOTs with predesigned emission features is dispersed homogeneously.[23] The polymer used should not have quenching effects on the fluorescence of the QDOTs and should satisfy other requirements, such as reasonable solubility in the selected solvent, long-term environmental stability, and good compatibility and miscibility with the QDOTs.

CdSe nanocrystals in the nanometer-size range can be well dispersed in toluene, which is also a good solvent for polystyrene (PS). Therefore, the info-ink prepared in our preliminary work consisted of CdSe nanocrystals (Evident Technologies[24]), high-molecular-weight PS, and toluene (both from Aldrich). According to their individual emission intensity, the CdSe QDOT ensembles were mixed with different ratios, together with a certain amount of PS and toluene to engineer the info-inks with required emission spectral features and viscosity for testing. Furthermore, in-house-synthesized CdSe-based nanocrystals and CdS nanocrystals were used in the present study.[25-29] The in-house CdSe QDOTs were synthesized with a slow growth rate for high-quality and large-scale production. This nonorganometallic approach involved the addition of a solution with a chalcogen source in tri-*n*-octylphosphine (TOP) to a solution of CdO in TOP at one temperature, with subsequent growth at a lower temperature. A slow crystal growth rate was achieved with the zero growth rate accomplished via tuning the Cd-to-Se precursor molar ratios, as assessed by the temporal evolution of the optical properties of the growing nanocrystals dispersed in nonpolar hexanes (Hex) and polar tetrahydrofuran (THF).

It has been acknowledged that a higher particle growth rate results in greater surface roughness, more surface defects, and lower PL efficiency. An approach for removing the surface defects is to control a small growth rate, ideally to achieve a zero growth rate, which means that the average rate of removal of atoms from the nanocrystal surface is equal to that of the addition of atoms to the nanocrystal surface. The nanoparticles with a size close to a zero growth rate in an ensemble possess the smoothest and defect-free surfaces and highest PL efficiency. In general, the in-house synthetic approach is excellent in terms of the control of the growth rate for the CdSe nanocrystals with high surface quality—a simple system but a rational choice for both fundamental understanding and tailoring applications. Also, such slow size growth is essential for large-scale production as well as for further in situ modification such as for core-shell materials. It is necessary to point out that the zero growth rate methodology leads to our magic-sized QDOT ensembles, the latest in our synthetic laboratories; these single-sized QDOT ensembles exhibit bandgap emission with a bandwidth as narrow as 8 nm are ideal for info-inks for security applications.[25,11]

13.3 Information Encoding

As mentioned above, wavelengths and intensities of the QDOTs span a 2D coding space. Figure 13.1 illustrates schematically the designed samples of info-inks consisting of three principal QDOTs with different emission wavelengths. Adjusting the amount of the QDOTs can produce a series of three-digit codes. The info-inks consisting of a polymer, solvent, multiple QDOTs, and other additives are prepared to label the objects that need to be coded.

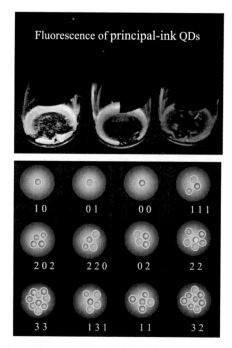

Figure 13.1 Information contained in info-ink.

It is the simplest information-coding method. As the info-ink is applied uniformly on the surface of the object, it is invariant to changes of position, rotation, and distance of the object, which greatly simplifies the data processing in the retrieval system. In the viewpoint of spatial coding, it is a zero-dimensional (0D) encoding method.

Further to the optical coding based on fluorescent semiconductor QDOTs, a concept of using mixtures of multiple single-color QDOTs for creating highly secret cryptograms based on their absorption/emission properties was proposed.[30] The key to readout of the optical codes is a group of excitation lights with the predetermined wavelengths programmed in a secret manner. The cryptograms can be printed on the surfaces of different objects such as valuable documents for security purposes. Monodispersed QDOTs have a very broad absorption spectrum, and their single emission wavelength (λ_{em}) is independent of excitation wavelengths (λ_{ex}) as long as λ_{ex} is shorter than the excitation threshold (the first absorption peak).

Figure 13.2 shows an emission spectra of a mixture of three single-color QDOTs in toluene excited at a λ_{ex} ranging from 350 to 510 nm with a constant interval of 10 nm. The excitation peaks (λ_{ex} = 510, 500, 490, 480, 470, 460, and 450 nm) and three diffraction-induced second-order peaks (from λ_{ex} = 350, 360, and 370 nm) are also recorded in the measurement range. The inset plot illustrates two sets of selected spectral data normalized to a 10-level (0–9) intensity scale.

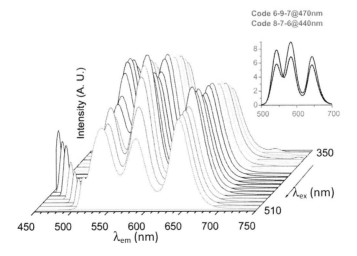

Figure 13.2 Emission spectra of a mixture of three single-color QDOTs.

A λ_{ex}-dependent spectral change can be simply used to increase the coding capacity. Assuming that a 10-level (0–9) intensity scale is used for coding, a dicolor QDOT system under the excitation of a single λ_{ex} can theoretically generate 99 codes, including those involving the fluorescence intensity level of 0. If multiple λ_{ex} can be applied to the system, the coding capacity is theoretically expanded to a multifold of 99. This should well accommodate all Latin characters, with a large redundancy. In the same way, thousands of the most frequently used Chinese characters could be coded with a tricolor QDOT system plus multiple λ_{ex} elements. Of course, given the number of single-color QDOTs that are available and the intensity levels that are distinguishable, using more spots spatially arranged in a particular way to represent a character could be a better way of increasing the coding capacity, though at the cost of reducing the readout speed.

This method is also a 0D coding method, but a time dimension for different exciting lights is needed, so it is a temporal 1D coding method.

The QDOT-based coding capacity can be easily expanded to a higher-dimensional coding space by deliberately arrange info-ink beans in a 1D sequence or 2D array. In this case, as each QDOT bean acts as a spectral 1D barcode, the QDOT-based info-ink has one more dimensional coding space than the ordinary 1D/2D patterns.

To further extend the coding capability of an info-ink to spatial three dimensions, that is, from a 2D surface to multiple layers, the two-photon exciting and three-dimensional (3D) information extraction method is proposed. The process of two-photon excitation is currently being examined for a variety of applications, for example, ultra-high-density optical data storage, biochemical imaging, and 3D microfabrication.[31]

We have investigated the two-photon excitation characteristics of QDOTs and have begun to explore their application for 3D information extraction. Theoretically, when laser light at a wavelength λ_p is tightly focused inside transparent QDOT materials so that its intensity in a small volume around the focus surpasses a specific threshold, two-photon absorption is triggered if $\lambda_p > \lambda_{em} > (\lambda_p/2)$, where λ_{ex} is denoted as any of fluorescent emission wavelength. Because the emission is localized in a part of the focal volume, 3D information could easily be extracted with the micron or submicron spatial resolution, depending on focal optics and laser intensity. Experimentally, we demonstrated two-photon excitation based on

CdSe, CdS, and/or CdSe/ZnS QDOTs and an 800 nm femtosecond (fs) laser with a repetition rate of 78 MHz and a pulse duration of 120–150 fs. Figure 13.3 shows the typical red fluorescence emission of soluble CdSe QDOTs by two-photon excitation. The volume of the focus spot can be manipulated by using various focusing objectives, laser beam diameter, and laser intensity.

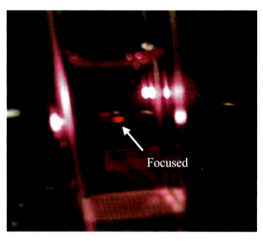

Figure 13.3 A typical red fluorescence emission of soluble CdSe QDOTs by two-photon excitation.

The fluorescence spectra of CdSe and CdSe/ZnS QDOTs by ultraviolet (UV) single-photon and 800 nm two-photon excitation are shown in Figs. 13.4a and b. Obviously, the two fluorescence spectra by two-photon excitation shown in 13.4a and b have low signal-to-noise ratios, with unambiguous recognition of fluorescence peaks and shapes. Moreover, the relationship between the emission intensity and the two-photon excitation pump intensity was investigated, with one CdS QDOT ensemble exhibiting an emission peak around 470 nm. While the fluorescent spectra of the CdS QDOT ensemble pumped by different pump power are shown in Fig. 13.5a, the fluorescent intensity versus the pump power shown in Fig. 13.5b was calculated as $y = 0.8515x^{1.4016}$. Such a relationship does not agree with that of theoretical prediction $y = \alpha \cdot x^2$ under the coherently driven two-photon resonant absorption. The discrepancy could be explained by the detuning of the 470 nm emission peak from the coherent two-photon pumping at around 800 nm, which is equal to 3723 cm^{-1}.[32]

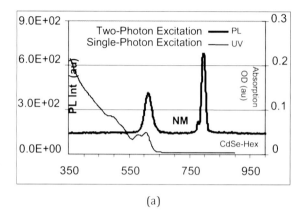

(a)

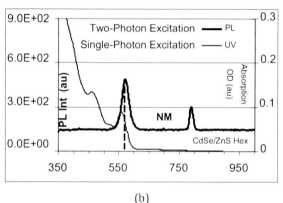

(b)

Figure 13.4 The fluorescence spectrums of CdSe-Hex QDs (a) and CdSe/ZnS-Hex QDs (b).

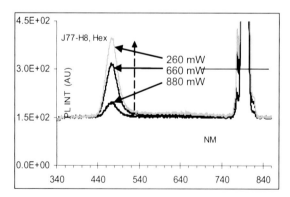

Figure 13.5a The fluorescent spectra of CdS QDOTs pumped by various excitation powers.

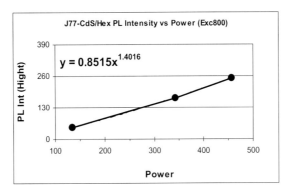

Figure 13.5b The relation between fluorescent intensity and pump power.

To prepare and apply an info-ink onto a cardboard object, such as a passport, an ID card, or a sticky label, is a crucial procedure for QDOT coding applications.

Here, four issues need to be addressed:

- The ability to accurately calculate the amount for each type of QDOT in the info-ink according to the encoding formula;
- The ability to mix and apply different types of QDOTs in the info-ink on an object;
- The ability to accurately apply a QDOT-ink to a predesigned 1D position; and
- The ability to accurately apply a QDOT-ink to a predesigned 2D area.

All the above functions can be easily reprogrammed and modified by auto/manual instruction.

To achieve these tasks, a practical solution is the use of a commercial inkjet color printer. Considering that a color printer has four cartridges for four basic inks—three principal colors and one black color—a color printer can be used for printing four different info-inks or QDOTs with different wavelengths.

However, all the four basic info-inks must be able to meet the requirement of the printer's nozzle: proper consistency, viscosity, and volatility, which is technically a chemical work of preparing the mixture of the info-ink.

For the four issues mentioned above, a computerized printing system can easily achieve those tasks. The main procedures are explained below:

- Applying amount control. A computer controls the applying amount of each info-ink—exactly the way like controlling the three principal and one black color for printing a color pattern. The depth resolution for each info-ink is typically 8 bits, 256 levels.
- Applying position control; and
- Applying area control. This can also be done by a computer, as easy as printing a color pattern in the predefined location and area.

In the case of more than four info-inks involved, more printers may be used in the printing system. All these printers can be controlled by one computer with multiple universal serial bus (USB) ports. All the encoding contents and patterns can be easily monitored and changed by a graphical user interface (GUI) in the application program. As the printing is a not heavy computing job, a digital signal processing (DSP)-based circuit could also be an option.

Figure 13.6 shows a scheme of a 16-info-ink printing system, in which four printers are employed.

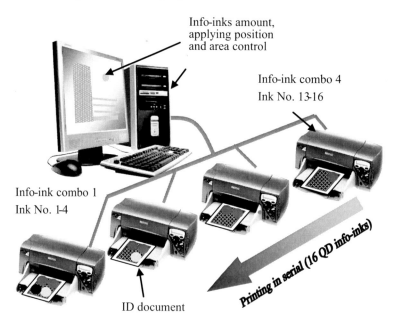

Figure 13.6 A four-printer, 16-info-ink, 2D encoding system.

13.4 Information Retrieval

Information retrieval basically consists of two parts, optical spectral signal detection and digital signal processing. The former detects and converts a QDOT fluorescence signal by a spectrometer, and the latter processes the signal by a computer with the following steps:

- Removing the noise by a digital filter;
- Separating the spectral center lines emitted by QDOTs from the overlapped spectra;
- Finding the wavelengths (Ws) and intensities (Is) of all the spectral center lines;
- Calibrating these Ws and Is; and
- Retrieving the original data according to a prior known code book.

Figure 13.7 shows a fluorescence spectrum measured from an info-ink containing only QDOTs with an emission peak at 535 nm. Because of the Gaussian-like profile, the neighboring spectral profiles may mutually affect the intensity of each other in an info-ink emitting multiple wavelengths, that is, introduce a spectral alias, as shown in the example in Fig. 13.8. The acquired spectrum is the top black curve, which is actually composed of spectra from two different QDOTs, represented by the light and dark gray curves in the figure. As the dark one is only about one-fifth of the light one in intensity, its peak could not be distinguished from the input spectrum. This effect will eventually result in a decoding error if no measure is taken.

A spectrum function of an info-ink can be described as

$$f(\lambda) = \sum_{i=1}^{N} k_i \cdot \delta(\lambda - \lambda_i) \otimes p(\lambda_i), \qquad (13.1)$$

where $\delta(\lambda)$ represents an impulse function, physically, a spectral line. k_i is the intensity of $\delta(\lambda)$ at λ_i. $p(\lambda_i)$ denotes the profile function centered at λ_i. $\otimes$ represents a convolution operation.

As described above, the broad wavelength profile of the info-ink is the main reason for the spectrum alias. To get rid of the alias effect, the spectrum line must be separated by means of de-convolution operation. Let $F(u)$ and $D(u)$ are Fourier transforms of $f(\lambda)$ and $\delta(\lambda)$, respectively. Provided all the $p(\lambda_i)$s have the same Fourier transformation $P(u)$, the separated spectrum line at λ_m is obtained by the inverse Fourier transform (IFT) of Eq. 13.1:

$$IFT[F(u)/P(u)] = \sum_{i=1}^{N} k_i \cdot \delta(\lambda - \lambda_i) \qquad (13.2)$$

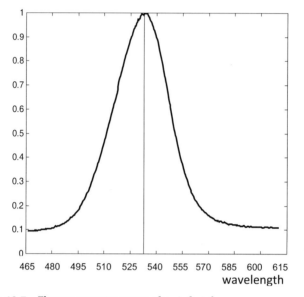

Figure 13.7 Fluorescent spectrum of an info-ink.

Equation 13.2 yields a series of $\delta(\lambda)$s, indicating that all the spectral lines are extracted and separated as individual impulses. However, as each spectrum profile of the info-ink is actually different from others, the de-convolution operation can only extract one narrow sharp impulse. To find all spectral lines, $k_i \delta(\lambda - \lambda_i)$, $i = 1 \ldots N$, N times operations are needed.

An experimental result of this procedure is illustrated in Fig. 13.8, in which the info-ink has center wavelengths at 611 nm and 632 nm and intensities 1 and 0.2, respectively. The spectrum of the info-ink is presented by the black bold line in the figure. Because the emission at 632 nm is weak, there is no noticeable peak at 632 nm in the spectrum. After two operations of spectrum line extraction, two spectral lines are obtained, as shown in Fig. 13.8. With all the extracted λ_is and k_is, the original data sequence can be retrieved eventually based on a prior decode book. A procedure for calibrating λ_is and k_is may be required before the decoding. In our experiments, we used a fiber optic spectrometer made by Ocean Optics.[33]

Compared to conventional spectrometers, this device dramatically reduces the size and cost and is easy to use.

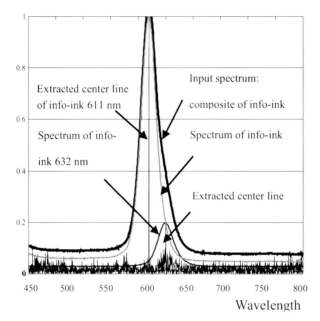

Figure 13.8 Illustration of spectral lines extraction.

For the 3D information extraction using two-photon exciting, a depth scanning device is required, with the collaboration of a 2D extraction system. Controlled by a computer, the information encoded in different depths will be retrieved layer by layer.

However in applying QDOT-based labeling technology to a sunshine-covered area, there is a critical issue to be addressed: the brightness of sunlight can overpower most optics-based solutions. Specifically, detecting a return fluorescent signal can be difficult as the return optical signal can get overwhelmed by sunlight.

As a fact of matter, the sun is not a perfectly continuous white source. There are many black lines seen in the solar spectrum—the Fraunhofer lines. As shown in Fig. 13.9, the solar spectrum has been split into pieces, from the blue (left) end to the red (right) end by these absorption lines—literally thousands, in fact, if one looks in fine-enough detail. These black spectrum lines provide the coding space. The info-ink will be prepared and mixed with the QDOTs whose fluorescent wavelengths are located in those black lines.

Coding by those wavelengths, the information emitted from the info-ink could not be covered by the reflection of the powerful daylight.

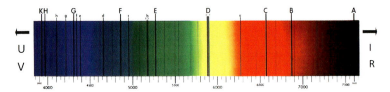

Figure 13.9 Fraunhofer lines in the solar spectrum.

Although the encoding information provided by these wavelength positions could not be covered by the reflection of the powerful daylight, in our retrieving device proposed, special filters are still used to extract the useful wavelengths, namely, Fraunhofer line signals, while screening the others.

13.5 Applications

Due to their flexible form, encoded QDOTs could be prepared as ink, paint, label, and powder. They can be applied to label varied objects easily. They could replace almost all the existing barcodes, either 1D or 2D technology. They can also be applied to objects where a barcode cannot be used, for example, a small area and a curved surface. QDOTs for biolabeling were proposed about 10 years ago.[10] Bawendi et al. in the MIT reported the design of using QDOTs for inventory control.[16,17] In this section, we will describe the applications of QDOTs for anticounterfeiting and friend/enemy discrimination.

13.5.1 Anticounterfeiting

Counterfeiting represents a major problem in the world community. It illegally results in economic losses of hundreds of billions of dollars to corporations and governments. It is also dangerous to the consumer by an increase of the risk of acquiring faulty, low-quality, or dangerous products, ranging from clothing, software, and medicine to currency and credit cards.

A direct application of QDOTs is ID card identification. A 2D information-retrieving system was reported by Chang et al.[34], which consists of an exciting light, a spectrum sensor, and a signal-processing unit. Figure 13.10 shows such a system for retrieving the information hidden in a tiny spot of an ID card. The exciting light is provided by a 370 nm light-emitting diode (LED) light source. A bunch of optical fibers guides the exciting light to the info-ink spot applied on the surface of any object. The fluorescence emitted from the QDOTs is collected by the detecting fiber in the fiber bunch and fed to a spectrometer. The data generated by the spectrometer is further delivered to an intelligent instrument, for example, a microprocessor or a personal computer (PC), which eventually extracts the information originally coded in the info-ink. To obtain an even exciting light, the exciting fibers are arranged to surround the detecting fiber evenly to form an optical fiber bundle, as shown in Fig. 13.10. A rubber cup is connected at the end of the fiber bundle to ensure that only the excited fluorescent light can enter the detecting fiber. Figure 13.11 provides a photo of the prototype of the system illustrated in Fig. 13.10.

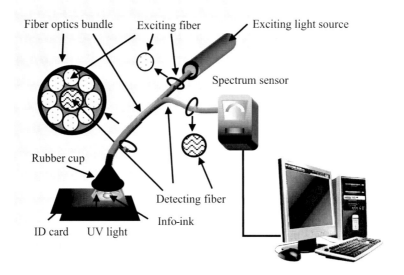

Figure 13.10 Scheme of a QDOT info-reader for an ID card.

To code the documents automatically and fast, principal inks should be prepared first. For an application using six wavelengths, six principal info-inks are needed. Controlled by a computer, these principal info-inks are mixed to form a coded info-ink and then applied onto the surface of an ID card. As the last step, a cover layer is applied to the top of the info-drop to prevent the info-ink from being scratched or damaged by any means.

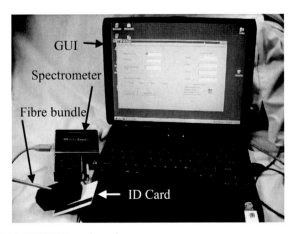

Figure 13.11 QDOT ID card reader.

13.5.2 Friend/Enemy Discrimination

The problem of identifying friendly fires from unfriendly forces becomes critical in modern battles. The rise of the so-called "smart weapons" has increased the accuracy of ordnance delivery but, sadly, has also increased the casualty rates of so-called "friendly fire" (FF) incidents. After World War II, the use of precision weapons and advances in intelligence technologies for air and space have drastically revolutionized air warfare. However, the percent casualties rose due to FFs. The percent casualties from 21% in World War II changed to 39% in the Vietnam War and 49% in the Persian War (estimated by the American War Library's report on FF casualties, both fatal and nonfatal, based on the historic war department, Department of the Navy, and Department of Defense casualty reports detailing various battle reports).[35]

After Desert Storm, the Pentagon conducted extensive research into developing ways for friend/enemy discrimination but ended up ditching many of them due to cost. One program—called the Battlefield Combat Identification System, or BCIS—had been introduced to prevent FF. Based on radar technology, such an automatic identification system was developed and used by aircraft. Under the BCIS, every combat vehicle in the US Army was equipped with a small transmitter and receiver, which would receive and send signals to identify itself between the firer and the fired upon. However, equipping the army with such automatic identification systems would have cost as much as $40,000 per vehicle.[36] To date, FF is still paying a role as a critical "enemy" in the battlefield. In April 2002, a US fighter pilot was involved in a FF incident that killed four Canadian soldiers in Afghanistan. In April 2003, US planes apparently fired by mistake on allied Kurdish guerrillas and US special-forces soldiers, killing at least 18 people and injuring more than 45.

Developing a simple and cheap friend/enemy discrimination device is becoming increasingly important. A solution is needed that is cost effective and easily deployable on individual soldiers. Such a solution should also be capable of being deployed on ordnance to avoid human error and misidentification by combatants such as pilots. One possible solution is the use of a QDOT information-coding and information-retrieving system. Instead of using expensively active radar technology, the QDOT system takes advantage of the passive fluorescent features of QDOTs. To avoid FF from attacking aircraft, the coded info-ink is painted or labeled on the top surfaces of vehicles and soldiers' helmets as ID labels, as shown in Fig. 13.12. By sending a probing beam and detecting the reflected spectrum, the aircraft can distinguish these labels before firing a missile or bomber.

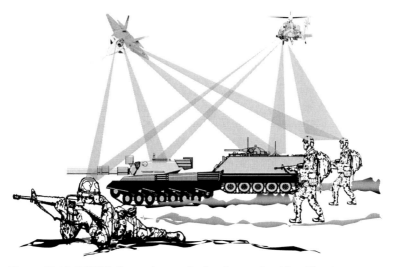

Figure 13.12 QDOT information retrieving for combat identification.

Figure 13.13 shows the QDOT info-reader and info-label applied on the helmet of a soldier. By adding a timing agency inside the info-label, this ID coding could expire at a setting date.

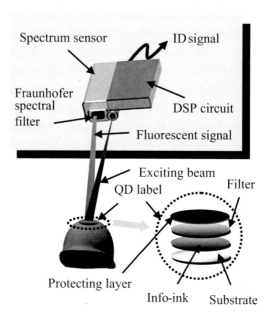

Figure 13.13 QDOT info-label and its reader.

In applying QDOT-based labeling technology to the battlefield, another critical issue to be considered is that in a cloudy or dusty environment, the exciting light from the aircraft and the emitted fluorescence from the ground objects may not be seen clearly.

To address the sunshine problem, Fraunhofer line coding may provide a solution. The second issue is also a common problem for aircraft using laser-guided missiles and bombers. It is difficult for a distant aircraft to detect the ID signal through a cloudy or dusty space. To receive more ID information, the aircraft has to lower the altitude, but it will increase the risk to be attacked by ground fires. Considering that the QDOT information retrieval device is relatively simple and cheap, such a device could be built into the missile or bomber rather than into the aircraft. When the missile/bomber approaches the target and the ID signal, emitted by QDOT labels, will become strong enough, the device will disable the explosion at the last moment if the target is found as a friend. Another more practical device, called probing bomber, can be developed. Similar to an illuminating projectile, it is fired by an aircraft and explodes at the near-ground area with strong UV light emission. After retrieval processing, the ID signal is converted to a radio signal and sent to the remote aircraft or battle control centre.

Figure 13.14 shows a scheme for an ID detection system mounted on warheads. Even after the firing, if the ID signal is detected as a friend, the intelligent device built into the warhead will disable the explosion at the last second. In the battlefield, the more the number of soldiers and vehicles involved, the stronger the ID label signals produced, which will further enhance the discrimination and reduce the large-scale casualties caused by FF.

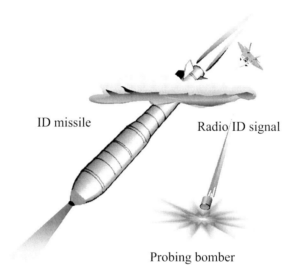

Figure 13.14 Warhead and probing bomber with QDOT ID.

13.6 Conclusions and Outlook

We have described methods for status identification using QDOT-based info-inks containing fluorescent substances with well-defined emission spectra. We also described the applications for document security and battlefield identification. PL QDOTs are suitable for security identification due to their intrinsic optical properties, such as narrow emission and broad absorption, good photostability, and easy tuning of their photoemission peak positions via control of their size, structure, and composition. Thus, QDOTs satisfy the main requirements in security identification, such as easy encoding and decoding, large coding space, invisibility, and ease of deployment.

There are some critical issues to be addressed:

(1) Long-term stability of QDOTs. The emission feature of QDOTs as a function of time and environment is yet to be established. A possible factor that leads to the change of the fluorescence spectra with time is the slow aggregation process of QDOT particles in a polymer matrix. However, it is possible to prevent this process by cross-linking the polymer (e.g., by UV) after the info-ink is applied to the targeted surfaces. A mixture of solvents and other additives may be used to improve the properties such as solubility, viscosity, volatility, storage stability, adhesion, etc., so that the info-ink can be applied to surfaces with different chemical natures.

(2) Detection of the spectrum signal from a distance. This issue is particularly important for the battlefield identification. After passing through the Fraunhofer line filter, the excited signal should be strong enough to identify itself. Two factors are involved in the performance of the detection. One is the exciting beam; the other is the sensitivity of the sensor. As both the exciting source and the sensor are built into the aircraft, the major effort will be focused on this most expensive device. On the other hand, research will be carried on concentrating the info-inks, to intensify the fluorescence, to enhance the fluorescent signal emitted from ID labels on the ground.

(3) Algorithms used for extracting and separating the spectrum data. The de-convolution procedure is adopted basically due to its ability to narrow and separate the overlapped neighboring QDOT spectra. However, if the QDOT spectrum changes its shape, this procedure will introduce some measurement error. Fortunately, QDOT-based info-inks have exhibited little changes in their individual emission shapes. Our experiments and computer simulations show that the spectrum profiles of info-inks have also little change before and after they blend with others. Although many algorithms of signal detection and spectrum evaluation are available[37], they are basically designed for the purpose of improving the signal-to-noise

ratio, which is at a lower priority compared to the spectrum overlap problem encountered in this application. However, when a useful signal in the input is weak, a calibrating procedure should be introduced.

This chapter provides only a basic idea and primary experimental results about the use of PL semiconductor nanocrystals, or QDOTs, for codification and authentification purposes. Theoretically, the proposed QDOT information-encoding and retrieval technology should have a bright future in many applications; however, they are not mature technologies at the present time. Additional experiments, such as sensitivity to nonuniform illumination, detection limitations, repeatability, and data-collecting time, will be continuously conducted in work.

References

1. A. P. Alivisatos, *J. Phys. Chem.*, **13226** (1996).
2. A. Empedocles, D. J. Norris, M. G. Bawendi, *Phys. Rev. Lett.*, **3873** (1996).
3. T. J. Gould, J. Bewersdorf, and S. T. Hess, *Zeitschrift fur Physikalische Chemie—Intern. J. Res.* in *Phys. Chem. Chem. Phys.*, **833** (2008).
4. K. Jacobs, D. Zaziski, E. C. Scher, A. B. Herhold, A. P. Alivistos, *Science*, **1803** (2001).
5. Z. Jakubek, J. de Vries, S. Lin, J. A. Ripmeester, K. Yu, *J. Phys. Chem. C*, **8153** (2008).
6. M. A. Hines, P. Guyot-Sionnest, *J. Phys. Chem.*, **468** (1996).
7. M. Danek, K. F. Jensen, C. B. Murray, M. G. Bawendi, *Chem. Mater.*, **173** (1996).
8. B. O. Dabbousi, J. Rodriguez-Viejo, F. V. Mikulec, J. R. Heine, H. Mattoussi, R. Ober, K. F. Jensen, M. G. Bawendi, *J. Phys. Chem. B*, **9463** (1997).
9. J. Ouyang, Md. B. Zaman, F. Yan, D. Johnston, G. Li, X. Wu, D. Leek, C. I. Ratcliffe, J. A. Ripmeester, K. Yu, *J. Phys. Chem. C*, **13805**(2008).
10. C. I. Ratcliffe, K. Yu, J. A. Ripmeester, Md. B. Zaman, C. Badarau, S. Singh, *Phys. Chem. Chem. Phys.*, **3510** (2006).
11. J. Ouyang, B. Wilkinson, D. Kingston, J. Kuijper, C. I. Ratcliffe, X. Wu, J. A. Ripmeester, K. Yu, *J. Phys. Chem. C*, **4908** (2008).
12. X. Zhong, Y. Feng, W. Knoll, M. Han, *J. Am. Chem. Soc.,* **13559** (2003).

13. M. Bruchez, Jr., M. Moronne, P. Gin, S. Weiss, A. P. Alivisatos, *Science*, **2013** (1998).
14. W. C. Chan, S. Nie, *Science*, **2016** (1998).
15. X. Gao, W. C. W. Chan, S. Nie, *J. Biomed. Opt.*, **532** (2002).
16. Bawendi et al., United States Patent: US6,602,671 B1. Aug. 5, 2003.
17. Bawendi et al., United States Patent: US6,617,583 B1. Sep. 9, 2003.
18. M. Zhou, S. Chang, C. P. Grover, United States Patent: US7,077,329. July 18, 2006.
19. S. Chang, M. Zhou, C. P. Grover, United States Patent: US7,202,943. April 10, 2007.
20. C. B. Murray, D. J. Norris, M. G. Bawendi, *J. Am. Chem. Soc.*, **8706** (1993).
21. Z. A. Peng, X. Peng, *J. Am. Chem. Soc.*, **123** (2001).
22. S. Nie et al., *Curr. Opin. Biotechnol.*, **40** (2002).
23. J. Lee, V. C. Sundar, J. R. Heine, M. G. Bawendi, K. F. Jensen, *Adv. Mater.*, **1102** (2000).
24. http://www.evidenttech.com.
25. K. Yu, J. A. Ripmeester, United States Patent: US20,050,238,568, April 21, 2007.
26. K. Yu, B. Zaman, J. A. Ripmeeste, *J. NanoSci. NanoTec.*, **669** (2005).
27. K. Yu, S. Singh, N. Patrito, V. Chu, *Langmuir*, **11161** (2004).
28. K. Yu, B. Zaman, S. Singh, D. Wang, J. A. Ripmeester, *Chem. Mater.*, **2552** (2005).
29. Y. Nien, B. Zaman, J. Ouyang, I. Chen, C. Hwang, K. Yu, *Mater. Lett.*, **4522** (2008).
30. Ming Zhou, Shoude Chang, C. P. Grover, *Opt. Exp.*, **2925** (2004).
31. W. Zhou, S. M. Kuebler, K. L. Braun, T. Yu, J. K. Cammack, C. K. Ober, J. W. Perry, S. R. Marder, *Science*, **1106** (2002).
32. B. R. Mollow, *Phys. Rev.*, **1969** (1969).
33. http://www.oceanoptics.com.
34. S. Chang, M. Zhou, C. P. Grover, *Opt. Exp.*, **12**, 143–148 (2004).
35. http://members.aol.com/amerwar/ff/ff.htm.
36. http://www.fas.org/man/dod-101/sys/land/bcis.htm.
37. C. C. Chan, W. Jin, M. S. Demokan, *Opt. Laser Tech.*, **299** (1999).

Chapter 14

Characterization Approaches for Blue and White Phosphorescent OLEDs

Brian W. D'Andrade
Exponent, 420 Lexington Ave, Suite 1740, New York, NY 10170, USA
bdandrade@exponent.com

The characterization of blue and white organic light-emitting diodes (WOLEDs) containing phosphorescent emitters are examined with a focus on the methods used for the characterization of new materials and device architectures. Early blue phosphorescent OLEDs had limited efficiency due to the lack of host materials with large triplet energies that could exothermically transfer energy to high-energy phosphorescent guests. Unfavorable efficiency was overcome with wide bandgap host materials that enabled internal quantum efficiencies near unity. One path to saturated deep-blue emitters favored the use of hosts that were electrically inert and required the use of phosphorescent guests capable of transporting charge. WOLED characterization is also discussed. White devices have been studied due to their potential for high-efficiency solid-state lighting; hence, their optical and electrical characterizations have been standardized for display and illumination engineers. Early

---

Luminescence: The Instrumental Key to the Future of Nanotechnology
Edited by Adam M. Gilmore
Copyright © 2014 Pan Stanford Publishing Pte. Ltd.
ISBN 978-981-4241-95-3 (Hardcover), 978-981-4267-72-4 (eBook)
www.panstanford.com

reports presented basic information on device performance, but more rigorous characterization methods have been developed and codified to foster the successful growth of the technology. Several tools for studying white device parameters and determining material properties are summarized in this chapter.

14.1 Introduction

Organic thin-film semiconductors are fascinating and promising materials for the development of new electronic and optoelectronic devices.[1] Traditional inorganic semiconductor analytical techniques are often not sufficient to fully characterize organic semiconductors, but significant progress has been made in modifying tools to accommodate the unique features of organic materials. In parallel, organic device characterization has steadily matured, and new international standards have been delineated for the reporting of device properties.

This chapter focuses on the discussion of blue and white phosphorescent organic light-emitting devices (PHOLEDs). Both devices contain phosphorescent emissive molecules, which imbue them with the ability to convert near 100% of injected current into light; however, devices need significant improvement in operational stability and high-yield-mass production methods to successfully compete with incumbent display and lighting technologies, so improvements in characterization and analyses of PHOLEDs are necessary to achieve the requisite device features.

Section 14.2 of this chapter describes the development of blue PHOLEDs and the characterization methods deployed in understanding the fundamental properties of these devices. Section 14.3 provides insight into white OLED (WOLED) optical and materials characterization methods.

14.2 Blue Electrophosphorescence

The development of PHOLEDs emitting blue wavelengths progressed at a slower pace than counterparts emitting in the red and green. Blue PHOLED emission efficiency and operating lifetime are two example characteristics that have had a longer development time than comparable green and red device characteristics. However, advances over the past decade continue to indicate that blue PHOLEDs remain the most promising path to high emission efficiency in displays and solid-state lighting. This section discusses some of the improvements made during the early development of blue PHOLEDs.

14.2.1 Device Architecture and Energy Transfer

In a typical PHOLED, a common approach to realizing high efficiency uses Förster energy from a fluorescent host material to excite a phosphorescent guest material. To ensure a suitable energetic alignment for this transfer (i.e., exothermic energy transfer from host to guest), the exciton energies of the host should exceed those of the guest. This ensures that energy transfer occurs from both the singlet and the triplet exciton state of the host and that the lowest excited energy state for the system is the guest triplet (i.e., the excited state is ultimately confined to the guest).

This favorable energy alignment was difficult to realize during the initial development of blue PHOLEDs. Phosphorescent dopants were synthesized with high triplet energy levels, but the singlet energies of the dopants were also necessarily high. As a result, to ensure efficient energy transfer and exciton confinement, both the singlet and triplet levels of a suitable host had to be even higher in energy than hosts previously employed for other emitters.

The lack of a suitable high-triplet-energy host was not a deterrent to the first demonstration of blue electrophosphorescence, which relied on endothermic energy transfer.[2] In an Adachi et al. paper[2] demonstrating the first blue electrophosphorescence, the host and guest materials were *N,N'*-dicarbazolyl-4-4´-biphenyl (CBP) and iridium(III)bis[(4,6-difluorophenyl)-pyridinato-*N*,$C^{2'}$]picolinate (FIrpic), respectively.

In that paper, the singlet energy level of CBP was shown to be larger than that of FIrpic; however, the CBP triplet level is below that of FIrpic. The result of the energy difference is an excited state that resides at the CBP triplet level prior to recombination on FIrpic. The endothermic energy transfer is not the most efficient method to produce emission, so new hosts with appropriate energies and transport materials were developed and new analytical tests were adapted to select guest-host combinations for device fabrication.

Transient temporal tests at low temperature were a key to understanding the energy transfer dynamics.

14.2.2 Identifying High-Triplet-Energy Host Materials

A significant improvement to the endothermic host-guest arrangement of Section 14.2.1 came with the use of new large-triplet-energy host materials capable of realizing exothermic energy transfer to blue phosphorescent guests. This favorable energetic alignment was first realized using *N,N'*-dicarbazolyl-3,5-benzene (mCP).[3]

The triplet energy of mCP (and of many fluorescent materials) can be estimated by examining cryogenic temperature phosphorescence from a host compound. At room temperature, the triplet states of fluorescent materials are nonradiative. The rate of nonradiative decay decreases as temperature decreases, so a fluorescent material cooled to cryogenic temperatures enables efficient phosphorescence and emission from the triplet state can be observed.

Using a streak camera and liquid nitrogen–cooled samples, one can determine that the energetic alignment between the triplet levels of mCP and FIrpic is suitable for triplet exciton confinement on the phosphorescent guest before fabricating a PHOLED.

The confinement of the triplet excited state on FIrpic can also be observed from measurements of the transient phosphorescent decay of FIrpic as a function of temperature. In this type of measurement, the decay from films of CBP:FIrpic and mCP:Firpic are measured as a function of temperature using a streak scope.

For the case of endothermic energy transfer (FIrpic in a CBP host), the lifetime of the FIrpic emission is extremely long. In fact, the natural lifetime of FIrpic is ~23 μs at 10 K and ~1 μs at 300 K in neat film. The long lifetime in CBP arises from the fact that the excited state spends significant time on the triplet of CBP before ultimately undergoing endothermic transfer to FIrpic and finally radiative recombination. Thus, the observed energy transfer dynamics between CBP and FIrpic reflect the energetic alignment between host and guest.

With a host matrix of mCP, the decay of FIrpic phosphorescence occurs on a much shorter time scale, consistent with the behavior of FIrpic in neat film. This suggests that the triplet excited state remains well confined to the guest triplet level and that the observed behavior reflects the intrinsic photophysics of FIrpic.[3]

Transient phosphorescence data together with measurements of the steady-state phosphorescence intensity as a function of temperature was used to determine if endothermic or exothermic conditions existed. In the case of CBP:FIrpic, significant temperature dependence was noted, and emission from FIrpic increased rapidly as the temperature was increased because of the thermal contribution required to complete the endothermic nature of the host-guest transfer. The mCP:FIrpic system showed very little dependence on temperature until very low temperatures, where a reduction in exciton diffusivity may have prevented exciton formation on FIrpic.[3]

Overall, the use of transient and low-temperature photoluminescence measurements, on thin films and solutions, has enabled more profound insight into energy transfer in PHOLEDs. These techniques have provided a means to screen potential host and guest materials prior to integration into an OLED.

14.2.3 A General Route to Deep-Blue Electrophosphorescence

While efficient triplet confinement on the emissive dopant remains a key design principle in PHOLEDs, the need for energy transfer is not a limiting factor. In fact, concomitant with the increase in the bandgap of blue emitters is an increase in the applied voltage across a device required to inject charge into the constituent material layers. As such, one advance in the design of blue PHOLEDs was the introduction of ultrawide energy gap host (UGH) materials.[4]

This approach uses the host only as an inert matrix to suspend the phosphorescent dopant, so charge transport and exciton formation occur primarily on the guest. In the methodology, carriers need not be injected into the relatively inaccessible highest occupied molecular orbital (HOMO) and lowest occupied molecular orbital (LUMO) levels of the UGH, permitting the excitation of deep-blue phosphorescent guests without corresponding increases in operating voltage.

In the initial demonstration of this approach[4], two UGHs, diphenyldi(o-tolyl)silane (UGH1) and *p*-bis(triphenylsilyly)benzene (UGH2), were paired with a blue phosphor iridium(III) bis(4´,6´-difluorophenylpyridinato)tetrakis(1-pyrazolyl)borate (FIr6). The performance of those devices marked the first demonstration of blue PHOLEDs' external quantum efficiencies that exceeded 10%.

In those high-efficiency structures, the key characterization activities sought to confirm that charge was carried primarily by the phosphorescent guest and that the host acted essentially as an inert matrix. An examination of device performance as a function of FIr6 doping confirmed the role played by the guest in the transport of charge.

The undoped host exhibited low current and recombination in the hole-transport layer 4,4′-bis[*N*-(1-napthyl)-*N*-phenyl-amino] biphenyl (NPD) because of the large barrier for hole injection into the UGH material. As FIr6 was added in significant quantities, the operating voltage of the device was reduced because FIr6 was shown to lower the resistive pathway for hole injection and transport into the emissive layer.

The ability of FIr6 to efficiently form excitons was also evidenced by the excellent performance obtained for neat-layer devices that lack the UGH material. Efficiencies of $\eta_{EXT} = (5.0 \pm 0.5)\%$ and $\eta_P = (7.0 \pm 0.7)$ lm/W were observed, among the highest ever reported for OLEDs employing a neat phosphorescent material.[4]

14.2.4 Application in White OLEDs

The use of wide-energy-gap hosts to realize efficient blue electrophosphorescence has since been exploited to demonstrate ultraviolet-emitting PHOLEDs, as well as high-efficiency WOLEDs. It is sometimes straightforward to modify the structure of the devices discussed in Section 14.2.3 to realize white emission by including red and green phosphorescent guests in the emissive layer of the device.

By tuning the relative concentration of blue, green, and red emitters, the spectral output and quality of the white can be adjusted. This is an attractive feature of WOLEDs, the ability to tailor device spectral output to customer specifications through tuning of the doping concentration.

Figure 14.1 shows the output from an FIr6-based WOLED that has a "triple doped" emissive layer consisting of FIr6, the green phosphor fac-tris(2-phenylpyridinato-N,C2′)iridium(III) (Ir(ppy)$_3$), and the red phosphor iridium(III) bis(2-phenylquinolyl-N,C2′) acetylacetonate (PQIr).[5] The characterizations of materials and architectures for WOLEDs are discussed in detail in Section 14.3.

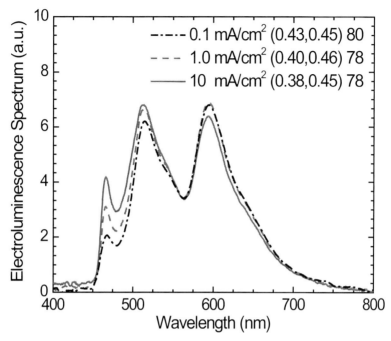

Figure 14.1 Normalized electroluminescence spectra of WOLED emission at various current densities. The corresponding CIE and CRI values for each spectrum are inset. *Abbreviations*: CIE, Commission Internationale de l'éclairage; CRI, coloring rendering index.

14.3 White Organic Light-Emitting Device

Solid-state light sources are the future of lighting, and WOLEDs[6,7] are being investigated to deliver exciting new illumination forms and functionalities. WOLEDs were introduced in the early 1990s[8-10], and substantial materials development led to device efficacies exceeding 100 lm/W in 2008.[11] This milestone delivered on the innate efficacy advantages of this technology, which can efficiently convert electronic energy into optical energy[12], operate at low voltage[13], and extract light from organic materials.[14]

Over time, the complexity of WOLED characterization developed to near fully describe optical, electrical[15], and material properties of these organic semiconductor devices. The growth in the number of types of measurements stems from the maturation of the technology from the lab to illumination designers. At each stage of development, the interests and insights of developers differed, so WOLED characterizations increased to support the demands of multiple groups of scientists, designers, and engineers.

14.3.1 Optical Characterization and Device Efficiency

The light emitted or electroluminescence from the WOLED surface normal can be minimally described by its 1931 CIE coordinates and luminance.[16] The CIE and luminance are easily captured by inexpensive luminance meters, and other photopic quantities may be derived. For example, luminous efficacy, luminous intensity, and power efficacy may be calculated from the luminance when the area of, operating voltage of, and current through a device are known and some assumption of luminance variation with viewing angle[17] and luminance variation[18] across the device active area are incorporated into the calculations.

Fundamental physical quantities required additional data that allowed conversion of photometric units to radiometric units.[19] For example, the external quantum efficiency can be calculated from the luminance information and the OLED spectral power distribution, which is measured separately using a spectrometer calibrated to measure relative intensities.

Researchers are often only interested in OLED forward emission into 2π steridians[20] because of intended flat-panel display applications. Hence, forward emission external quantum efficiency was determined with a calibrated flat silicon photodetector[21] that provided an inaccurate estimate of the emitted optical power, which together with the current was used to calculate the quantum efficiency. However, a cosine collector[22] is required to obtain a more accurate determination of the OLED forward optical flux, and these collectors are often attached to integrating spheres that are used to measure total flux.[17]

Spectroradiometers capture the absolute spectral power distribution of a light source, and this single instrument can be used to calculate parameters such as external quantum efficiency, CRI, and correlated color temperature (CCT);[16] however, one still required assumptions about the variation of the spectral power distribution with viewing angle and luminance variation across the device active area to support calculations. As a note, spot meters, such as the spectroradiometers, accurately measure the device emission from a fraction of the device active area, so awareness of measurement errors caused by variation of luminance across the surface of a device is important. These color or intensity variations arise from intentional color variations, or pixilation and resistive losses.

The existence of metameric sources of light, that is, light sources with a similar CIE and different spectral power distributions, requires complete knowledge of the emission characteristics of sources to assist in OLED design. If the emission characteristics of red, yellow, and blue emitters are individually known, linear combinations of the emission can be created to determine potential CIE and CRI combinations, as shown in Fig. 14.2. The information from such calculations is exploited to optimize device color characteristics.

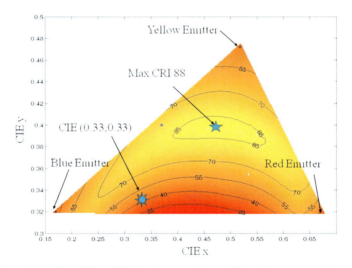

Figure 14.2 The CIE vs. CRI for three emitters. The unique spectral power distributions determine the CRI contour pattern. Emitters with the same CIE can have very different CRI contours. The maximum CRI is displaced from the white point (0.33, 0.33). Incandescent sources have a CIE near to (0.42, 0.41).

Spectroradiometers are versatile instruments, and their usage extends beyond OLED optical emission. This instrument can be used to measure the photoluminescence, reflectivity, and absorption of OLED materials. Typically, fluorimeters were used to record photoluminescence, but the spectra of a source obtained from a spectroradiometer and from a fluorimeter tend to have noticeable differences. Therefore, one may choose a spectroradiometer for both electroluminescence and photoluminescence measurements to ensure consistency of results. The reflectance of metal electrodes and the absorbance of thin films may be obtained by using a spectroradiometer and a light source with emission at all wavelengths under investigation; however, other specialized instruments are typically engaged for these types of measurements.

To fully capture the WOLED spectral output characteristics, a spectroradiometer coupled with a goniometer would be required to record the spectral power distribution over 4π steridians. This is not trivial, is time consuming, and is costly. Hence, more expedient methods of characterization, such as imaging spheres, are exploited.

For example, the total power efficacy[23] of a WOLED can be ascertained by using an integrating sphere with a photopic or a spectroscopic detector instead of integrating the spectral power distributions over 4π steridians. An integrating sphere measurement has several advantages over other measurements. There are no assumptions about the luminance distribution versus angle or about the luminance uniformity[24] of the OLED active area, and the measurement can be completed in minutes.

Integrating spheres are also essential in determining the photoluminescence quantum yield[25] of thin films (~100 nm) of organic materials employed in OLEDs. The photoluminescence quantum yield is a fundamental property of OLED materials, and over time more reliable and consistent measurement systems have been developed. Current tools are portable, so they can be used in confined spaces where OLEDs are processed to avoid deleterious effects of oxygen and water on the photoluminescence of the organic materials.

An imaging sphere[26] is another tool that has recently been used to plot the luminance, luminous intensity, and CIE versus the viewing angle over 2π steridians or over the forward emitting direction of the WOLED. These advanced imaging systems are necessary to generate more complete WOLED emission characteristics that are desired by illumination designers.

Presently OLED characterization has been formalized by the Illuminating Engineering Society of North America (IESNA) with its LM-79-08 and LM-80-08 procedures.[23,27] These documents provide insight into the type of measurements systems and practices that are necessary for the consistent presentation of WOLED data and for assessing the quality of products.

14.3.2 Characterization of Organic Semiconductor Materials

Possibly, the most difficult aspect of OLED characterization is accurately measuring the optical output. The current-voltage characteristics of OLEDs can be rigorously determined from a semiconductor parameter analyzer or a number of simpler digital multimeters. However, some other intrinsic characteristics are usually not trivially determined.

For example, mobility of organic materials has been extensively studied, and they vary significantly upon the method used to determine this quantity. Time-of-flight and electrical pulse measurements are a couple of techniques used to calculate mobility.[28–30]

Capacitance measurements are sometimes employed in the analysis of devices, but these measurements have not found widespread acceptance. Organic materials dielectric constants are approximated from the refractive index, and significant tabulation of dielectric constants of organic materials based on capacitance measurements is difficult to find in the literature. This is due to the nature of the development of WOLEDs. Materials are rapidly developed, and little time is spent on fully characterizing any single material.

However, there are some quick measurements that provide the basic properties of organic materials that constitute WOLEDs. This section will focus on a few that find practical application in the design of WOLEDs: complex refractive index, molecular orbital energies, radiative decay rates, morphology, and purity.

Material complex refractive index spectra $[n(\lambda) + ik(\lambda)]$ are required for detailed modeling of all WOLED optical cavities. There are numerous WOLED materials, and their indexes can significantly vary over the visible range of wavelengths, so spectroscopic ellipsometers[31] are valuable tools to determine n and k over visible wavelengths. Transmission and absorbance measurements can also be used to determine n and k. Both methods yield valuable n and k data that can be directly used to design WOLEDs and organic solar cells.[32] Examples of n and k data are shown in Fig. 14.3.

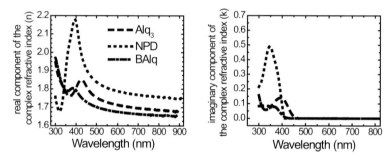

Figure 14.3 The complex refractive index components for three common organic thin films are show in the above plots. These values were determined using ellipsometry.

Knowledge of both the HOMO and LUMO energies of organic thin films is essential for understanding and for designing organic devices. For example, an OLED usually consists of several layers of various stacked organic thin films, and offsets in the HOMO and LUMO energies between layers act as potential energy barriers to the flow of charge and molecular excited states (or excitons).

Two conventional methods to ascertain HOMO energies (E_{HOMO}) are ultraviolet photoemission spectroscopy (UPS) and cyclic voltammetry (CV). UPS experiments determine the ionization energy (E_i) of a molecule on the surface of a thin film, where E_i = $-E_{HOMO}$. Solution-based CV experiments determine the relative molecular oxidation potentials (V_{CV}), which are indirectly related to E_i. It is desirable to determine E_{HOMO} from UPS data; however, the high cost and complexity of UPS systems tend to favor the use of CV in many laboratories. A linear relationship between the HOMO energy found using UPS and the oxidation potential found from CV was demonstrated[33], so either CV or UPS can be used to estimate HOMO energies. Figure 14.4 is a plot of ionization energy versus electric potential for several organic thin films.

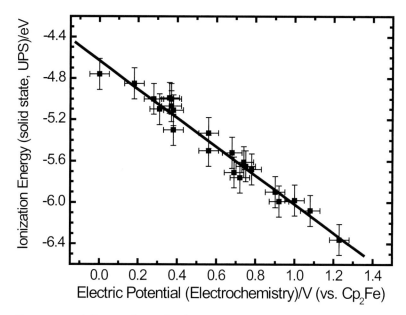

Figure 14.4 A linear relationship between ionization energy of organic thin films and the solution electric potentials is shown in the above plot.

LUMO energies are calculated by approximating the organic material optical bandgap as the energy equivalent to the onset of absorption in absorbance spectra and by using the HOMO energy derived from the methods described in the paragraph above. Inverse photoemission spectroscopy can more accurately determine the energy of the LUMO, and it is used in more fundamental studies of organic materials.[34]

Two other fundamental properties of WOLED constituent materials are the radiative and nonradiative decay rates, which are determined by measuring the photoluminescence quantum yield[25] and photoluminescence lifetime. Lifetimes are determined using a streak camera[35,36] or a time-correlated single-photon counter[37] or a flashlamp system, and this data finds applications in any number of OLED material studies.

The morphology of amorphous organic materials is often too complex to fully consider in the evaluation of materials incorporated into WOLEDs. Atomic force microscopy, X-ray crystallography, and transmission electron microscopy are examples of specialized techniques that have been used to analyze the very thin material films that are common in WOLEDs. Morphological stability is typically inferred from glass transition temperatures[38,39], and the glass transition temperature of an organic material is often chosen to be high to ensure the operational longevity of devices.

WOLED efficacy increased from <1 lm/W to over 100 lm/W in the span of 16 years. This improvement was achieved through the development and optimization of measurement techniques, understanding of device operational physics, and novel materials. The operational longevity[34,40,41] of devices has also been extensively investigated with highly specialized equipment, which was not discussed in this chapter. Intense interest in deploying WOLEDs as illumination sources ensures the continued introduction of novel characterization methods and advancement of presently esoteric analysis methods. Light sources permeate our world, and new, commercially available, efficient WOLED illumination systems are near to the present and are certainly in the future.

References

1. S. R. Forrest, *Nature*, **428**(6986), 911–918 (2004).
2. C. Adachi, R. C. Kwong, P. Djurovich, V. Adamovich, M. A. Baldo, M. E. Thompson, S. R. Forrest, *Appl. Phys. Lett.*, **79**(13), 2082–2084 (2001).
3. R. J. Holmes, S. R. Forrest, Y. J. Tung, R. C. Kwong, J. J. Brown, S. Garon, M. E. Thompson, *Appl. Phys. Lett.*, **82**(15), 2422–2424 (2003).
4. R. J. Holmes, B. W. D'Andrade, S. R. Forrest, X. Ren, J. Li, M. E. Thompson, *Appl. Phys. Lett.*, **83**(18), 3818–3820 (2003).
5. B. W. D'Andrade, R. J. Holmes, S. R. Forrest, *Adv. Mater.*, **16**(7), 624–628 (2004).
6. B. W. D'Andrade, S. R. Forrest, *Adv. Mater.*, **16**(18), 1585–1595 (2004).
7. D. Gupta, M. Katiyar, Deepak, *Opt. Mater.*, **28**(4), 295–301 (2006).
8. M. Berggren, G. Gustafsson, O. Inganas, M. R. Andersson, T. Hjertberg, O. Wennerstrom, *J. Appl. Phys.*, **76**(11), 7530–7534 (1994).
9. A. Dodabalapur, L. J. Rothberg, T. M. Miller, *Appl. Phys. Lett.*, **65**(18), 2308–2310(1994).

10. J. Kido, M. Kimura, K. Nagai, *Science*, **267**(5202), 1332–1334 (1995).
11. B. W. D'Andrade, J. Esler, C. Lin, V. Adamovich, S. Xia, M. S. Weaver, R. Kwong, J. J. Brown, *SPIE Proc.*, **7051** (2008).
12. M. A. Baldo, D. F. O'Brien, Y. You, A. Shoustikov, S. Sibley, M. E. Thompson, S. R. Forrest, *Nature*, **395**(6698), 151–154 (1998).
13. K. Walzer, B. Mannig, M. Pfeiffer, K. Leo, *Chem. Rev.,* **107**(4), 1233–1271 (2007).
14. M. H. Lu, J. C. Sturm, *J. Appl. Phys.,* **91**(2), 595–604 (2002).
15. B. Ruhstaller, T. Beierlein, H. Riel, S. Karg, J. C. Scott, W. Riess, *IEEE J. Sel. Top. Quantum Electron.*, **9**(3), 723–731 (2003).
16. G. Wyszecki, W. S. Stiles, *Color Science: Concepts and Methods, Quantitative Data and Formulae*, 2nd ed., New York: Wiley-Interscience, 2000.
17. R. Meerheim, R. Nitsche, K. Leo, *Appl. Phys. Lett.,* **93**(4) (2008).
18. X. Zhou, J. He, L. S. Liao, M. Lu, X. M. Ding, X. Y. Hou, X. M. Zhang, X. Q. He, S. T. Lee, *Adv. Mater.,* **12**(4), 265–269 (2000).
19. Illuminating Engineering Society of North America, *The IESNA Lighting Handbook: Reference & Application*, 9th ed., New York: Illuminating Engineering Society of North America, 2000.
20. M. Agrawal, S. Yiru, S. R. Forrest, P. Peumans, *Appl. Phys. Lett.,* **90**(24), 241112–241111 (2007).
21. S. R. Forrest, D. D. C. Bradley, M. E. Thompson, *Adv. Mater.,* **15**(13), 1043–1048 (2003).
22. Optical Society of America, *Handbook of Applied Photometry*, Springer, 1997.
23. Illuminating Engineering Society of North America, *LM-79–08: Approved Method: Electrical and Photometric Measurements of Solid-State Lighting Products*, (2008), 1–16.
24. Y. Tomita, C. May, M. Törker, J. Amelung, M. Eritt, F. Löffler, C. Luber, K. Leo, K. Walzer, K. Fehse,. Q. Huang, *SID Symp. Digest Tech. Papers*, **38**, 1030–1033 (2007).
25. A. Endo, K. Suzuki, T. Yoshihara, S. Tobita, M. Yahiro, C. Adachi, *Chem. Phys. Lett.,* **460**(1–3), 155–157 (2008).
26. H. Kostal, R. Rykowski, *Laser Focus World*, **44**(8) (2008).
27. Illuminating Engineering Society of North America, *LM-80–08: IES Approved Method for Measuring Lumen Maintenance of LED Light Sources*, (2008), 1–16.

28. H. H. Fong, K. C. Lun, S. K. So, *Chem. Phys. Lett.*, **353**(5–6), 407–413 (2002).
29. S. W. Tsang, K. L. Tong, S. C. Tse, S. K. So, *Proc. SPIE, Org. Light Emitting Mater. Devices X*, **6333**, 633313 (2006).
30. S. C. Tse, H. H. Fong, S. K. So, *J. Appl. Phys.* **94**(3), 2033–2037 (2003).
31. B. W. D'Andrade, J. J. Brown, *Appl. Phys. Lett.*, **88**(19), 192908 (2006).
32. P. Peumans, A. Yakimov, S. R. Forrest, *J. Appl. Phys.*, **93**(7), 3693–3723 (2003).
33. B. W. D'Andrade, S. Datta, S. R. Forrest, P. Djurovich, E. Polikarpov, M. E. Thompson, *Org. Electron.*, **6**(1), 11–20 (2005).
34. I. G. Hill, A. Kahn, J. Cornil, D. A. dos Santos, J. L. Bredas, *Chem. Phys. Lett.*, **317**(3–5), 444–450 (2000).
35. M. A. Baldo, M. E. Thompson, S. R. Forrest, *Nature*, **403**(6771), 750–753 (2000).
36. B. W. D'Andrade, S. R. Forrest, *Chem. Phys.*, **286**(2–3), 321–335 (2003).
37. N. C. Giebink, B. W. D'Andrade, M. S. Weaver, P. B. Mackenzie, J. J. Brown, M. E. Thompson, S. R. Forrest, *J. Appl. Phys.*, **103**(4), 044509–044501 (2008).
38. B. W. D'Andrade, S. R. Forrest, A. B. Chwang, *Appl. Phys. Lett.*, **83**(19), 3858–3860 (2003).
39. Y.-J. Lee, H. Lee, Y. Byun, S. Song, J.-E. Kim, D. Eom, W. Cha, S.-S. Park, J. Kim, H. Kim, *Thin Solid Films*, **515**(14), 5674–5677 (2007).
40. D. Y. Kondakov, W. C. Lenhart, W. F. Nichols, *J. Appl. Phys.*, **101**(2), 24512–24511 (2007).
41. R. Meerheim, S. Scholz, S. Olthof, G. Schwartz, S. Reineke, K. Walzer, K. Leo, *J. Appl. Phys.*, **104**(1), 014510–014511 (2008).

Index

absorption   3–5, 12, 17, 79, 140, 150–53, 155, 164, 182–85, 244, 246, 349, 363, 375
   cross-polarized   11, 13, 15
acceptor photobleaching (AP)   263
acceptor tubes   141, 143, 148, 152, 155–56
AFM *see* atomic force microscope
anisotropy decay   235, 237, 240, 245, 247
anthracene   170, 331–32
AP *see* acceptor photobleaching
atomic force microscope (AFM)   63, 68, 118, 164

bandgap modification for visible emission   328–29, 331
batteries   84, 87
bioimaging   323–26, 332, 334, 337
blue electrophosphorescence   368–69, 371

cancer cells   184, 299, 302, 308
cancers   291–92, 312–14
capacitors   87–88, 220
capping agents   326, 329
carbon   1, 19, 26, 35–36, 40, 47–50, 54–55, 61–62, 102, 163–64, 230–31
carbon atoms   36–37, 40, 43, 47, 62
carbon caps   47, 50–51
carbon nanotubes (CNTs)   23–26, 32, 55–56, 62–63, 66, 88, 90, 102, 125–26, 163–65, 169–71, 177–79, 181, 183–85, 189
carbon precursors   43–45, 54

catalysts   35, 37–38, 40–44, 54
CdSe nanocrystals   347
cell viability   306, 326, 337–38
cells
   plant   325, 329, 332, 336–37
   single   306
chemical vapor deposition (CVD)   2, 24, 36, 62–63
chemistry, nanomaterial   302
chloroform   83, 118, 126, 186–88, 295–96, 300–2
CNT sidewalls   169–71
CNTs
   photothermal conversion of   184
   soluble   189
CNTs *see* carbon nanotubes
Co clusters   44–45
Co-doped ZnO   327, 329–30, 332, 337
Co-Mo catalysts   38, 41, 45–46
conductivity   74–76, 82, 84, 92, 101, 116–17, 119–20, 125
conjugated polymers   69, 116, 125–29, 141, 156
CVD *see* chemical vapor deposition
cytoplasm   306, 325–26, 332–33
cytotoxicity   309, 323, 326, 334, 337–38

density gradient ultracentrifugation (DGU)   81–82, 137–38, 145–46, 148–49
devices
   photovoltaic   84–85
   supercapacitor   88–89

DGU *see* density gradient ultracentrifugation
diazonium salts 81–82
dielectric environment 150–51, 154
digital signal processing (DSP) 354–55
diodes 98, 100
dipoles 12–14, 277
discriminators 217, 220
dispersal agents 163–82, 184, 186, 188
donor excitation 263, 275, 282
donor lifetimes 264
donor nanotubes 143, 149, 155
donor s-SWNTs 134
doped ZnO 328–30, 339
doping 81–83, 121, 301, 324–25
doping concentration 118–19, 372
doxorubicin 312–13
drug delivery 312
DSP *see* digital signal processing
dyes 209, 229, 232, 238–40, 244–46, 266, 324

EET *see* exciton energy transfer
EET efficiency 143–44
electrodes 54, 61, 64, 86–88, 91, 116, 118, 126, 177, 251
　drain 94, 96, 98
　thin-film 85, 88–89
electroluminescent performance 117, 119, 121, 123
electron transfer, photo-induced 86, 172–73, 177
electron transport 116, 120, 123–25
electronic confinement 77
electronics 64, 84, 206, 222, 224–25, 297
emission
　dual-color 332–33
　forward 374
　midgap 139
emission dipole 11–12, 246
emission energies 5, 14, 136
emission peaks 10, 134–35, 232, 299, 351, 355
emission polarizer 12
emission sidebands 144–45, 156
emission spectra 1, 6, 244, 268, 273, 276, 278, 282–83, 324, 336, 346, 349
emission wavelengths 28, 31, 39, 135, 138–41, 146, 298, 301, 346, 348
emissive layer 119, 372
energy
　exciton 4, 148, 369
　photon 77–78, 155
energy separation 6, 10, 145, 147
energy transfer 137, 143, 261, 269–70, 369, 371
　endothermic 369–70
energy transfer lifetime 142–43
EQE *see* external quantum efficiency
excitation energy 6, 14–15, 18, 137, 139, 151
excitation light 29, 264, 349
excitation matching 136
excitation pulse 213, 222
excitation wavelengths 7, 9, 18, 28, 135, 138–39, 146, 272, 280, 331, 344–45, 349
exciton dissociation 85, 116, 126, 128–29
exciton energy transfer (EET) 133–34, 136–48, 150–52, 154–56
exciton–exciton resonances 135–37, 144, 153, 156
exciton relaxation 142–43, 148
exciton states 3, 7, 15, 18, 324
excitons 2, 4, 6, 11, 13, 15–17, 119–20, 126, 134, 137, 139, 141–44, 148–49, 155–56, 378

longitudinal 13, 15, 17, 19
photoexcited 143, 149
quasi-dark 15, 17, 19
transverse 3, 13, 15–17, 19
external quantum efficiency (EQE) 120–22, 374

FESEM *see* field-emission scanning electron microscope
FETs *see* field-effect transistors
field-effect transistors (FETs) 43, 53–54, 63, 125
field-emission scanning electron microscope (FESEM) 326
FLIM-FRET (fluorescence lifetime imaging/Förster resonance energy transfer) microscopy 264, 276–77
fluorescein 241, 324, 331–32
fluorescence 204, 213, 226, 231–34, 236, 238, 246, 248, 250–52, 278, 330, 334, 347, 359, 364
  size-dependent 234, 251–52
fluorescence anisotropy decays 229, 232, 239, 241–42
fluorescence decay 211, 237, 244
fluorescence emissions 217, 219, 264, 334–35, 337
fluorescence lifetime 209, 213, 218, 232, 247, 249, 264, 276
fluorescent materials 370
fluorescent proteins (FPs) 261, 265, 272–73, 278, 280, 284
fluorophores 232, 234, 247, 259–61, 264–65, 269, 285
Förster resonance energy transfer (FRET) 133–34, 141, 144, 259–62, 265, 269, 272, 279–82, 312
FP donor–acceptor FRET pair 272–73, 275, 277

FPs *see* fluorescent proteins
FRET *see* Förster resonance energy transfer
FRET donors 262, 267, 280–81
FRET microscopy 260–61, 263, 284
FRET pairs 260–61, 265–66, 268, 272–73, 276, 279, 282–83, 285
FRET reactions 267, 280–81
FRET signals 263, 274–75, 280–81

gold, bulk 232–33, 249–50
gold nanoclusters 230, 234, 243, 250–52
gold nanoparticles 179, 229–30, 242, 253
gold nanorods 230, 233, 243, 249
graphene sheet 36, 62
graphite 10–11, 37, 77–78, 184

hexane 296, 300–2
highest occupied molecular orbital (HOMO) 120, 128, 250, 371
hole-conducting layer 118, 124–25
HOMO *see* highest occupied molecular orbital
HOMO energies 378–79
honeycomb structures 186–87

IFT *see* inverse Fourier transform
indium tin oxide (ITO) 64, 79–80, 83–87, 93, 115, 127, 177
inverse Fourier transform (IFT) 355–56
inverters 98, 100–1
ITO *see* indium tin oxide

laser ablation 24, 36, 62
laser diodes 204, 210–11, 237

lasers 17, 27, 29, 83, 184–85, 209, 212–15, 276, 351
  pump 155, 209, 213
  supercontinuum 207, 213–14
LEDs
  see light-emitting diodes
  composite 127–29
light-emitting diodes (LEDs) 61, 84, 87, 125, 129, 204, 207, 210, 215, 236–37, 359
light-emitting materials 116–17
live cells 263–65, 270, 272–77, 280, 285, 306, 338
lowest unoccupied molecular orbital (LUMO) 120, 128, 250, 371, 379
luminance 118, 374–76
luminance variation 374–75
luminescence 1, 35, 61, 88, 115, 133, 143, 153, 155, 163, 203, 229, 233, 250, 311–12
LUMO see lowest unoccupied molecular orbital

membranes 71–72, 246
metal clusters 40, 47–51, 54–55
methanol 45, 71–72, 83, 187, 252, 295, 326, 331
microviscosity 235, 245
monochromator 23–27, 219, 237

nanoclusters 47, 55, 233, 242–43, 249
nanocrystals 294–95, 299, 301–2, 313, 324, 326, 330, 333–34, 337, 345
  semiconductor 259–60, 278, 298–99
nanomaterials 115–16, 180, 204, 250
nanoparticle loading 337–38
nanoparticles, semiconductor 292–93, 302
nanotube bundles 144, 155–56

nanotube dispersions 139, 148, 156
nanotube/polymer composites 182–83, 185, 187
nanotube species 137, 147, 152, 154
nanotubes 1, 6, 16, 19, 26, 35–36, 61–63, 66, 85, 93, 125–26, 137–38, 142, 151–55, 163–64
  individual 4, 64–65, 69
  short 46
nickel 326, 329
nickel-doped ZnO 326–30, 337

OLED see organic light-emitting diode
OLED devices 87–88
OLED materials 375–76
one-photon excitation 151–52, 234
optical absorbance spectra 81–82
optical conductivities 76–77
optical transitions 2, 11–12, 15, 150–51
optoelectronic devices 125, 128, 368
  organic 115–16
organic 115–16
organic dye donor–acceptor FRET pair 267, 269, 271
organic dyes 259, 261, 265, 267, 278, 280, 284, 293, 308, 325, 344
organic light-emitting diode (OLED) 85, 87–88, 116–18, 120–23, 371–72, 374–78
organic materials 368, 373, 376–77, 379–80
organic thin films 378

percolation threshold 74–75, 79, 97

PHOLEDs *see* phosphorescent organic light-emitting devices
PHOLEDs, blue 368–69, 371
phonon energies 5–6
phonon sideband peaks 6
phonon sidebands 3, 6, 10, 18, 136–39, 141, 143–45, 147–49
    excitonic 3, 5–6, 137
    resonant excitation of 138–39
phonons 5–6, 10, 137, 144–45
phosphorescent organic light-emitting devices (PHOLEDs) 368, 370–71
photocurrent 128–29
photomultiplier 217, 219, 224
photon exchange 141, 144
photons 149, 204–5, 212, 264, 330
polymer, hole-conducting 117, 121, 123
polymer electrolytes 88–89
porphyrins 86, 163, 170, 172–73, 175, 177

QD *see* quantum dot
QD donors 280–81, 285
QD formulations 305, 308, 313
QD–organic dye FRET pairs 278–79, 281, 283
QD–protein conjugates 279, 281
QDs
    commercial 338
    functionalized 293, 304, 306–8
QR *see* quantum rod
quantum dot (QD) 156, 230–31, 234, 252, 259–61, 264–65, 278–82, 284–85, 291–313, 323–24, 333–34, 336–39
quantum rod (QR) 302, 305, 308
quantum yields (QYs) 260, 265, 269, 278, 326, 331–32

QYs *see* quantum yields

radial breathing mode (RBM) 9, 52, 171
RBM *see* radial breathing mode
resonance Raman spectroscopy (RRS) 52–53
RRS *see* resonance Raman spectroscopy

SANS *see* small-angle neutron scattering
sapphire 44–45, 212
sapphire laser 24–26, 213
scanning electron microscopy (SEM) 183, 245, 247
scanning tunneling microscopy (STM) 52, 178
Schottky barriers 75, 93
SDS *see* sodium dodecyl sulfonate
SEM *see* scanning electron microscopy
semiconducting tubes 38, 43, 45, 63
semiconductor devices 63, 217, 219
silica 70, 73, 230–31, 235, 240, 326
silica colloids 229, 232, 241–42
silicon QDs 300–1
single-walled carbon nanotubes (SWNTs) 1–15, 17–19, 26, 35–46, 48–56, 61–69, 71–77, 79–94, 96, 98, 100–1, 136–38, 154–56, 163–75, 177–87
small-angle neutron scattering (SANS) 165, 231
sodium cholate 25, 82, 137, 145–46, 166–68
sodium dodecyl sulfonate (SDS) 2, 25, 67–68, 82, 134–35, 165, 167–68

spectral FRET microscopy 263, 273–76, 282, 285
spectral power distributions 374–76
spectroradiometers 374–76
STM *see* scanning tunneling microscopy
supercapacitors 84, 87–89
surface capping 326, 330
surface chemistry 238, 251
surface coatings 304–6, 339
surface defects 323, 325, 347
SWNTs *see* single-walled carbon nanotubes
  aligned 45, 69, 94, 96, 101
  applications of 55, 66
  as-grown 45, 66
  clone 46
  electronic properties of 62, 67
  enriched 155
  intrinsic 4, 11
  large-diameter 53, 81
  large-gap 139, 155
  low-density 94–95
  metallic 52, 63–64, 74–75, 79, 81–83, 93–94, 96, 98, 165, 179
  optical properties of 133–34
  optical transitions of 150, 152
  optoelectronic properties of 62, 65, 74–75, 77, 79
  purification and dispersion of 65
  selective synthesis of 35–38, 40, 42, 44, 46, 48, 50, 52, 54
  semiconducting 2, 15, 53, 62–63, 74–75, 82, 92–94, 96–97, 175, 178
  small-bundle 86
  small-diameter 18, 81
  solubilization of 166–67, 174–75, 181
  vacuum-filtrated 74, 77

TCSPC *see* time-correlated single-photon counting
TCSPC timing electronics 220, 222
TDCs *see* time-to-digital converters
TEM *see* transmission electron microscope
TFT *see* thin-film transistor
thin-film chemicapacitor 90–91
thin-film deposition processes 69, 71, 73
thin-film transistor (TFT) 61, 81, 83–84, 93–94, 100–1, 183
thin films 23–25, 27–31, 61–62, 64–65, 67, 69–74, 76–80, 83–95, 97–98, 101–2, 126, 371, 376, 378
time-correlated single-photon counting (TCSPC) 203–4, 206–8, 211, 213–14, 216–18, 221, 236, 276
time domain luminescence instrumentation 203–4, 206, 208, 210, 212, 214, 216, 218, 220, 222, 224, 226
time-to-digital converters (TDCs) 222
timing device 219–21
transit time spread (TTS) 216–17
transmission electron microscope (TEM) 39, 48, 52, 65, 164, 231
transmittance 76, 79, 88
triplet dark excitons 145, 147
triplet exciton states 10, 369
TTS *see* transit time spread
tubes, photomultiplier 216–17
two-photon excitation 151, 230, 350–52

ultraviolet photoelectron spectroscopy (UPS) 85, 378–79

UPS *see* ultraviolet photoelectron spectroscopy

vacuum filtration 71–72

water-dispersible QDs 294, 302–4
white organic light-emitting device 373, 375, 377, 379
white organic light-emitting diodes (WOLEDs) 367–68, 372–73, 376–77, 379–80

WOLEDs *see* white organic light-emitting diodes

YAG *see* yttrium aluminum garnet
yttrium aluminum garnet (YAG) 29, 209, 215

ZnO nanocrystals 325–26, 336, 338
ZnO nanoparticles in biosensing applications 323–24, 326, 328, 330, 332, 334, 336, 338